Vicious
Circles

CSLI Lecture Notes Number 60

Vicious Circles

On the Mathematics of Non-Wellfounded Phenomena

Jon Barwise and Lawrence Moss

CSLI Publications

CENTER FOR THE STUDY OF
LANGUAGE AND INFORMATION
STANFORD, CALIFORNIA

Copyright © 1996
CSLI Publications
Center for the Study of Language and Information
Leland Stanford Junior University
Printed in the United States
00 99 98 97 96 5 4 3 2 1

Library of Congress Cataloging-in-Publication Data

Barwise, Jon.
Vicious circles / Jon Barwise and Lawrence Moss.
p. cm. — (CSLI lecture notes ; no. 60)
Includes bibliographical references and index.
ISBN 1-57586-009-0 (alk. paper).
ISBN 1-57586-008-2 (pbk. : alk. paper).
1. Hypersets. 2. Cycles. I. Moss, Lawrence Stuart, 1959– . II. Title.
III. Series.
QA248.B376 1996
511.3'22–dc20 96-809
CIP

CSLI was founded early in 1983 by researchers from Stanford University, SRI International, and Xerox PARC to further research and development of integrated theories of language, information, and computation. CSLI headquarters and CSLI Publications are located on the campus of Stanford University.

CSLI Lecture Notes report new developments in the study of language, information, and computation. In addition to lecture notes, the series includes monographs, working papers, and conference proceedings. Our aim is to make new results, ideas, and approaches available as quickly as possible.

∞The acid-free paper used in this book meets the minimum requirements of the American National Standard for Information Sciences—Permanence of Paper for Printed Library Materials, ANSI Z39.48-1984.

The painting on the cover of the paperback edition of this book is *La reproduction interdite*, 1937, (Portrait of Edward James), by René Magritte, from the collection of the Museum Boymans-van Beuningen, Rotterdam. Oil on canvas: 81.3 x 65 cm.

to
Yiannis Moschovakis

Contents

Part I Background   1

1   Introduction   3
    1.1   Set theory and circularity   5
    1.2   Preview   6

2   Background on set theory   11
    2.1   Some basic operations on sets   11
    2.2   Sets and classes   15
    2.3   Ordinals   17
    2.4   The Axiom of Plenitude   21
    2.5   The Axiom of Foundation   24
    2.6   The axioms of set theory   27

Part II Vicious Circles   31

3   Circularity in computer science   33
    3.1   Streams   34
    3.2   Labeled transition systems   35
    3.3   Closures   40
    3.4   Self-applicative programs   42
    3.5   Common themes   45

4   Circularity in philosophy   47
    4.1   Common knowledge and the Conway Paradox   47
    4.2   Other intentional phenomena   49
    4.3   Back to basics   50
    4.4   Examples from other fields   51

5   Circularity and paradox   55
    5.1   The Liar Paradox   55
    5.2   Paradoxes of denotation   57

5.3 The Hypergame Paradox   58
5.4 Russell's Paradox   59
5.5 Lessons from the paradoxes   60

Part III Basic Theory   65

6   **The Solution Lemma**   **67**
    6.1 Modeling equations and their solutions   70
    6.2 The Solution Lemma formulation of *AFA*   72
    6.3 An extension of the Flat Solution Lemma   74

7   **Bisimulation**   **77**
    7.1 Bisimilar systems of equations   77
    7.2 Strong extensionality of sets   81
    7.3 Applications of bisimulation   83
    7.4 Computing bisimulation   87

8   **Substitution**   **91**
    8.1 General systems of equations   92
    8.2 Substitution   92
    8.3 The general form of the Solution Lemma   97
    8.4 The algebra of substitutions   100

9   **Building a model of** *ZFA*   **103**
    9.1 The model   104
    9.2 Bisimulation systems   107
    9.3 Verifying *ZFC⁻*   109
    9.4 Verifying *AFA*   111

Part IV Elementary Applications   117

10   **Graphs**   **119**
    10.1 Graphs and the sets they picture   119
    10.2 Labeled graphs   125
    10.3 Bisimilar graphs   128

11   **Modal logic**   **131**
    11.1 An introduction to modal logic   132
    11.2 Characterizing sets by sentences   137
    11.3 Baltag's Theorems   142
    11.4 Proof theory and completeness   145
    11.5 Characterizing classes by modal theories   149

12   **Games**   **159**
    12.1 Modeling games   159

12.2 Applications of games   165
12.3 The Hypergame Paradox resolved   170

13 The semantical paradoxes   177
13.1 Partial model theory   177
13.2 Accessible models   181
13.3 Truth and paradox   183
13.4 The Liar   187
13.5 Reference and paradox   191

14 Streams   197
14.1 The set A^∞ of streams as a fixed point   197
14.2 Streams, coinduction, and corecursion   200
14.3 Stream systems   205

Part V Further Theory   209

15 Greatest fixed points   211
15.1 Fixed points of monotone operators   212
15.2 Least fixed points   214
15.3 Greatest fixed points   216
15.4 Games and fixed points   220

16 Uniform operators   223
16.1 Systems of equations as coalgebras   224
16.2 Morphisms   228
16.3 Solving coalgebras   230
16.4 Representing the greatest fixed point   233
16.5 The Solution Lemma Lemma   235
16.6 Allowing operations in equations   239

17 Corecursion   243
17.1 Smooth operators   244
17.2 The Corecursion Theorem   248
17.3 Simultaneous corecursion   255
17.4 Bisimulation generalized   258

Part VI Further Applications   265

18 Some Important Greatest Fixed Points   267
18.1 Hereditarily finite sets   268
18.2 Infinite binary trees   272
18.3 Canonical labeled transition systems   275
18.4 Deterministic automata and languages   277

18.5 Labeled sets   281

19   Modal logics from operators   283
19.1 Some example logics   284
19.2 Operator logics defined   286
19.3 Characterization theorems   294

20   Wanted: A strongly extensional theory of classes   301
20.1 Paradise lost   301
20.2 What are *ZFC* and *ZFA* axiomatizations of?   303
20.3 Four criteria   307
20.4 Classes as a *façon de parler*   309
20.5 The theory SEC_0 311
20.6 Parting thoughts on the paradoxes   319

21   Past, present, and future   323
21.1 The past   324
21.2 The present   325
21.3 The future   326

Appendix: definitions and results on operators   335

Answers to the Exercises   337

Bibliography   381

Index   385

Part I

Background

1

Introduction

All these tidal gatherings, growth and decay,
Shining and darkening, are forever
Renewed; and the whole cycle impenitently
Revolves, and all the past is future.

Robinson Jeffers, "Practical People"

Mathematics is frequently described as "the science of pattern," a characterization that makes more sense than most, both of pure mathematics, but also of the ability of mathematics to connect to a world teeming with patterns, symmetries, regularities, and uniformities.

A major source of pattern in the world stems from the cyclical movements of the earth, sun and moon. It is not that these bodies move in circles, but that their movement results in the "shinings and darkenings" of day and night, the waxing and waning of the moon with its "tidal gatherings," and the seasons of the year, marked by "growth and decay." The cycles help establish the patterns which make life possible, and are celebrated in our division of time into hours, days, months, seasons, and years. We might characterize this cyclical nature of time by means of unfolding "streams" of weeks and seasons; they unfold without end but with a cyclic pattern to their nature.

$$week \quad = \quad (Su, (M, (Tu, (W, (Th, (Fr, (Sat, week)))))))$$
$$seasons \quad = \quad (spring, (summer, (fall, (winter, seasons))))$$

Cycles occur not only in the physical world around us, but also in the biological and psychological world within. Our heart beats to a regular, repetitive rhythm, our digestive system works in cycles, and our emotional life has its own patterns of ups and downs. Our interactions with each other are marked by the give and take of reciprocal activity. The basic structure of a two person conversation might be characterized as follows:

$$conversation \quad = \quad (1^{st} \; speaker, (2^{nd} \; speaker, conversation))$$

The philosopher David Lewis uncovered a deep source of circularity in human affairs, described in his famous study of convention (Lewis 1969). All social institutions, from language to laws to customs about which side of the sidewalk to use, are based on conventions shared by the community in question. But what does it mean for a society to share a convention? Certainly, part of what it means is that those who accept some convention, say C, behave in the given way. But Lewis also argues that another important part of what makes C a convention is that those who accept C also accept that C is a shared convention.

For example, suppose that C is the convention that people walk or drive on the right side of the street. To accept C means that you know that normally you should walk on the right. But it also means that you expect others to do the same thing, and indeed that you expect others to expect you to do the same thing, etc. In short, you accept that the C is a shared convention. Note that this is circular, since part of your understanding of C is that others understand the same thing.[1]

Circularity is also important in the design and understanding of technology: most notably in the computer systems that surround us. Hardware designers make use of a notion of the state of a system, laying out the design of the system in a way that insures that the system returns to a given state in regular ways. This notion of state is embodied in Turing's original notion of a Turing machine, and in later notions of finite state machine and labeled transition system.

Over the past twenty years or so, ever more intrinsically circular phenomena have come to the attention of researchers in the areas of artificial intelligence, computer science, cognitive science, linguistics, and philosophy. For example, mathematical accounts of circularity have been pursued by computer scientists interested in the notions of processes and streams, computational linguists working on feature structures, philosophers studying theories of truth and reference, and workers in artificial intelligence studying terminological cycles, among others. These different fields have been working with what can be seen as a common set of ideas, the notions that constitute the themes of this book.

---

[1] Another aspect of conventions is that they are conventional and so can be violated. There is nothing in them that compels their acceptance. For example, we are going to use some standard conventions of mathematical discourse and hope that you, our reader, share these conventions with us, but there is nothing that forces this to be the case.

1.1 Set theory and circularity

In certain circles, it has been thought that there is a conflict between circular phenomena, on the one hand, and mathematical rigor, on the other. This belief rests on two assumptions. One is that anything mathematically rigorous must be reducible to set theory. The other assumption is that the only coherent conception of set precludes circularity. As a result of these two assumptions, it is not uncommon to hear circular analyses of philosophical, linguistic, or computational phenomena attacked on the grounds that they conflict with one of the basic axioms of mathematics. But both assumptions are mistaken and the attack is groundless.

Set theory has a dual role in mathematics. In pure mathematics, it is the place where questions about infinity are studied. Although this is a fascinating study of permanent interest, it does not account for the importance of set theory in applied areas. There the importance stems from the fact that set theory provides an incredibly versatile toolbox for building mathematical models of various phenomena.

Moreover, most parts of classical mathematics can be modeled within set theory. Just because set theory *can* model so many things does not, however, mean that the resulting models are the best models. Even more, the successes of set theory do not compel us to take it as "the foundations" for mathematics. That is, knowing that things like real numbers, functions, relations, and the like can be represented faithfully in set theory does not mean that they *are* sets, any more than the planes we ride are the scale models once tested in wind tunnels.

The universe of sets may or may not be a suitable tool for modeling some particular phenomena we find in the world. Some other mathematical universe may well be better in some given instance. Choosing the best mathematical model of a phenomenon is something of an art, depending both on taste and technique. However, circularity is not in and of itself any reason to despair of using sets, as has sometimes been assumed. Indeed, this book is concerned with extending the modeling capabilities of set theory to provide a uniform treatment of circular phenomena.

The most familiar universe of sets in the century now drawing to a close, known as the iterative (or cumulative) universe, is indeed at odds with circularity. For example, if we attempt to model the "streams" displayed above, treating pairs (x, y) in the standard way, then none of the streams displayed has a set-theoretical analogue in the iterative universe! The basic reason is that in this universe you can't have a stream $s = (a, s')$ unless s' is "constructed before" s, which is impossible if s is nested within s', as in our examples.

There is, however, a richer universe, first explored by Forti and Honsell (1983) and Aczel (1988), that gives us elegant tools for modeling all sorts of circular phenomena. One of our aims here is to show that these tools unify much of the work going on in the fields mentioned above. This universe of sets contains what are called *non-wellfounded* sets, or *hypersets*.[2]

1.2 Preview

The core notions that have arisen in the study of various sorts of circular systems are the following:

1. *observationally equivalent* or *bisimilar* phenomena,
2. *canonical representations* for various kinds of objects,
3. *solving equations* to find canonical objects,
4. *largest fixed points* of monotone operators defining universes in which we can find canonical systems,
5. *coinduction*, a method of proof about canonical systems, and
6. *corecursion*, a method for defining operations into canonical systems.

If you look in a standard textbook on set theory written before 1990 you will find none of these topics. Within the universe of hypersets, however, these notions in combination with more standard ideas from set theory, give rise to a rich universe of sets. We also get powerful mathematical methods for using the extended universe in modeling circular and otherwise non-wellfounded phenomena.

We have had two objectives in writing this book. One is to make it easy for computer scientists, linguists, philosophers, cognitive scientists, and the like, to *use* hypersets in modeling real world phenomena. We hope our readers will end by knowing when the universe $V_{afa}[\mathcal{U}]$ of hypersets is a natural place to look for solutions to problems,[3] what the easiest way to show that some phenomena can be modeled in $V_{afa}[\mathcal{U}]$ might be, and what tools are at hand for proving things about the resulting models. This leads us to present some older material in this new setting. Thus much of the material in this book draws on work of the researchers who have pioneered the mathematical study of circular phenomena.

Our other objective has been to contribute to the study of the universe of hypersets. Thus a fair amount of material in the book is original. We will do our best to acknowledge the work of others at the end of each chapter. We will not try to explain the history of hypersets themselves, though. The serious scholar will want to look at Aczel (1988), both for different ways of presenting

---

[2]The first of these names comes from the fact that such sets are ruled out by the iterative conception. On that conception, all sets are wellfounded. We prefer the second name because the change is conceptual and so a new name seems appropriate. Also, we like the term "hyperset" because of an analogy to the hyper-real numbers, another extension of a standard mathematical universe. Out of habit, we will use the two terms interchangeably.

[3]The subscript "afa" stands for the Anti-Foundation Axiom, the name Aczel gave to the basic axiom of hypersets. The $\mathcal{U}$ stands for "urelements," and we discuss these at length in Section 2.4.

much of the material presented here, but also for a discussion of the history of hypersets. Additional historical material can be found in Felgner (1971).

Prerequisites

As we said above, the intended reader of this book is a researcher in one of several fields, someone who might want to use hypersets is typically not a mathematician interested in the theory for its own sake. Nevertheless, you will need some background in mathematics to read it. We review the basic concepts of set theory in Section 2.1, but we assume you already have some facility with this material. You should be able to learn or review this material from any book on elementary set theory. The recent books of Devlin (1993) and Moschovakis (1994) contain this material, for example. (These books are also among the first textbooks to present hypersets.) Certainly we assume that the reader has enough mathematical familiarity to follow proofs and to work out exercises.

But this is really the only prerequisite. At times we discuss applications in various fields, and we trust that when we do so, we have given all the background one would need.

How to read and use this book

It is the style of some mathematics books to aim for a high level of abstraction, because this often leads to a more elegant theory with more powerful results. Since we are not aiming at a purely mathematical audience, we did not want to write that kind of book. Instead, we have tried hard to discuss examples along with theory, and to study simpler versions of ideas before moving on to more abstract forms.

We have also tried to make this book accessible to people with widely differing backgrounds and interests. As a result, there is more than one way to proceed. Chapter 2 is primarily for reference, and you may choose to study its sections when they are needed. But if your background in set theory is not strong, you should read through it to get a quick look at some aspects of the subject that will come up in the book. We have tried to point out where more advanced topics like ordinals, transfinite recursion, and consistency proofs are used in the book, to enable readers to skip those discussions without missing too much.

Part II of the book contains a catalog of the kinds of phenomena that people have used hypersets to model. In discussing these examples, we have tried to stay away from specialized vocabulary. Also, if you are not familiar with some of the issues in some particular field, you need not feel stuck. You should read Section 3.1, since the notion of streams is used repeatedly through the book to provide examples of more general theory.

The heart of the book is Part III, where the universe of hypersets is explained. The main notions here are those of canonical system, solving a system of equations, and bisimulation. If you read Part III you will have the background to follow most discussions about hypersets and their applications in the literature.

Part IV of the book gives several fairly simple applications of the basic theory. Three of the chapters in this part cover new ground: our discussion of modal logic and sets in Chapter 11 leads to new results in that area. The work on the hypergame paradox in Chapter 12 presents new work, as does Chapter 13 on the semantical paradoxes. We would like to think that these applications are just a sample of what will be done in each area.

Part V of the book uses the applications in Part IV as motivation for extending the general theory developed in Part III. We develop the methods of greatest fixed points, coinduction, and corecursion in this part. The material in Part V is more abstract than the first half of the book. Although we have tried to make the more difficult mathematics easier to absorb, we know that those chapters will be rougher going than the earlier parts of the book. We hope you will find them to be rewarding, nonetheless. In Part VI, we give some additional applications as a way of illustrating the methods developed in Part V.

We have included many exercises, with solutions at the end of the book. The exercises should guide one through concrete examples of the theory and stimulate one to think about new results. They range from working out concrete versions of abstract ideas, to routine problems, and on to some harder ones.

The graph on page 10 shows the logical dependencies among the chapters of this book.

Acknowledgments

One of the exciting things about hypersets is that it is part of a general re-vitalization of logic generated by its applications to new areas, especially in computer science. The origins of this book are in courses on hypersets that we have given in two computer science summer schools. Barwise presented a course at the 1992 Banff Workshop on Process Algebra and Moss gave one at the 1993 Summer School in Logical Methods in Concurrency held at Aarhus University. We would like to thank the organizers and participants in both of these courses for their encouragement and enthusiasm.

The book borrows from Chapter 3 of Barwise and Etchemendy (1987) and from the expository article Barwise and Moss (1991). We have presented seminars on some of the new material in this book to the Indiana University Logic Program over the past few years. We would like to thank our colleagues and students in, and visitors to, this program for their interest and encouragement.

The penultimate draft of the book was used in a joint computer science, philosophy, and mathematics seminar on the Mathematics of Circularity at Indiana University during the fall of 1995. Many improvements, major and minor, resulted, and some new results were discovered during this seminar.

Those results that can be attributed to specific individuals are so attributed in the text. Others are attributed to the seminar as a whole, as the MOC Workshop. The members of this seminar were Alexandru Baltag, Axel Barceló, Albert Chapman-Leland, Jefferson Davis, David Ritterskamp, Pragati Jain, Hossein Mehdizadeh, Yuko Murakami, Matthias Scheutz, and Gerry Wojnar. In addition, a number of other people read the book at various stages and made useful comments. We are grateful to Johan van Benthem, Venkatesh Choppella, Andrew Dabrowski, Jean-Yves Marion, Paul Miner, Piero Ursino, Xuegang Wang, and Glen Whitney in this regard.

We thank Mary Jane Wilcox for her help, especially in proofreading the book. We also want to thank the publication staff at CSLI, especially Dikran Karagueuzian and Tony Gee, for their patience and help in producing this book.

Finally, of course, we acknowledge the great debt we owe to Peter Aczel. It was he, after all, who introduced us to the universe of hypersets. This universe has reawakened our love of set theory. Beyond the two goals articulated above, our overarching hope is that this book will help some readers discover the beauty and usefulness of set theory.

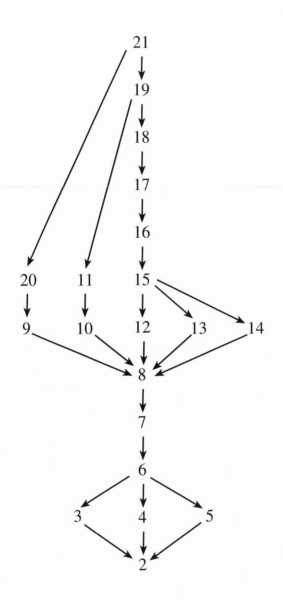

2

Background on set theory

In this chapter, we gather together some background material from set theory. Except for Section 2.4 on the Axiom of Plenitude, our background is quite standard. However, our presentation of some topics might differ from what you have seen before. So even if you have a good knowledge of elementary set theory, you should look through this chapter.

2.1 Some basic operations on sets

The importance of set theory rests in the fact that sets provide one of the most flexible and all-encompassing toolboxes for constructing models of structured objects. These include both mathematical structures (like the natural numbers, rationals, reals, etc.) and non-mathematical structures (models of computational devices, biological systems, human language, etc.). We are consistently impressed that a single concept, that of a set and its elements, is able to model such a wide variety of structure.

Relations, ordered pairs, and relational structures The first step in working out the modeling of mathematical objects is the treatment of functions and relations. After all, the basic concept of a set includes no structure whatsoever. For any sets a and b, we can model the informal notion of the *ordered pair* (a, b) by finding an operation $\langle a, b \rangle$ defined in terms of sets alone, so that

$$\langle a, b \rangle = \langle c, d \rangle \quad \text{iff} \quad a = c \text{ and } b = d.$$

The standard way to do this is to define

$$\langle a, b \rangle = \{\{a\}, \{a, b\}\}.$$

The reader who is unfamiliar with this definition should convince himself or herself that it does indeed have the desired property, and that the proof of this does not make any use of the Foundation Axiom. (If you are unfamiliar with the Foundation Axiom, you might return to this point after reading Section 2.5.) Having ordered pairs, we can use these in other definitions.

A set R is a *relation* if every element of R is an ordered pair. In this case, instead of writing $\langle a, b \rangle \in R$ we usually write aRb, or $R(a, b)$. Once we have defined ordered pairs, we can define ordered triples by $\langle a, b, c \rangle = \langle a, \langle b, c \rangle \rangle$, and so on.

Suppose that A is a set and R is a relation with the property that for all a and b such that $R(a, b)$, a and b belong to A. Then we say that R is a *relation on A*.

Frequently in mathematics one finds *relational structures*. These are defined as pairs $\langle A, R \rangle$, such that A is a set, and R is a relation on A. Usually, one is interested in relational structures which satisfy additional axioms. Some of the most common of these are listed on page 18 below.

Functions One of the most important notions in mathematics is that of a function. In set theory, it is customary to define a *function* (or *map*) to be a set f of ordered pairs with the property that if $\langle a, b \rangle$ and $\langle a, c \rangle$ belong to f, then $b = c$. In this case, we would write $f(a) = b$ (or $f_a = b$ in situations when we want to avoid parentheses). The *domain* of f is the set

$$dom(f) \quad = \quad \{a \mid f(a) = b \text{ for some } b\},$$

and the *range* of f is the set

$$rng(f) \quad = \quad \{b \mid f(a) = b \text{ for some } a\}.$$

f is said to be a function *from c to d* if $c = dom(f)$ and $rng(f) \subseteq d$. For any sets c and d, there is a set $c \to d$ of all functions from c to d. There is also a set $c \rightharpoonup d$ of all *partial functions* from c to d, that is, functions f with $dom(f) \subseteq c$ and $rng(f) \subseteq d$. Notice that $c \to d$ is a subset of $c \rightharpoonup d$. It is customary to write $f : c \to d$ for $f \in c \to d$ and similarly for $f : c \rightharpoonup d$. Given a function f and a subset $c_0 \subseteq dom(f)$, we define the *image of c_0 under f* to be

$$f[c_0] \quad = \quad \{f(a) \mid a \in c_0\},$$

and the *restriction of f to c_0* to be

$$f \restriction c_0 \quad = \quad \{\langle a, b \rangle \mid \text{for some } a \in c_0, f(a) = b\}.$$

A function $f : c \to d$ is *one-to-one* (or *injective*) if whenever $c_1 \neq c_2$ are elements of c, then $f(c_1) \neq f(c_2)$. f *maps onto d* (or f is *surjective*) if for every $d_1 \in d$ there is some $c_1 \in c$ such that $f(c_1) = d_1$. A function which is one-to-one and onto is called *bijective* ; in this case, f itself would be called the *bijection*.

Natural numbers Another important step in working out the way to model structure in set theory is modeling the natural numbers. The standard way is to take certain sets, the *finite von Neumann ordinals*, and to identify them with the natural numbers. (Ordinals in general are discussed in Section 2.3.) In this way of doing things

$$
\begin{aligned}
0 &= \emptyset \\
1 &= \{0\} \\
2 &= \{0,1\} = \{\emptyset, \{\emptyset\}\} \\
&\vdots
\end{aligned}
$$

(The symbol $\emptyset$ denotes the empty set, the unique set with no elements at all.) In this way, every number is the set of its predecessors. We also define ω to be the set of natural numbers. So $\omega = \{0, 1, 2, \ldots\}$.

Product and union A number of operations on sets are fundamental to the use of set theory in modeling. We review some of them here. First is the operation of *cartesian product*. For all sets A, B there is a set

$$ A \times B = \{\langle a, b \rangle \mid a \in A, b \in B\} $$

of all ordered pairs whose first element is drawn from A and whose second element is drawn from B. The *union* of A and B is the set

$$ A \cup B = \{a \mid a \in A \text{ or } a \in B\}. $$

The *intersection* of A and B is the set

$$ A \cap B = \{a \mid a \in A \text{ and } a \in B\}. $$

The *difference* of A and B is the set

$$ A - B = \{a \mid a \in A \text{ and } a \notin B\}. $$

The *disjoint union* of A and B is the set

$$ A + B = (\{0\} \times A) \cup (\{1\} \times B) $$

This is a trick which is needed when you want a set which contains a copy of each of A and B but where you keep track of which set things come from, in case some things are in both sets. If $a \in A \cap B$, then the object $\langle 0, a \rangle$ is the surrogate of a when you are thinking of a as a member of A, while $\langle 1, a \rangle$ is the surrogate of a when you are thinking of a as a member of B.

There is a *unary* (one-place) operation of union defined as follows:

$$ \bigcup a = \{x \mid x \in y \text{ for some } y \in a\} $$

This operation generalizes the binary union operation since $a \cup b = \bigcup\{a, b\}$.

The power set operation For any set a, the *power set* of a, $\mathcal{P}(a)$, is the set of all subsets of a. Every subset of a is an element of $\mathcal{P}(a)$, and every element of $\mathcal{P}(a)$ is a subset of a. For example, if $a = \{\emptyset, p\}$, then

$$\mathcal{P}(a) \quad = \quad \{\emptyset, \{\emptyset\}, \{p\}, a\}.$$

Note that $\emptyset$ and $\{\emptyset\}$ are different because the second of these has an element (namely $\emptyset$) and is therefore non-empty.

For every set a, the empty set $\emptyset$ and a itself are elements of $\mathcal{P}(a)$. Among the other elements of $\mathcal{P}(a)$ are the singletons (one-element sets) determined by the elements of a.

We sometimes have occasion to consider the finitary version of $\mathcal{P}(a)$, the set $\mathcal{P}_{fin}(a)$ of all finite subsets of a. Notice $a \in \mathcal{P}_{fin}(a)$ iff a is finite.

The transitive closure operation A set a is *transitive* if every set b which is an element of a has the property that all of *its* elements also belong to a. There are a number of equivalent ways to say this, including that every set which belongs to a also belongs to $\mathcal{P}(a)$; and also that $\bigcup a \subseteq a$.

For any set a, $TC(a)$ is the smallest transitive set including a. It is called the *transitive closure* of a. That is $TC(a)$ is transitive, $a \subseteq TC(a)$, and if b is any transitive set such that $a \subseteq b$, then $TC(a) \subseteq b$. The existence of this set can be justified using $\bigcup$ as follows:

$$TC(a) \quad = \quad \bigcup\{a, \bigcup a, \bigcup\bigcup a, \ldots\}$$

The smallest transitive set that has a as an *element* is just $TC(\{a\})$.

Example 2.1 If x and y are not sets and $a = \{x, \{y\}\}$ then $TC(a) = \{x, \{y\}, y\}$ since $\{x, \{y\}, y\}$ is clearly transitive, contains a as a subset, and is the smallest such set. (Notice that if x or y were a non-empty set, then we would have to include its members, its members' members, and so on, into the transitive closure of a.)

Exercise 2.1[1] Suppose that a and b are sets satisfying:

1. $a = \{b\}$, where b is as in (2), and
2. $b = \{\emptyset, a\}$, where a is as in (1).

Prove that $a = \{\mathcal{P}(a)\}$. (Later on, we'll see that such sets exist, assuming the Anti-Foundation Axiom.)

Exercise 2.2 Suppose a, b, c, and d satisfy the system:

$$\begin{aligned} a &= \{b\} \\ b &= \{\emptyset, \{\emptyset\}, c, d\} \\ c &= \{\emptyset, a\} \\ d &= \{a\} \end{aligned}$$

Show that $a = \{\mathcal{P}(\mathcal{P}(a))\}$.

---

[1] The answers to the exercises are in the back of the book, beginning on page 337.

Exercise 2.3 Compute the transitive closures of the sets a and b:

$$
\begin{aligned}
a &= \{b, \{b, \emptyset\}\} \\
b &= \{\{\{a\}\}\}
\end{aligned}
$$

Exercise 2.4 The Axiom of Extensionality says that two sets a and b with the same elements are equal:

$$(\forall a)(\forall b)\left[(\forall c)(c \in a \leftrightarrow c \in b) \to a = b\right].$$

Suppose a and b just happen to have the property that $a = \{a\}$ and $b = \{b\}$. Can we use the Axiom of Extensionality to prove that $a = b$?

2.2 Sets and classes

In this section we review an important distinction in set theory, the distinction between sets and classes. Intuitively, one might think that any predicate determines a fixed set, the set of all things satisfying that predicate. It turns out, however, that this is an incoherent idea, as we will see in Chapter 5. Some predicates have an extension which is just too "large" to be considered a set. Examples are things like: being a set, being a set which is not a member of itself, being an ordinal number, being a cardinal number, being a vector space, having exactly one element. The list goes on. However, we still like to speak of the collection of things determined by a predicate, and we call this the *extension* of the predicate. A *class* is the extension of some predicate.

Most of the familiar operations used in mathematics do not lead to classes which are too large, and so do not cause any problem. It is now the practice in mathematics to distinguish between small classes, called sets, and large classes, called *proper classes*. The basic intuitions are the following:

1. Every predicate determines a class, namely, the class of all sets that satisfy the predicate.
2. Any subclass of a set is a set. Another way of saying this is that for any set a and any predicate P, there is a set of those $x \in a$ such that $P(x)$. (This makes sense on the intuition that a subclass of a cannot be any larger than the original set a.)
3. The sets are closed under basic operations like pairing, union, and powerset.
4. The sets are closed under taking images of operations. That is, if f is an operation on sets and if a is a set, then $\{f(x) \mid x \in a\}$ is a set. (This makes sense on the intuition that the image of a cannot be any larger than the set a.)
5. If a class is a member of a class, then it is a set.

Of these principles, (5) requires comment. Like the others, it is part of the standard conception of set these days, but it is the least intuitive based on intuitions about size. It makes good sense under the cumulative conception of set, since in order to form a set under that conception, all of its members already have to be formed so that there can't be too many of them. But unlike the other principles, there is nothing in the intuition of size alone that would seem to justify it. We will follow tradition and assume (1)–(5) as working axioms in our theory of sets.[2] In fact, though, we will seldom use (5) and we return to re-examine it and related issues in Chapter 20.

It follows from (2) that the class V of all sets is large. For if V were small (that is, a set), then by (2), the class $R = \{a \in V \mid a \notin a\}$ would be small, and so a member of V. But then it follows that $R \in R$ if and only if $R \notin R$. This shows that V cannot be small. Apply (5), we see that V cannot be a member of any class.

A very common method of showing that some class is large is to show that if it were small, then the class V of all sets would be small. Here is an example of this method.

Example 2.2 Consider the predicate

x is an ordered pair $\langle a, b \rangle$ and $b = \mathcal{P}(a)$

By (1), this predicate determines a class, those sets x of the described form. We identify this class with the power set operation $\mathcal{P}$ itself. This class is large. To see this, note that the union of its domain (or range, for that matter) is the class V of all sets, itself a proper class.

Parallel to the distinction between sets and proper classes is the distinction between functions and those *operations* (like $\mathcal{P}$) that are too large to be modeled by *sets* of ordered pairs. We will have occasion to use large operations frequently in this book.

We'll go into the set-class distinction and its history further in Chapter 20. For now it is enough to remember that a set is a small class and that a proper class is a large class.

Exercise 2.5 Show that the class C of all singleton sets is a proper class.

Exercise 2.6 Fix a natural number n. Define an operation f_n on sets by

$$f_n(a) \quad = \quad a \cup \bigcup a \cup \bigcup\bigcup a \cup \cdots \cup \bigcup^n a$$

Let C_n be the class of all sets a such that $f_n(a)$ is finite. Prove that C_n is a proper class.

Later we will prove a contrasting result. Assuming either the Foundation

---

[2]In terms of our axiomatization, (5) follows from the Union Axiom, the assumption that if b is a set, so is $\bigcup b$. But that principle also does not seem justified solely on the basis of intuitions about size. That is, it seems unclear why a small class b would have to have a small class as its union $\bigcup b$. For example $\{X\}$ (if it exists) is small, for any class X, regardless of its size, but $X = \bigcup\{X\}$.

Axiom or the Anti-Foundation Axiom, $\bigcap_n C_n$ is a set. (See Exercise 18.3 on page 271.)

2.3 Ordinals

At several points in this book, we shall need the notion of an *ordinal*. This concept is very important in set theory, but it is typically not central to applications of set theory to other fields. Accordingly, many experts in the fields of applications of this book are less comfortable with ordinals than with the topics reviewed above.

While it is possible to read this book and omit all references to the ordinals, to have a really good grasp of the material presented, you need to have a mastery of the ordinals and transfinite recursion. The subject is quite standard, and you can learn it from many books. Accordingly, this section is a glimmer of an introduction to the ordinals for those who have not seen them. For those who have seen them long ago, it should also be a reminder of what ordinals and wellorderings are all about.

Basic definitions concerning ordered structures

The basic motivation for the ordinals is concerned with the notion of the *stage* of an infinite process. The examples of such a process which lead to the theory of the ordinals has to do with a certain operation on sets of real numbers. This operation takes a set A or real numbers and returns another set $A' \subseteq A$, called the *Cantor-Bendixson derivative* of A. The details of this operation need not concern us here. What is important is that it is possible to start with some A, and get a sequence A', A'', ..., A^n, ... These sets are a decreasing sequence, and it is possible that the sequence does not stabilize at any point. It is then possible to consider $A^\omega = \bigcap_n A^n$. After this, we might want to take derivative again, to get a sequence $A^{\omega+1}$, $A^{\omega+2}$, ... Once again, this sequence will be decreasing, but it might not stabilize anywhere.

When we continue like this for long, we run out of notations for the *stages* of this kind of process. What kind of structure is the set of stages? It is clear that the stages should be some kind of ordered set, but what kind of order do we want? How "long" could a set of stages be? How do we define functions on such stage-sets, like the iteration of derivatives that we have seen?

The answers to these questions constitute some of the most fundamental work in set theory. The appropriate abstract structure for talking about the stages is that of a wellorder, and we review the definition below. But we begin with a review of some of the basic order-theoretic concepts which we need at various places in this book.

Definition Let X be any set, and let R be a relation on X.

R is *reflexive* if for all $x \in X$, $x \ R \ x$.

R is *irreflexive* if for all $x \in X$, $\neg(x \ R \ x)$.

R is *symmetric* iff whenever $x \ R \ y$, then also $y \ R \ x$.

R is *antisymmetric* iff whenever $x \ R \ y$ and $y \ R \ x$, then $x = y$.

R is *transitive* if whenever $x \ R \ y$ and $y \ R \ z$, then also $x \ R \ z$.

An *equivalence relation* is a pair $\langle A, \equiv \rangle$ such that $\equiv$ is a reflexive, symmetric, and transitive relation on a set A.

A *preorder* is a pair $\langle P, \leq \rangle$ such that $\leq$ is a reflexive and transitive relation on a set P.

A *partial order* is a preorder $\langle P, \leq \rangle$ whose relation $\leq$ is antisymmetric.

A *linear order* is an irreflexive, transitive preorder $\langle L, < \rangle$ with the additional property that for all $x, y \in L$, either $x < y$, or $x = y$, or $y < x$.

A *wellorder* is a linear order $\langle W, < \rangle$ with the additional property that every non-empty $S \subseteq W$ has a least element.

We are primarily interested in wellorders in this section. An equivalent definition is that a wellorder is a linear order with the property that whenever $f : \omega \to W$ has the property that $f(n+1) \leq f(n)$ for all n, then there must be some N so that for all $n > N$, $f(n) = f(N)$.

The condition on f is connected to the idea of stages in an interesting way. If we want to compute A^w for some $w \in W$, we probably want to use A^v, for some $v < w$; in fact, we sometimes want to use all of the previous values. To compute A^v, we will want to use some A^{v_1} with $v_1 < v$, etc. The idea is that our retracing things like this should not go on forever. Eventually, after some finite number of steps, we should reach a "bottom" value, and f would be defined directly on this, by some "base case" in its definition.[3] This is what the condition above gives us.

Here is an example of a wellorder: Consider

(1) $\langle W, < \rangle$,   where $W = \{-2, 3\}$ and $<\ = \{\langle -2, 3 \rangle\}$.

Canonical wellorders

In a sense, all of the information in a wellorder is contained in the ordering relation $<$. So in our example, the important thing is that we have a first element, and then a second. The exact nature of those elements is pretty much irrelevant.

---

[3]This is part of the intuition behind recursion as it is usually presented. You might be interested to know that the perspective of this book shows that the "bottoming-out" idea is not really needed to do recursion. In Chapter 17, we present some interesting "co-recursions" which do not have a "base case" and yet define functions in a perfectly satisfactory way. This does not mean that the theory of ordinals is in any way defective or superfluous. It just means that in the context of hypersets, it does not account for all of the possible ways to define a function using some sort of recursion.

Definition Let $\langle A, R \rangle$ and $\langle B, S \rangle$ be two relational structures. An *isomorphism* between $\langle A, R \rangle$ and $\langle B, S \rangle$ is a bijection $f : A \to B$ (see page 12) with the property that $R(a, a')$ iff $S(f(a), f(a'))$, for all a and $a' \in A$.

We say that $\langle A, R \rangle$ and $\langle B, S \rangle$ are *isomorphic* if there is some isomorphism between them. We write $\langle A, R \rangle \cong \langle B, S \rangle$ in this case.

With this vocabulary in mind, we can say that although the example in (1) is a wellorder, we would prefer an isomorphic copy that was more "natural."

Definition Let A be any set. The *membership relation on A*, $\in_A$, is

$$\{\langle a, b \rangle \in A \times A \mid a \in b\}.$$

A *canonical wellorder* is a wellorder $\langle \alpha, R \rangle$ such that α is a transitive set and R is $\in_\alpha$. We call α an *ordinal* (or *ordinal number*) in this case.

The basic idea behind this definition is if that $\langle W, < \rangle$ is a wellorder and $w \in W$, then each $v < w$ is a kind of *component* of w. In a wellorder, no w can be a component of itself. And unlike the example in (1), there should be a close relationship between w and its components. The closest thing would simply be for w to *be* the set of its components. If α is an ordinal, then it is easy to check that α is precisely the set of its components.

Consider the set $2 = \{\emptyset, \{\emptyset\}\}$. The relation $\in_2$ is $\{\langle \emptyset, \{\emptyset\} \rangle\}$. If we call this relation $<$, then $\langle 2, < \rangle$ is a wellorder. Also, 2 is a transitive set. So 2 is an ordinal. The wellorder $\langle 2, < \rangle$ is isomorphic to the wellorder of (1). For more examples, note that if α is an ordinal, then so is $\alpha \cup \{\alpha\}$. This gives us infinitely many examples. Indeed, it gives the finite ordinals from page 13.

Next, we mention some basic properties of the ordinals:

Proposition 2.1 *Let α, β, and γ be ordinals.*

1. *Every element of α is itself an ordinal.*
2. $\alpha \notin \alpha$.
3. *If $\alpha \in \beta \in \gamma$, then $\alpha \in \gamma$.*
4. *Exactly one of the following holds: $\alpha \in \beta$, $\alpha = \beta$, or $\beta \in \alpha$.*
5. *If S is any non-empty set of ordinals, then S has a least element.*

Proposition 2.2 *Every ordinal α satisfies exactly one of the following three conditions:*

1. $\alpha = \emptyset$. *(In this case, we often write $\alpha = 0$.)*
2. *For some unique ordinal β, $\alpha = \beta \cup \{\beta\}$. (Here we write $\alpha = \beta + 1$, and we call α a* successor *ordinal.)*
3. $\alpha \neq 0$, *and for all $\beta \in \alpha$, $\beta + 1 \in \alpha$ as well. (Here we call α a* limit *ordinal.)*

Proposition 2.3 *Let $(W, <)$ be a wellorder. Then there is a unique function f defined on W so that for all $w \in W$,*

$$(2) \qquad\qquad f(w) \quad = \quad \{f(v) \mid v < w\}.$$

The image of f is an ordinal, say α, and f establishes an isomorphism between W and $\langle \alpha, \in_\alpha \rangle$.

Proofs of these assertions may be found in any book which develops the ordinals. Proposition 2.1 shows that on the class of all ordinals, $\in$ has the properties of wellorder. For this reason, when dealing with ordinals, one usually writes $\alpha < \beta$ instead of $\alpha \in \beta$.

Proposition 2.3 is noteworthy in that it uses the Collection Scheme (or Replacement Scheme) of set theory. It was in proving this result that von Neumann first noticed the need for axioms of set theory beyond the basic axioms of Zermelo.

A final important result ties together the ordinals with the definitions like that of the iterated Cantor-Bendixson derivative that we mentioned at the beginning of this section.

Proposition 2.4 (Transfinite Recursion) *Let $\varphi(a)$ be any definable operation taking functions to sets. Then there is a definable operation F from ordinals to sets, so that for all α,*

$$F(\alpha) \quad = \quad \varphi(F \restriction \alpha).$$

We say that F is defined by *transfinite recursion* on the ordinals. There are a number of other versions of transfinite recursion, including one where a function F is defined on a wellordered set instead of on the ordinals. Yet another version allows for parameters in the definition of F. We will see a few uses of transfinite recursion in this book.

Exercise 2.7 Use Proposition 2.1 to show that the class *On* of ordinals is a proper class. This result is called the *Burali-Forti Paradox*.

Cardinals

The theory of wellorders solves the problem of giving a workable theory of the notion of the stage of an infinite process. Furthermore, the ordinals give us a yardstick for measuring the lengths of wellorders. This is because every wellorder W is isomorphic to a unique ordinal α, so α can serve as the length of W.

One application of the theory of ordinals is to solve another, even more important, foundational problem in set theory. This is the matter of establishing a yardstick for the sizes of sets.

Given two sets or classes a and b, we say a *is at most as big as* b if there is a one-to-one function $f : a \to b$. It is a basic theorem of set theory that if a

and b are sets such that a is at most as big as b, and vice-versa, then there is a bijection from a onto b. In this case, we say that a and b are *equinumerous*.

In more detail, the problem of size is to find a class C of sets so that:

1. For every set a, a is equinumerous to a unique element $|a|$ of C, called the *cardinality of* a.
2. C should have an order $\leq$, and it should be the case that for all sets a and b, $|a| \leq |b|$ iff a is at most as big as b.
3. For all a, the cardinality of $|a|$ should again be $|a|$. That is, $||a|| = |a|$.
4. The class C, its order $\leq$, and the operation $a \mapsto |a|$ should be definable in set theory.

To solve this problem, we define the *cardinals* to be those ordinals α with the extra property that for all $\beta < \alpha$, there is no bijection between β and α. Every ordinal α is equinumerous to some cardinal, namely the smallest ordinal β to which it is equinumerous.

We use letters like κ and λ for cardinals, and we use C for the class of all cardinals. So $\alpha \in C$ iff $|\alpha| = \alpha$. The order on C is just the reflexive version of the ordering on ordinals restricted to C; that is, $\kappa \leq \lambda$ iff $\kappa \in \lambda$ or $\kappa = \lambda$.

The Axiom of Choice implies that every set is equinumerous to some ordinal, and hence to some cardinal. This cardinal is easily seen to be unique, and in this way the cardinals serve as the measures of the sizes of sets.

Exercise 2.8 Assume principles (1) – (4) concerning sets and classes above. Also assume the results on ordinals and cardinals from earlier in this section. Prove that a class a is a set iff a is equinumerous to some cardinal $|a|$.

2.4 The Axiom of Plenitude

Intuitively, we know that there are many sorts of things in the world. Some of them are sets, but many aren't. In spite of this, it is not uncommon to find set theory formulated in such a way that it assumes that everything is a set.

Now you might wonder how one would ever get started. What are there sets of, if everything is a set? The answer is that there is, at least, the empty set $\emptyset$ and sets that can be built up from it. It turns out that one can model an astonishing amount of mathematics starting this way.

However, starting with nothing but the empty set is rather tedious and unnatural, especially when you want to apply set theory outside mathematics. We want to be able to have sets of sticks, stones, and broken bones, as well as more abstract objects like numbers and whatnot. So in this book we allow ourselves a proper class $\mathcal{U}$ of *urelements*, objects in our universe which are not sets or classes and which have no members. The reason that we want a proper class of urelements, not just a set, is a bit technical. Suffice it to say for now that we don't ever want to worry that we've run out of urelements in the middle of an argument, so we're going to take an axiom that insures us that there are plenty of urelements.

Notation In this book, we reserve letters like x, y, and z for urelements. Sets or classes of urelements are usually denoted by upper-case letters such as X or Y. Objects which are either sets or urelements are denoted by p, q. All other letters denote sets or classes. We usually use upper-case letters for proper classes, but if we have a class which is not known to be proper, we might use a lower-case letter.

We started out this chapter with a look at some operations defined on all sets. These operations are not functions, since their domains are proper classes. What makes them operations is that their action is definable. Although we will not go into the precise formulation of definability in this book (it uses more logic than we otherwise need), you should be able to see that operations like the power set are far from random operations. In this book, all of the classes and operations may be taken to be definable.

Definition For each set a, the *support of* a, *support*(a) is $TC(a) \cap \mathcal{U}$. The elements of *support*(a) are the urelements which are "somehow involved" in a. A set a is *pure* if *support*$(a) = \emptyset$. Finally, for all subsets $A \subseteq \mathcal{U}$, we define

$$V_{afa}[A] \quad = \quad \{a : a \text{ is a set and } support(a) \subseteq A\}.$$

This is always a proper class, of course. If $A = \emptyset$, we omit it from the notation and just write V_{afa}. So V_{afa} is the class of all pure sets.

Note that $V_{afa}[A]$ is a collection of sets, so no urelements belong to $V_{afa}[A]$; in particular $A \cap V_{afa}[A] = \emptyset$ for all $A \subseteq \mathcal{U}$.

$V_{afa}[\mathcal{U}]$ is the collection of all sets. The urelements do not belong to it, but they do belong to $V_{afa}[\mathcal{U}] \cup \mathcal{U}$. The axioms of set theory are logical statements about $V_{afa}[\mathcal{U}] \cup \mathcal{U}$. In standard set theory, these axioms use the membership symbol $\in$, but here we also allow $\mathcal{U}$ as a unary relation symbol. For example, we define "x is a set" to mean that $\neg \mathcal{U}(x)$. We abuse notation a bit and write $x \notin \mathcal{U}$ in this case. We have already been doing something similar when we wrote "$A \subseteq \mathcal{U}$" above.

The axioms we will be using most often tell us that the universe of sets and urelements is closed under unordered pairs, unary unions of sets, and the power set operation. $V_{afa}[\mathcal{U}]$ also satisfies some additional axioms which will not be of central interest in this book. And since that material is standard, we focus only on what might be unfamiliar: the urelements.

One important fact is that although the urelements are members of the universe, they themselves have no members. (We take this as an axiom.) After all, the idea is that they are bereft of set-theoretic structure and that all of the structure comes in the sets. In the same way, the usual axioms of set theory must be re-cast in order to accommodate urelements. For example, we state

AXIOM OF EXTENSIONALITY For all $a, b \notin \mathcal{U}$, if $c \in a \leftrightarrow c \in b$ for all c, then $a = b$.

The main axiom we use concerning urelements is a principle which guarantees that there are always fresh urelements around:

AXIOM OF PLENITUDE For every set S there is a one-to-one function $f : S \to \mathcal{U}$ whose image $f[S]$ is disjoint from S.

Typically when we use this axiom we'll have some set S around. We'll want some new urelements x_s to be in one-to-one correspondence with the elements $s \in S$. We might also want to be sure that the urelements do not appear in anything we are considering. In that case, we would let T be all the urelements at hand, and apply the axiom to $S \cup T$.

The class form of the Axiom of Plenitude

Some of the results in this book define operations on proper classes, such as the class of all sets. For this, we'll need large collections of urelements. Our Axiom of Plenitude as it stands does not guarantee that the collections we will need really do exist. We therefore adopt the following stronger form.

STRONG AXIOM OF PLENITUDE There is an operation $\mathsf{new}(a, b)$ so that

1. For all sets a and all $b \subseteq \mathcal{U}$, $\mathsf{new}(a, b) \in \mathcal{U} - b$.
2. For all $a \neq a'$ and all $b \subseteq \mathcal{U}$, $\mathsf{new}(a, b) \neq \mathsf{new}(a', b)$.

The idea is that $\mathsf{new}(a, b)$ gives an urelement which is new in the sense that it does not belong to b. And for fixed b, the operation of giving new urelements for b is injective. Although the axiom only applies when b is a set of urelements, we can really use it in other cases: if b were not a subset of $\mathcal{U}$, then we would only need to consider $\mathsf{new}(a, support(b))$ to get urelements which not only didn't belong to b but also to any of the elements of b, etc.

Since this axiom posits the existence of a definable operation, what we mean here is that the basic vocabulary of our set theory must be increased by taking new as a function symbol.

Exercise 2.9 (Only for those who have seen consistency proofs in set theory.) Sketch a proof that the Strong Axiom of Plenitude is consistent relative to standard ZF set theory.

Assuming all our axioms, we have that for all $A \subseteq \mathcal{U}$, $V_{afa}[A]$ is a collection of sets which again satisfies all our axioms. The only exception to this is the Axiom of Plenitude: if A is small (i.e., a set), then of course $V_{afa}[A]$ will not satisfy this axiom, even in its weak form.

Transitivity and transitivity on sets One point which we should clarify has to do with urelements and the notion of transitivity. A transitive set may contain urelements. In fact, for every set a, $support(a) \subseteq TC(a)$. In some cases we want a modified notion of transitivity. We say that a class C is *transitive on sets* if whenever b and c are sets and $b \in c \in C$, then $b \in C$. For example, each class $V_{afa}[A]$ is transitive on sets but is transitive only if $A = \emptyset$.

2.5 The Axiom of Foundation

One of the main points of this book is that a number of circular phenomena may be modeled naturally using sets, provided that non-wellfounded sets are used. Put another way, the restriction to wellfounded sets makes the modeling of some phenomena awkward. In order to make this point clear, we need to discuss the Foundation Axiom of set theory, and to mention some of its more prominent consequences.

A binary relation R on a set S is *wellfounded* if there is no infinite sequence $b_0, b_1, b_2, \ldots$ of elements of S such that $b_{n+1} R b_n$ for each $n = 0, 1, \ldots$. If there is such a sequence, then R is said to be *non-wellfounded*, and such a sequence is called a *descending sequence* for R. A bit more generally, we say that the *wellfounded part of* R is the set of all $b \in S$ such that there is no descending sequence starting with b. So R is wellfounded if the wellfounded part of R is S. One important fact is that b is in the wellfounded part of R iff every b' such that $b' R b$ is also in the wellfounded part of R.

R is said to be *circular* if there is a finite sequence $b_0, \ldots, b_k$ such that $b_0 = b_k$ and $b_{n+1} R b_n$ for each $n = 1, \ldots, k$. Such a sequence is called a *cycle* in R. If there are no such cycles, then R is called *noncircular*. We begin with the following simple but important observation.

Proposition 2.5 *If a relation R is circular then it is non-wellfounded. In other words, if R is wellfounded then it is noncircular.*

Proof Let $b_0, b_1, \ldots, b_k$ be a cycle for R. We can get an infinite descending sequence by simply repeating this cycle indefinitely:

$$\ldots \quad b_0 \ R \ b_1 \ R \ \ldots \ R \ b_k \ (= b_0) \ R \ b_1 \ R \ \ldots \ R \ b_k.$$

$\dashv$

Example 2.3 The order relation $<$ on natural numbers is wellfounded, but the same relation, as a relation on the integers (positive and negative) is non-wellfounded, since $-1, -2, -3, \ldots$ is a descending sequence. However, $<$ is noncircular on the integers.

For every set a, we have a structure $\langle a, \in \rangle$, where "$\in$" here is the membership relation on the members of a. The Foundation Axiom (FA) states that for every set a, the structure $\langle a, \in \rangle$ is wellfounded.

PRINCIPLE OF PROOF BY $\in$-INDUCTION Suppose that $\varphi(a)$ is a property of sets, and that for all a,

If $\varphi(b)$ for all $b \in a$, then also $\varphi(a)$.

Then (using FA): for all a, $\varphi(a)$.

The idea is that any counterexample a to φ would be forced to contain an element a_0 which also did not satisfy φ. This a_0 would contain some a_1 with the same property, etc. In this way we would generate an infinite descending sequence, contradicting *FA*.

This principle depends crucially on *FA*, an axiom we will not have available to us in this book. But even if we drop *FA*, we still do have an induction principle, but instead of being able to conclude that a property holds for all sets, we must restrict attention to the wellfounded sets. We call a set *a wellfounded set* if the membership relation restricted to $TC(a)$ is wellfounded. The Axiom of Foundation is equivalent to the statement that all sets are wellfounded.

Even without the Axiom of Foundation, there are many wellfounded sets. Indeed, the same proof that shows that the class of all sets is a proper class shows that the class of all wellfounded sets is a proper class. (Verify this. The main observation you need, in addition to the earlier proof, is that any set of wellfounded sets is itself a wellfounded set.)

PRINCIPLE OF PROOF BY $\in$-INDUCTION ON WELLFOUNDED SETS Suppose that $\varphi(a)$ is a property of sets, and that for all wellfounded a,

$$\text{If } \varphi(b) \text{ for all } b \in a, \text{ then also } \varphi(a).$$

Then: for all wellfounded a, $\varphi(a)$.

Exercise 2.10 In addition to the principle above, there is a principle of Proof by $\in$-Induction (on all sets). To state it, one takes the principle above and drops the two occurrences of "wellfounded." Be sure you see the difference between the two principles. Show that the Principle of Proof by $\in$-Induction implies *FA*. (In contrast, the Principle for Wellfounded Sets does not imply *FA*.)

As we'll see in Chapters 3–5, there are a number of cases where one would like to model some circular phenomenon by means of equations. Unfortunately those equations can be shown to have no solutions, using *FA*. What we want to do here is to foreshadow those examples by mentioning a number of ways in which *FA* acts to prohibit circularity.

Proposition 2.6 *Assuming the Axiom of Foundation:*

1. *For all a, $a \notin a$.*
2. *There is no finite sequence $a_1, a_2 \ldots, a_n$ so that*

$$a_1 \in a_2 \in \cdots \in a_n \in a_1 .$$

3. *There are no a, b so that $a \in TC(b) \in a$.*
4. *There are no a, b so that $a \in TC(b)$ and $b \in TC(a)$.*
5. *If $c = \langle a, b \rangle$, then $c \neq a$, $c \neq b$, $c \notin a$, and $c \notin b$.*
6. *For all A, there are no non-empty X so that $X = A \times X$.*

7. *The only solution of $X = X \times X$ is $X = \emptyset$.*
8. *There are no functions f so that f belongs to the domain of f.*
9. *In fact, it is impossible to find a finite sequence of functions*

$$f_1 : A_1 \to A_2 \qquad f_2 : A_2 \to A_3 \qquad f_n : A_n \to A_1$$

and an element a of A_1 so that

$$f_n f_{n-1} \cdots f_2 f_1(a) \quad = \quad f_1 .$$

10. *For every A, there are no solutions to $X = X \to A$.*

Proof Most of the parts of this are easy. The last part follows from the next proposition. ⊣

In Chapter 8 we shall return to Proposition 2.6 to see how the situation changes when we adopt the Anti-Foundation Axiom. As it happens, there are some failures of circularity which are even more basic because they are forbidden by the axioms of set theory without using *FA*.

Proposition 2.7 *In set theory,* even without the Axiom of Foundation*:*

1. *If F is a function then $\{x \in dom(F) \mid x \notin F(x)\}$ is not in the range of F.*
2. *If F is a function from a set X to its power set, then consider the relation R on X defined by xRy iff $x \in F(y)$. Then*

$$Y \quad = \quad \{x \in X \mid x \text{ is in the wellfounded part of } R \}$$

is not in the range of F.

3. *There are no solutions to $X = \mathcal{P}(X)$.*
4. *The only solution to $X = \emptyset \to X$ is $X = \{\emptyset\}$.*
5. *$X = X \to A$ has no solutions unless A is a singleton $\{a\}$ and X is a singleton $\{f\}$ such that $f(f) = a$.*
6. *Every solution to $X = X \to X$ is a singleton $\{x\}$ satisfying $x = \{\langle x, x \rangle\}$.*

Proof Part (1) is a basic result of set theory known as Cantor's Theorem. If F is the identity function, this is really the Russell paradox argument given earlier to show that the class of all sets is large. We leave the similar argument here to you.

For (2), suppose that Y were in the range of F. Say that $Y = F(y)$. We claim that $y \in Y$. To see this, suppose that xRy. Then by definition of R, $x \in F(y) = Y$. So R is wellfounded below x. But if R is wellfounded below all such x's, it is wellfounded below y, too. We therefore know that $y \in Y$. This means that yRy. And we get a descending sequence $yRyRy\ldots$. So $y \notin Y$ after all.

Part (3) follows immediately from (1) and from (2). The standard proof is via (1), but the argument of (2) is one we will see several times in the book.

For part (4), note that the empty set is a function, and indeed for all X, $\emptyset$ is the only function from $\emptyset$ to X.

Turning to part (5), if A has at least two elements, then there is a map of $X \to A$ onto $\mathcal{P}(X)$. There can be no map from X onto $X \to A$ since composing the two would give us a map of X onto $\mathcal{P}(X)$, contradicting (1). Also, (4) shows that $X = \emptyset$ is not a solution to $X = X \to A$. So any solution would have to be a singleton, say $X = \{f\}$. Then A is in one-to-one correspondence with $X \to A$, so A must be a singleton as well.

Finally, (6) follows from (5).   ⊣

Exercise 2.11 Sometimes it is sufficient to have a natural isomorphism instead of equality. Fix a set A. Show how to solve $X \cong A \times X$ where $\cong$ is as natural a bijection as you can make it. [Hint: the idea is that if $X \cong A \times X$, then also $X \cong A \times A \times X$, etc. So we would like to think that

$$X \quad \cong \quad A \times A \times \cdots$$

At least we would like to formalize something like this.]

Part of the technical thrust of our work is to show how one can turn $\cong$ into identity. See Chapter 14.

2.6   The axioms of set theory

Although we will be working informally in this book, we cannot in good conscience proceed without at least stating the axioms of set theory and making a few comments about them.[4]

An important feature of the axioms of set theory is that they are entirely *first-order*. This means that they are statements which can ultimately be written out in logical notation, using the propositional connectives $\neg$, $\wedge$, $\vee$, $\to$, and $\leftrightarrow$; the equals sign $=$ and the non-logical relation symbols $\in$ and $\mathcal{U}$; the function symbol new; quantifiers $\forall$ and $\exists$; and variables ranging over objects. This contrasts with other famous theories like (second-order) Peano Arithmetic, where we need quantifiers over arbitrary subcollections of the domain.

The basic axioms of set theory, listed in the box, constitute the system ZFC^- (Most studies in set theory do not use urelements, so the Axioms of Urelements and Strong Axiom of Plenitude would be dropped, and some of the others would change a bit.) The system ZFC adds the Foundation Axiom to this list, and the system ZFA which we will study adds the Anti-Foundation Axiom; we'll get to this in Chapter 6.

One feature of the Collection and Separation Schemes is that they are not single sentences at all, but rather infinite schemata. This means that for every

---

[4]That is, we will not be giving formal proofs from the set of first-order axioms.

The Axioms of ZFC^-

Urelements $(\forall p)(\forall q)[\mathcal{U}(p) \to \neg(q \in p)]$.

Extensionality $(\forall a)(\forall b)[(\forall p)(p \in a \leftrightarrow p \in b) \to a = b]$.

Pairing $(\forall p)(\forall q)(\exists a)[p \in a \wedge q \in a]$.

Union $(\forall a)(\exists b)(\forall c \in a)(\forall p \in c)\, p \in b$.

Power Set $(\forall a)(\exists b)(\forall c)[c \subseteq a \to c \in b]$

Infinity $(\exists a)[\emptyset \in a \wedge (\forall b)[b \in a \to (\exists c \in a)\, c = b \cup \{b\}]]$.

Collection
$(\forall a)(\forall p \in a)(\exists q)\varphi(a, p, q) \to (\exists b)(\forall p \in a)(\exists q \in b)\varphi(a, p, q)$.

Separation $(\forall a)(\exists b)(\forall p)[p \in b \leftrightarrow p \in a \wedge \phi(p, a)]$.

Choice $(\forall a)(\exists r)[r$ is a wellorder of $a]$.

Strong Plenitude
$(\forall a)(\forall b)\,[\,\mathcal{U}(\mathsf{new}(a, b))\ \wedge\ \mathsf{new}(a, b) \notin b$
$\wedge\ (\forall c \neq b)[\mathsf{new}(a, b) \neq \mathsf{new}(c, b)]]$.

formula ϕ in the language we get a new axiom. In our case, the formulas are permitted to have not only the logical symbols mentioned above, and $\in$, but also $\mathcal{U}$ and new. They are also permitted to have variables besides the ones shown below (except for b).

There is a scheme called the Replacement Scheme which is often found in the standard axiomatizations of set theory. It differs from the Collection Scheme in that the existential quantifier $(\exists q)$ is replaced by a unique-existential quantifier. In ZF, the Replacement Scheme implies the Collection Scheme. But the proof of this uses the Foundation Axiom. So our basic set theory includes the (apparently) stronger Collection Scheme.

There is a strong form of the Axiom of Choice called the *Global Axiom of Choice*: there is a definable bijection $G : On \to V_{afa}[\mathcal{U}]$. This strengthening is a parallel to what we saw with the Axiom of Plenitude in that it asserts the existence of a single predicate which globally does the job that a proper class of sets would have to do. It is known that the AC does not imply the Global AC. Assuming the consistency of the other axioms, it is consistent to add the

Global AC to our set theory. We do not need the Global AC for the results of this book, but we mention it in connection with Exercise 2.12 below.

While the axioms of set theory could be written out fully in logical notation, it is much more convenient to use abbreviations. For example, we write $(\forall x \in a)\phi$ to mean $(\forall x)(x \in a \to \phi)$. Another source for abbreviations comes from our conventions on variables (see page 22). For example, $(\exists x)\phi$ actually means $(\exists x)(\mathcal{U}(x) \wedge \phi)$. In the axioms below, we also use abbreviations that come from set theory itself. For example, $a \neq \emptyset$ is abbreviation for $(\exists p)p \in a$.

This is a fairly simple abbreviation; a more complicated one would be the use of "r is a wellorder relation on a" for its rather long official definition.

Exercise 2.12 This exercise gives equivalent formulations of the Axiom of Plenitude and the Strong Axiom of Plenitude.

1. Prove that the Axiom of Plenitude is equivalent to the assertion that the class $\mathcal{U}$ of urelements is a proper class. You will need to use the Axiom of Choice.

2. Prove that the Strong Axiom of Plenitude implies that there is a proper class $C \subseteq \mathcal{U}$ and a proper class $<$ of pairs of elements of C which is a wellorder of C.

3. Assume the Global Axiom of Choice. Prove that the Strong Axiom of Plenitude is equivalent to the assertion that there is a proper class of urelements.

Historical Remarks

The material of this chapter is quite standard, with the exception of our formulation of the Axiom of Plenitude. We read Proposition 2.7.2 in d'Agostino and Bernardi (to appear). Claudio Bernardi has informed us that the result is actually due to Smullyan, who presented it in seminars several years ago.

Part II

Vicious Circles

3

Circularity in computer science

The sensible practical man realizes that the questions which he dismisses may be the key to a theory. Furthermore, since he doesn't have a good theoretical analysis of familiar matters, sometimes not even the concepts needed to frame one, he will not be surprised if a novel situation turns out to be genuinely problematic.

G. Kreisel, "Observations on popular discussions of foundations."

This part of the book (Part II) describes a number of phenomena which are essentially circular, or otherwise non-wellfounded, and so are natural candidates for mathematical modeling via non-wellfounded sets.

In this chapter, we begin with examples from computer science. Pedagogically, the most important example is that of streams, since we use it throughout the book as a source of examples and motivation. The example of labeled transition systems is also important. The other examples in this section are here mainly to give a feeling for other uses to which the theory of hypersets is being put in computer science. We begin with examples from computer science because they have been quite important sociologically. Some logicians explored set theories with non-wellfounded sets over the past century, but the resulting work was outside the mainstream until these theories were used to develop elegant mathematical treatments of circular phenomena in computer science.

In Chapter 4 we turn to circular phenomena in philosophy and related disciplines. Our goal is to suggest uses to which hypersets might be put in developing rigorous accounts of philosophically central notions. We turn briefly to linguistics, economics, and mathematics to point out some circular phenomena in some different but related fields.

Finally, in Chapter 5 we turn to consider the paradoxes that led logicians to hold circular phenomena to be highly suspicious. This is important in understanding why the theory of hypersets has been so long in coming. We return to apply the theory of hypersets to the paradoxes in Chapters 12 and 13.[1]

3.1 Streams

Let A be some set. By a *stream* over A we mean an ordered pair $s = (a, s')$ where $a \in A$ and s' is another stream.[2] As the name suggests, we think of a stream as being an element of a followed by another stream. Two important operations on streams s are taking its first element $1^{st}s$, which gives an element of A, and taking its second element $2^{nd}s$, which yields another stream.

Here is a simple procedure which we can think of as defining some streams: Define f by:
$$f(n) = (n, f(n + 1)).$$
Then for any n, $f(n)$ is a stream. For example, $f(0)$ is the stream
$$(0, (1, (2, \dots))).$$
It would be easy enough for a programmer to write a program which had this stream as its (ideal) output.

We would like to model streams by using the ordinary pairing operation of set theory to represent pairs: $(x, y) = \langle x, y \rangle$. But there is a problem. Consider the constant stream c_a which is a followed by c_a itself. Then in our modeling, $c_a = \langle a, c_a \rangle$. But then c_a would be non-wellfounded, by Proposition 2.6(3).

The intuitive concept of a stream, then, when modeled in a mathematically simple way, leads to non-wellfounded sets. At this point, one could re-work the modeling; the natural choice is probably to take streams over A to be functions $f : N \to A$. Doing things that way would run counter to the intuition that streams are pairs, however. So in this book, we want to explore the possibility of *directly* modeling streams as pairs, and we'll have a set theory that makes this possible.

Once again, we want to think of a stream s as having *two* parts, $1^{st}s$ and $2^{nd}s$, and not as a function from N to A. Thus, if we let A^∞ be the streams over A, then we should have $A^\infty \subseteq A \times A^\infty$. Actually, the converse should also be true, since any element of A followed by a stream would be a perfectly good stream. So we really want A^∞ to satisfy the equation

(1) $$A^\infty \quad = \quad A \times A^\infty.$$

---

[1] Actually, most of these chapters could be read following Chapter 6.

[2] We use round brackets for the intuitive operation for forming ordered pairs, and angle brackets for the set theoretic operation modeling this operation.

Equations like this will be a central focus of our work in Chapter 15.

One issue we'll have to face is how to define operations on streams. For example, consider the map

$$zip : A^\infty \times A^\infty \to A^\infty$$

such that all $s, t \in A^\infty$,

(2) $zip(s, t) \quad = \quad \langle 1^{st}(s), \langle 1^{st}(t), zip(2^{nd}(s), 2^{nd}(t)) \rangle \rangle.$

As its name suggests, zip takes two streams s and t and returns the stream u which is the first element of s followed by the stream v which is the first element of t followed by .... One test of our modeling is whether functions such as zip can be defined, and whether we can prove the things about them that are intuitively true.

There is something of a recursive character to (2) despite the fact that there is no "base case." When we model streams as functions on N, this feature is largely lost. As we'll see in Chapter 14, it turns out that ZFA provides some tools that make working with streams quite natural. For example, there is a direct way in which (2) turns out to define zip.

3.2 Labeled transition systems

It is always hard to say just where an idea begins, but some important parts of the theory of hypersets can be traced at least as far back as the early days of automata theory, as will be seen in what follows.

Let *Act* be a set of *atomic actions*. We want to model the notion of a system $\mathcal{T}$ which has a state at each moment, and which changes its state according to evolution by the atomic actions. Whether these changes of state are due to factors inside or outside of $\mathcal{T}$ is not captured in this model. What will be captured is that in certain states, certain types of actions are possible, and each of those actions has certain possible outcomes. The evolution in our models will be non-deterministic. This means that if the system is in a certain state s and an action a is performed, then there might be more than one possibility for the next state s' of the system.

Here is our formal definition of a model of this kind of system:

Definition A *labeled transition system (lts)* over *Act* is a set pair $\mathcal{T} = \langle S, \delta \rangle$, where S is a set, and $\delta : S \to \mathcal{P}(Act \times S)$ is a *(nondeterministic) transition function.*

As you see, for each state s, $\delta(s)$ is a set of pairs. If the pair $\langle a, t \rangle$ belongs to $\delta(s)$, this means that under the action a, the system can evolve to state t. To show this more graphically, we often write $s \xrightarrow{a} t$ for $\langle a, t \rangle \in \delta(s)$. These systems are called *nondeterministic* since for a given s and a, there might be more than one t such that $s \xrightarrow{a} t$. There might also be no such t.

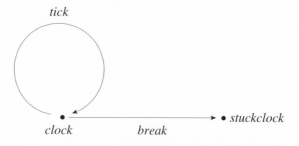

FIGURE 1 A picture of a labeled transition system.

A labeled transition system is often depicted by means of a diagram of the sort shown in Figure 1. This depicts a device C with two states, called *clock* and *stuckclock*, with two actions, *tick* and *break*. The idea is that C is a device which could either be a normal clock or a broken clock; these are the states of C. As a clock, C could undergo two actions: it could tick, after which it would again be a normal clock, or it could break and turn into a stuck clock. As a stuck clock, C can do nothing. Formally the transition function δ of C is given as follows:

$$\delta(clock) \quad = \quad \{\langle tick, clock\rangle, \langle break, stuckclock\rangle\},$$

and $\delta(stuckclock) = \emptyset$.

It is frequently convenient to specify a device D by its behavior alone. What this amounts to is the identification of devices with the same behavior. When defining the clock in Figure 1, for example, one might write

(3) $\qquad clock \quad = \quad tick.clock + break.stuckclock$

The equals sign in (3) is both suggestive and challenging. It suggests that the clock's state *clock* just *is* its behavior. At this level of modeling, any two clocks that work in this way, however different their internal mechanisms, are seen as different instances of the same abstract device.

To push this point further, suppose we also had a device C' behaving as follows:

(4) $\qquad \begin{aligned} clock' \quad &= \quad tick.clock'' + break.stuckclock' \\ clock'' \quad &= \quad tick.clock' + break.stuckclock' \end{aligned}$

Notice that this clock has three states, not two. A *tick* takes it back and forth between two states, and a *break* takes it to the state of being *stuckclock*. (Draw the graph of this device.)

Suppose you are a scientist who wants to tell the difference between C and

C' (or for that matter, that you wanted to test whether C' is in the *clock'* or *clock"* state). Since you are not a watchmaker, you cannot open C or C' up. Your only acts of observation are to observe their actions, that is watch them tick, or to break them. There is a sense in which you would not be able to tell the difference between C and C'. This basic idea is captured by the following important definition.

Definition Let $\mathcal{T}_1$ and $\mathcal{T}_2$ be labeled transition systems with the same set of actions. A *bisimulation* of $\mathcal{T}_1$ and $\mathcal{T}_2$ is a binary relation $R \subseteq S_1 \times S_2$ between the state sets so that:

1. If $s_1 R s_2$ and $s_1 \xrightarrow{a} t_1$ in S_1, then there is some t_2 so that $s_2 \xrightarrow{a} t_2$ in S_2 and $t_1 R t_2$.
2. If $s_1 R s_2$ and $s_2 \xrightarrow{a} t_2$ in S_2, then there is some t_1 so that $s_1 \xrightarrow{a} t_1$ in S_1 and $t_1 R t_2$.

We say that $\mathcal{T}_1$ and $\mathcal{T}_2$ are *bisimilar* if there is a bisimulation relation between the two which relates every $s_1 \in S_1$ to some $s_s \in S_2$, and which also relates every $s_s \in S_2$ to some $s_1 \in S_1$.

The idea is that if s_1 and s_2 are related by a bisimulation, then an outside observer cannot tell the two apart because every state that one device could evolve into could be matched by an action of the other device. Returning to our clock devices C and C', here is bisimulation relating the *clock* state of C with the *clock'* state of C':

$$\{\langle clock, clock'\rangle, \langle clock, clock''\rangle, \langle stuckclock, stuckclock'\rangle\}$$

If we want to characterize a device by its behavior under actions, then a desirable goal in modeling labeled transition systems is that bisimilar systems should be identified. An elegant way to do this is to find a *canonical form* for devices so that bisimilar devices have identical canonical forms. Then the devices which are in canonical form capture the essence of the behavioral view of the device.

Carrying out this project requires us to say exactly what canonical devices are, and this means that we must specify the states as fixed, determinate, mathematical objects. In our clock example, order to get bisimilar objects to be identical, not just similar in some sense (isomorphic), we need *clock* and *clock'* to be the very same object.

Further, and here is where (3) is a challenge, if we are working in set theory, we would like to take the dot in (3) as a notation for forming ordered pairs, and the plus sign is a shorthand for disjoint union. So the right-hand side of (3) becomes

$$\{\langle tick, clock\rangle\} \cup \{\langle break, stuckclock\rangle\}$$

We would like *clock* to be exactly this set. This is not a *necessary* part of the definition of labeled transition system, but does seem like a good candidate for a natural way to model the the canonical *clock*.

Definition A labeled transition system $\mathcal{T}$ is *canonical* if for each state $s \in S$,

$$s \;=\; \{\langle a, t \rangle \in Act \times S \mid s \xrightarrow{\;a\;} t\}.$$

In other words, $\mathcal{T}$ is canonical if its transition function δ is the identity function on the state set S. If S is a canonical labeled transition system which is bisimilar to $\mathcal{T}$, then we say that S is a *canonical form* of $\mathcal{T}$.

Exercise 3.1 An lts $\mathcal{T}$ is *deterministic* if for each state s and each $a \in Act$ there is a unique state t such that $\langle a, t \rangle \in \delta(s)$. Let $\mathcal{T}$ be a canonical, deterministic labeled transition system. Prove that each state of $\mathcal{T}$ is a function from Act to the states of $\mathcal{T}$. Is this true in general? [There is really nothing to prove here. We state it as a problem so that you can check to see if you understand the discussion above.]

Let Can be the set of canonical states over a fixed set Act. At this point, it is useful to put down an equation like (1) for Can. This would be:

$$(5) \hspace{3cm} Can \;=\; \mathcal{P}(Act \times Can)$$

Once again, the intention is that while the states of a labeled transition system are arbitrary, in a canonical labeled transition system the states are just their essential mathematical content. We would like our modeling to have the feature that every labeled transition system is bisimilar to a canonical one. However, this is clearly not going to be possible unless we adopt some alternative to the Axiom of Foundation. Our canonical *clock* state, for example, is clearly not a wellfounded set. More generally, whenever we had a device that could undergo an infinite sequence of actions, the device could not possibly have a canonical form, since this would be necessarily non-wellfounded. More generally, we have the following negative result:

Proposition 3.1 *Assume the Axiom of Foundation. Let $\mathcal{T}$ be a labeled transition system. Then the following are equivalent:*

1. *$\mathcal{T}$ has a canonical form.*
2. *There are no infinite sequences $a_0, a_1, a_2, \ldots$ of acts and $s_0, s_1, s_2, \ldots$ of states so that in $\mathcal{T}$,*

$$s_0 \xrightarrow{\;a_0\;} s_1 \xrightarrow{\;a_1\;} s_2 \xrightarrow{\;a_2\;} \cdots$$

Proof Suppose first that $\mathcal{T}$ contains infinite sequences of acts and states as in (2) but that $\mathcal{T}$ were bisimilar to a canonical labeled transition system $\mathcal{T}'$ via the relation R. Suppose $s_0 R s_0'$. Then by the bisimulation condition, there would be some s'_1 so that $s'_0 \xrightarrow{a_0} s'_1$. Since $\mathcal{T}'$ is canonical, this means that $\langle a_0, s'_1 \rangle \in s'_0$. In particular, the s'_1 belongs to the transitive closure of s'_0. And we can continue this way indefinitely, building a sequence of states, each of which belongs to the transitive closure of its predecessors. This contradicts the Axiom of Foundation.

Going the other way, suppose (2) holds. Let $\rightarrow$ be the union of the relations $\xrightarrow{a}$. Then $\rightarrow$ is well-founded. We define a function f on S by recursion on $\rightarrow$, so that[3]

$$f(s) = \{ \langle a, f(t) \rangle \mid s \xrightarrow{a} t \}.$$

Then we get a canonical $\mathcal{T}'$ by taking as states the sets $f(s)$ for $s \in S$. Moreover, f, considered as a set of ordered pairs, provides the needed bisimulation. This proves (1). ⊣

A theme of this book is to find canonical versions of all the objects we study. Assuming AFA, we'll be able to prove that every labeled transition system, whether it has infinite sequences of actions or not, is bisimilar to a unique canonical labeled transition system.

Exercise 3.2 (For those familiar with automata.[4]) An *automaton (over Act)* is a tuple

$$\mathcal{A} = \langle S, \delta, q_0, F \rangle$$

such that $\langle S, \delta \rangle$ is a labeled transition system over Act, $q_0 \in S$, and $F \subseteq S$. The state q_0 is called the *initial state* and F is called the set of *accepting states*.

1. Define a notion of bisimulation for finite automata over Act. In doing this, keep in mind that part of the observable behavior of a finite automaton is whether or not it is in an accepting state. Show that bisimilar automata accept the same sequences of actions.

2. Recall (or look up) the standard proof of the Myhill-Nerode Theorem on minimizing finite automata. Verify that the proof is by constructing a bisimilar automata to the automata to be minimized.

3. Define a *canonical automaton* to be one such that each state s is a set of triples $\langle i, a, t \rangle$, where $i = 1$ (for accepting states) or $i = 0$ (for non-accepting states), $a \in Act$, and t is a state, and $s \xrightarrow{a} t$. Give an example of a finite automaton that is not bisimilar to any canonical one, under the assumption of the Axiom of Foundation.

4. State the analogue of (5) for automata.

---

[3] If you have not studied definition by recursion on wellfounded relations, you may skip this proof with no loss of continuity.

[4] We assume familiarity with the notion of *recognition of a finite sequence from Act* in this exercise. Usually in automata theory one studies *finite* automata only. However, in this book we will not discuss any results which need the finiteness of automata, and so we allow our automata to have infinite state sets.

If you are interested in previewing our later results on automata and hyper-sets, see Section 18.3.

3.3   Closures

The material of this section is based on Milner and Tofte (1991) which uses hypersets to model the semantics of a simple functional programming language. It will not be referred to later in the book and so can be skipped without loss of continuity. It is included here since it is a nice application of non-wellfounded sets in computer science with a different flavor than the other examples in this section.

Let L be a language which is rich enough to contain expressions such as

(6)
```
fix factorial(n) =
    (1 if (n = 0)
            (else times(n, factorial(pred n)))))
```

which allow one to define the factorial function by recursion. The expression (6) says that to compute the factorial of a number n, first see if $n = 0$. If $n = 0$, the output is 1. If $n \neq 0$, then the output is n multiplied by the factorial of $n - 1$. This *recursive call* is a central feature of programming languages. The language has constants such as 0 and `pred`, and it also has variables like n. Also, `factorial` is a variable symbol of the language. Intuitively, it should range over functions, but this is getting ahead of the story. The constants and variables are examples of expressions, as is a function application such as `pred(n)`. The main syntactic feature is that the if `exp` is an expression and `f` is a function variable, then `fix f = exp` is again an expression.

Suppose we want to give a semantics for L, that is, a rigorous mathematical account of the meanings of its expressions. To do this, we need to consider the meanings not just of the constants, but also of the variables. In fact, we'll also want some meaning for the expression (6) even without any arguments to apply it to. Let's see what we need. We might as well interpret the constants as elements of a set $Const$ which contains the natural numbers and the predecessor function. We are going to need to interpret expressions, and $Const$ is not the right set for this. Following standard usage, we'll call the set of interpretations of expressions *closures*, and we denote this by $Clos$. We'll have more to say about closures below. The set of possible values of all expressions is therefore $Val = Const + Clos$, where $+$ again denotes disjoint union.

Variables, whether for numbers or functions, cannot be interpreted on their own. It takes an *environment*, a finite partial function from variables to values to get this going. For example, to interpret $(f\ n\ m)$, we would need to know the interpretations of f, n, and m. In an environment like

$$\{\langle \mathtt{f}, \times \rangle, \langle \mathtt{n}, 5 \rangle, \langle \mathtt{m}, 3 \rangle\}$$

f is assigned the multiplication operation, and n and m are assigned particular numbers. Using this environment, we could evaluate the expression (f n m) to get the number $5 \times 3 = 15$.

To interpret expressions, let's consider (6) again. The interpretation of (6) is going to have to be an environment E which in turn will give the meanings of 0, pred, etc. The interpretation will also have to tell us what to do to the environment, and for that it would be natural to give the expression exp on the right hand side of the identity sign. Finally, the variable n has to be mentioned somewhere. The upshot is that we would like for the closure which interprets (6) to be a triple $\langle \mathtt{n}, exp, Env \rangle$, where Env is itself an environment.

To summarize, we would like for the semantic functions to satisfy the following equations:

$$
\begin{array}{rcl}
Val & = & Const + Clos \\
Env & = & Var \rightharpoonup Val \\
Clos & = & Var \times Exp \times Env
\end{array}
$$

Now here's the rub. The closure c which interprets (6) in the empty environment is going to have to be the triple $\langle \mathtt{n}, exp, Env \rangle$ for some environment Env. But Env is going to have to tell us the meaning of everything involved in exp, including factorial. And the way we are setting things up, the meaning of factorial should be c itself. So we see that

$$c \quad = \quad \langle \mathtt{n}, exp, \langle \mathtt{factorial}, c \rangle \rangle.$$

Of course, c is not well-founded.

Milner and Tofte (1991) use the theory of hypersets to develop a rigorous mathematical semantics for a language of this kind. We will not go into it here, hoping that this short introduction will alert the interested reader so he or she can pursue the topic after reading Part III of this book.

The subject of functional programming suggests the untyped lambda calculus, a field where, intuitively, there is a universe D in which every object is a function f defined on the whole universe D. So $f \in D$ can apply to all of the functions in D, including itself. The standard modeling of functions as pairs in set theory means that every $f \in D$ would be non-wellfounded.

As far as is currently understood, the theory of hypersets will not be much help in studying the lambda calculus. It is possible that some of the ideas might be suggestive, but it is nevertheless true that even if one adopts the axiom *AFA* which we study, there still are no non-trivial sets D which are literally subsets of $D \rightarrow D$ (Exercise 14.3).

3.4 Self-applicative programs

Is there a computer program P with the property that when run, P produces itself as output? We call such a program *self-producing*. At first glance, it seems improbable that there are any self-producing programs. It also seems that whether or not these programs exist might depend on the exact programming language L used. In this section we discuss a surprising result: provided that L is "reasonable," it will contain self-producing programs.

We are not so much interested in self-producing programs as in reviewing the background results of computability theory that are needed to study the issue. These results include the S_n^m-Theorem and the Recursion Theorem. This review will lead us to a related topic: programs designed to be applied to themselves. This might seem like yet another topic which is theoretically fascinating but of no practical value. But here, too, there are surprises. As we'll see, there are issues in *compiler generation* which are specifically framed in terms of a certain kind of *self-applicative* program.

The material of this section is not needed in the rest of this book. This is because our work does not depend on any results of computability. Conversely, computability theory does really not depend on set theory in any essential way.[5] We are presenting it for a few reasons: some of the issues below, like fixed-points and self-application, are going to come up in the book. On a technical note, a version of the Recursion Theorem will reappear in the very last part of the book, in Chapter 20. Finally, we feel that there may be applications of the techniques of this book to the semantical issues raised by self-application. We hope that some ambitious readers decide to pursue this topic.

In order to speak about self-producing or self-applicative programs, we must have a programming language which can manipulate expressions written in that language. It would not do to have a language whose commands were written in Roman letters while the actual programs only dealt with numerals. We'll make a few assumptions at this point. All of the languages we consider will use the same symbol set. We'll assume that this contains both numerals and letters of the alphabet, though nothing much hinges on this choice. The objects in each language will be called *programs*; this means that numerals count as programs for us. Each program P has a *semantics* $[\![P]\!]$. This semantics will be a partial function of zero or more programs. When defined, its output is always another program.

We also assume that our languages have the following properties:

(Completeness) For every computable partial function on programs, say,

---

[5]This is not true of the study of the *semantics of computation*, where one might choose to use non-wellfounded sets to model some phenomenon or other. This is the point of the preceding sections of this chapter. But the classical theory of computability is not explicitly concerned with issues of modeling.

$f(P_1, \ldots, P_k)$ there is a program Q of L which computes f in the sense sense that $[\![Q]\!](P_1, \ldots, P_k) \simeq f(P_1, \ldots, P_k)$. Conversely, the meaning of every program is a computable operation.

$(S_n^m$ property) For all $m, n \geq 0$ there is a program S_n^m of L such that for all $R, P_1, \ldots, P_m, Q_1, \ldots, Q_n$,

$$[\![[\![S_n^m]\!](R, P_1, \ldots, P_m)]\!](Q_1, \ldots, Q_n) \simeq [\![R]\!](P_1, \ldots, P_m, Q_1, \ldots, Q_n).$$

The completeness condition is obviously not precise. What it means is that L is sufficiently expressive to write all possible computer programs. You may assume that your favorite programming language is complete. The S_n^m property is also quite standard: it basically says that the operation of plugging in constants for variables is computable. A complete language with the S_n^m property will be called *Turing-complete*. From the assumption of Turing-completeness, we derive one of the most important results of recursive function theory, Kleene's Recursion Theorem.

Theorem 3.2 (Recursion Theorem) *Let L be Turing-complete. Let $n \geq 0$, and suppose that $P(Q, R_1, \ldots, R_n)$ is a program of $n + 1$ inputs. Then there is a fixed program Q^* such that for all $R_1, \ldots, R_n$,*

$$[\![P]\!](Q^*, R_1, \ldots, R_n) \quad \simeq \quad [\![Q^*]\!](R_1, \ldots, R_n).$$

Proof We only work with the case $n = 1$. Consider the function $f(Q, R) \simeq [\![P]\!]([\![S_1^1]\!](Q, Q), R)$. This is a computable function of two arguments, so by completeness there is some program Q_0 that computes it. Let Q^* be the program $[\![S_1^1]\!](Q_0, Q_0)$. Then for any program R,

$$
\begin{aligned}
[\![P]\!](Q^*, R) \quad &\simeq \quad [\![P]\!]([\![S_1^1]\!](Q_0, Q_0), R) && \text{by definition of } Q^* \\
&\simeq \quad [\![Q_0]\!](Q_0, R) && \text{by definition of } Q_0 \\
&\simeq \quad [\![[\![S_1^1]\!](Q_0, Q_0)]\!](R) && \text{by the } S_n^m \text{ property} \\
&\simeq \quad [\![Q^*]\!](R) && \text{by definition of } Q^*
\end{aligned}
$$

This rather tricky reasoning completes the proof. ⊣

Among the many consequences of the Recursion Theorem is the solution to the problem at the beginning of this section.

Proposition 3.3 *Let L be Turing-complete. There is a program P^* of L such that $[\![P^*]\!]_L(\ ) \simeq P^*$. That is, P^* is a program of no inputs which outputs itself.*

Proof Let $f(P)$ be the partial recursive function $f(P) \simeq P$. By the Recursion Theorem, there is a program P^* such that

$$[\![P^*]\!]_L(\ ) \simeq f(P^*) \simeq P^*,$$

as desired. ⊣

So our solution means that to write a self-producing P, all you have to do is write S_0^1. In fact, this is usually not a trivial matter.

Exercise 3.3 Consider the following relation on programs of L:

$$H(Q, R) \quad \text{iff} \quad [\![R]\!](Q) \text{ is defined.}$$

$H(Q, R)$ means that if R is a program of one input, and R is run on Q, then R will eventually halt.

1. Prove that there is a program Q so that $H(Q, R)$ iff $Q = R$. That is, there is a program halts only on itself. [Hint: Use the Recursion Theorem.]
2. Prove that the wellfounded part of H is not a recursively enumerable set. That is, there is no program Q so that for all R, $[\![Q]\!](R)$ is defined iff R is in the wellfounded part of H.

We now come to the other topic of this section: the application of the S_n^m property to give the theoretical background behind compiler generators.

Definition Let L and M be languages.

1. *int* is an *L-interpreter for M* if for all P and Q,
   $[\![int]\!]_L(P, Q) \simeq [\![P]\!]_M(Q)$
2. *comp* is an *L-compiler of M* if for all P and Q,
   $[\![[\![comp]\!]_L(P)]\!]_L(Q) \simeq [\![P]\!]_M(Q)$.
3. *mix* is an *L-partial evaluator for M* if for all P,Q, and R,
   $[\![[\![mix]\!]_L(P, Q)]\!]_L(R) \simeq [\![P]\!]_M(Q, R)$.
4. *cogen* is an *L-compiler generator for M* if whenever *int* is an *L*-interpreter for M, then $[\![cogen]\!]_L(int)$ is a compiler for M.

The idea of an interpreter *int* for M is that although it works as a program of L, it pretends to be a program of M. A compiler is a translator, so it does more work than an interpreter. The idea behind a partial evaluator is contained in the S_n^m property. A compiler generator is the best of all: it turns an interpreter into a compiler. The interest in compiler generators is that for real programming languages, interpreters are much easier to write than compilers. So writing one efficient compiler generator would be an easy way to automatically get compilers.

Exercise 3.4 Using only the definition of Turing-completeness, prove that interpreters, compilers, and partial evaluators exist.

Theorem 3.4 (Futamura) *Let L and M be Turing-complete. There is an L-compiler generator cogen for M. In fact, given a partial evaluator mix, we can use cogen $= [\![mix]\!](mix, mix)$.*

Proof We'll show more than what is stated. First, let *int* be an *L*-interpreter for M and P be a program of M. For convenience, we drop the "L" subscripts.

We see that for all P and Q,

$$[[\![mix]\!](int, P)]\!](Q) \quad \simeq \quad [\![int]\!](P, Q) \quad \simeq \quad [\![P]\!]_M(Q).$$

Second, we claim that $[\![mix]\!](mix, int)$ is an L-compiler for M. The reason is that for any program P,

$$[[\![mix]\!](mix, int)]\!](P) \quad \simeq \quad [\![mix]\!](int, P),$$

and as we have seen, applying this to Q gives $[\![P]\!]_M(Q)$. Finally, *cogen* is a compiler generator because for all L-interpreters int for M,

$$[\![cogen]\!](int) \simeq [[\![mix]\!](mix, mix)]\!](int) \simeq [\![mix]\!](mix, int).$$

As we have just seen, this is an L-compiler for M. ⊣

Futumura's Theorem first appeared in print in 1971. It is the very beginning of a field of study known as *partial evaluation*. However, the result is not of practical value by itself: unless one works hard, the program *mix* will be very inefficient (and thus so will *cogen*). The usual ways to establish the S_n^m property for a real programming language would not give an efficient *mix*. Most of the work in partial evaluation is concerned precisely with cleverness in writing *mix* programs. These programs are designed to be self-applicative. In fact, to get *cogen* we need to apply *mix* to itself and then apply *that* program to *mix*. So partial evaluators are important examples of self-applicative programs.

3.5   Common themes

Though the examples in this chapter differ with regard to motivation, with the exception of self-applicative programs they do have a couple of important features in common. Conceptually, they all posit a domain of structured objects. The objects have the ability to change or evolve in such a way that they sometimes return to the original state. And the objects are all capable of referring, pointing, or evolving, to *themselves*.

Mathematically, each of the examples call for a solution to a *fixed point equation* involving sets. The model of streams over A was a set A^∞ satisfying

$$A^\infty \quad = \quad A \times A^\infty.$$

Similarly, the set of all canonical labeled transition systems over *Act* would be a set C satisfying

$$C \quad = \quad \mathcal{P}(Act \times C).$$

Strictly speaking, in both of these examples our discussions only lead us to seek sets satisfying slightly weaker conditions. For streams we asked for A^∞ so that $A^\infty \subseteq A \times A^\infty$, and similarly for C we had the analogous inclusion. However, a case could be made that the intuitive notion in each case would be the *largest possible* set satisfying the appropriate inclusion, and such a set would satisfy the equation, as we'll see in Chapter 15.

The closure example was a bit different. Assuming that sets $Const$ and Var are fixed collections, then the set $Clos$ would satisfy the equation

$$Clos = Var \times Exp \times (Var \rightharpoonup Const + Clos).$$

The study of "fixed point equations" of this kind is another of the main themes of the book. The observations in this chapter show that the Foundation Axiom implies that all of these equations, and many others like them, will not have any solutions. We'll study the Anti-Foundation Axiom, prove that it is consistent and that it leads to solutions to many useful fixed point equations.

Historical Remarks

The material of section 3.3 is from Milner and Tofte (1991). Every book on the basics computability should have the Myhill-Nerode Theorem; one standard source would be Section 11.4 of Rogers (1967). Similarly, the Myhill-Nerode Theorem may be found in books on automata theory. Exercise 3.3.2 is from d'Agostino and Bernardi (to appear). For more on self-applicative programs and their applications, see Jones, Gomard and Sestoft (1993).

4

Circularity in philosophy

Circularity comes up in many parts of philosophy. The most dramatic form is in the various logical and semantical paradoxes, to which we devote a separate chapter. In this chapter we discuss some non-paradoxical examples of circularity in philosophy and related disciplines.

4.1 Common knowledge and the Conway Paradox

A good way into the issue of circularity in philosophy is to think of the circular nature of the awareness of the world given by our perception of it. This circularity is brought out by what is known as the Conway Paradox. This "paradox" has many forms, from stories of unfaithful spouses, to dirty children, to card players. Here is a rather innocuous form.

Three logicians, Bob, Carol, and Ted, are having dinner together on the day that the nominations for the Nobel Prize in Logic are known to have been announced to the nominees. Unknown to the others, each of the three has been told they have been nominated for the prize. Their friend Alice comes into the restaurant and joins them. She asks them whether they know whether either of the others has been nominated for the prize. They all answer "no." "Now do you know whether either of the others has been nominated for the prize?" Again they all answer "no" since the earlier answers give them no new information to go on.

Contrast this with a slightly different story. When Alice sits down, she says "At least one of you has been nominated for the prize. Do you know whether either of the others has been nominated for the prize?" Each answers

"No." She asks again, "Now do you know whether either of the others has been nominated for the prize?" This time, each answers "Yes, both of the other two has been nominated for the prize."

Why? Look at it from Bob's point of view. Bob realizes that if Carol had not been nominated, she would know it. So from Alice's first remark she would have concluded that one of Bob and Ted had been nominated. Thus Carol would have answered "yes" to the first question. Since she answered "no" it can only be that she herself has been nominated. Similarly for Ted. Hence both of them must have been nominated. For this reason, Bob answers "yes" to the second question. And the others do likewise.

Why might this seem paradoxical? At first sight, it seems paradoxical because the only difference between the first scenario and the second scenario is that in the second, Alice told them something they already knew: that at least one of them was nominated. How could telling them something they already knew possibly make a difference? Yet doing that somehow radically changed the situation. How can that be?

The solution is quite simple. Initially the fact in question, that at least one of the three had been nominated, was known to each of the logicians individually, but it was private knowledge. Once Alice publically announced the fact, it became common or public knowledge. That's what makes the difference. Not only did each know it, they also knew that it was public knowledge. What made it public knowledge? A public announcement. To see this, notice that if Alice had whispered the same fact to each of them privately, it would not have had the same effect.

This example makes it look like a rather special sort of phenomenon, but it is not. Consider the difference between two forms of five card poker, draw and stud. In draw poker, each player is supposed to see his own cards but no one else's. In stud poker, each player has three of his five cards face up on the table for all the players to see. Clearly the situation is a lot different. But let's imagine a draw game in which each player secretly cheats by looking in a mirror and seeing three of his opponent's cards. The question is this: is stud poker equivalent to draw poker with this kind of cheating?

Anyone who has ever played poker will realize there is an enormous difference. In the case of stud poker, the values of the "up" cards is public information, everyone knows the values of these cards and they know that this information is public. By contrast, in draw poker with cheating, it may be that the players have the same knowledge about the cards, but their knowledge about each other is much more limited. For example, suppose that your opponent knows the values of some of your cards but you don't know that she knows this. Your betting in this situation would be quite different from stud poker.

CHAPTER 4

Treatments of common knowledge The best analysis of how public, or common, knowledge arises is due to David Lewis, in his book (Lewis 1969), and has been taken up by Clark and Marshall (1981). On Lewis's account, common knowledge is necessary for something to be a convention among a group of people. On his analysis, common knowledge among b, c and t of some fact σ arises from a situation or event s that each of them knows about, and that clearly demonstrates that σ is a fact, and that each of b, c, and t knows the facts of s. We might symbolize this as:

$$s \models \sigma$$
$$s \models (b \text{ knows } s)$$
$$s \models (c \text{ knows } s)$$
$$s \models (t \text{ knows } s)$$

Here $s \models \tau$ is intended to indicate that τ is one of the facts supported by the situation s. So the idea is that common knowledge arises from the perception of some event which supports the fact in question, but also supports the very awareness of the other agents of the situation in question.

This analysis is very appealing, but its very coherence has been doubted because of an apparent circularity. Thus, for example, if you want to model situations using sets in the standard way, then s is going to have to be a element of its own domain and so a non-wellfounded set. Some who have thought that this notion of non-wellfounded set was incoherent have concluded that the analysis was similarly incoherent. However, once one has the tools provided by *AFA*, it is easy to make this model into a rigorous one and study its properties.

There are two other analyses of common knowledge. One is circular and is due to Harman (1986). The other is based on iterated statements in epistemic logic. We will not go into these here, but rather refer the reader to Barwise (1989), where they are studied in detail, and where hypersets are used to model and compare all three notions. In addition, Lismont (1995) has further results on models of common knowledge, using hypersets.

4.2 Other intentional phenomena

"Intentionality" is the name philosophers give to the phenomena of semantic reach whereby some things seem to be about other things. The most standard examples of intentional phenomena are language and belief, both of which can be about things other than themselves. More generally, intentionality arises whenever the question of "meaning" rears it head. Various philosophers, most notably Paul Grice and John Searle, have come to the conclusion that there is something intrinsically circular in intentional phenomena.

Let us consider the meaning of stop-signs and the way they function at a four-way stop. When we learn to drive and pass our driving test, we learn the rules that govern these signs: that we have to stop, that the first person to stop has the right of way, and that in the case of ties, the person on the right has the right of way.

There are two ways to think about these signs. One way is just to suppose that everyone goes out and follows the rules, not thinking at all about the other drivers. If we all follow the rules, then everything works out fine. The other way to think about it, though, is to take into account what the other driver is supposed to know. For example, if we see a stopped car on our left with a huge "Student driver" sign on top, we will be less likely to go without careful inspection of that car even if it is our turn.

The point here is that it is not enough for each of us to know the rules of the road. Rather, the conventions of the road, and our acceptance of these conventions, must be common knowledge if we are going to be able to rely on each other to behave in the expected manner. When one of us violates these conventions, we are not just breaking the law, we are also violating the expectations of the other drivers around us. This is part of what makes it dangerous.

Grice, in his essay "Meaning" (Grice 1966), makes this same observation about the meaning of language. When we use English, there is more going on than each of us knowing the same language. Part of what is going on is that the conventions of language are public information within the English language community. We rely on this when we talk to each other. Grice proposes to analyze "a means something by S," (where a is a speaker and S is a sentence) as follows:

> a intends the utterance of S to produce some effect in an audience by recognition of this very intention.

The idea is that the speaker wants his audience to be aware of, and respond to, his intentions in saying what he does. We won't defend this analysis here, referring the reader instead to Grice's famous article. For us the interesting thing about his proposal is that on this account, a's intention I is circular, it is an intention to do something that crucially involves reference to I itself. John Searle (1983) has gone on to find similar circularity in much intentional activity. Rather than go into this here, we look at one of the earliest implicit uses of intentional circularity in modern philosophy.

4.3 Back to basics

Let us consider Descartes' famous dictum, "Cogito, ergo sum." Recall Descartes' project: to question all that could be questioned until he could find something certain on which to build. The one thing he could not question was the fact of his own thinking. From that he concluded that he existed, and from that he attempted to build up a coherent philosophy.

So Descartes' "Cogito, ergo sum" was intended to be an irrefutable argument from undeniable premises. Descartes could not doubt the fact that he thought. But why? You and I can certainly doubt that Descartes thought. We might wonder if he never even existed; maybe "Descartes" was a pseudonym for some group of French philosophers the way "Bourbaki" is a pseudonym for a group of French mathematicians. Why can we doubt that Descartes thought while he could not?

The reason is that Descartes' act of doubting itself requires thinking and Descartes was aware of this. Basically, Descartes' famous dictum is shorthand for something more like: I am thinking this thought, and this I cannot doubt because my doubting requires my thought.

This thought of Descartes is circular in an important way. It is this circularity which gets Descartes' ball rolling. In a similar way, Grice's attempt to characterize non-natural meaning is circular, as is Searle's characterization of other intentional activities. To provide mathematical models of these characterizations one needs convenient tools for modeling circularity.

Exercise 4.1 Let us consider how circularity can be implicated in belief. Suppose we model the notion of possible world as a first-order structure M, and a proposition p as a set of possible worlds, intuitively, those worlds in which the intuitive proposition is true. Let us further model belief as a relation a bel p between individuals a and propositions p. If $M \models (a \ bel \ p)$ then a's belief is true in M if $M \in p$. Show that if $M \models (a \ bel \ p)$ and a's belief is true in M, then M is a non-wellfounded structure. (This is one of the reasons that possible worlds have traditionally been taken to be primitives rather than structured objects.)

4.4   Examples from other fields

From linguistics

There are several well known examples of non-wellfounded phenomena in the linguistics literature. We will discuss some examples concerning reference in Section 5.2. Here we discuss an example concerning anaphora.

In a sentence like

Professor Hill denounced the judge who had harassed her

we think of *her* as having as antecedent the noun phrase *Professor Hill*, and so has to refer to the same person as this noun phrase. Furthermore, the reference of the noun phrase *the judge who had harassed her* depends on the referent of *her*. We are tempted to think of the semantic interpretation of this sentence as assigning individuals first to *Professor Hill*, next to *her*, and finally to *the judge who had harassed her*.

The order *Professor Hill, her, the judge who had harassed her* is not quite left-to-right, but it might be something like the depth-first order of noun phrases. The order could work differently. Consider

> The law school professor who had worked for him denounced Judge Thomas.

Here the order is right to left. The referent for *the law school professor* depends on the referent of *him*, which depends on the referent of *Judge Thomas*. But in both cases, we have a wellfounded process of assigning referents.

But now consider the following example, known in the linguistics literature as a Bach-Peters sentence:

> The law school professor who had worked for him denounced the judge who had harassed her.

If we try to use the same technique in the above examples, we find ourselves in a vicious cycle, since the referent of *the law school professor* depends on the referent of *him*, which depends on the referent of *the judge who had harassed her*, which depends on the referent of *her*, which depends on the referent of *the law school professor*. Still, people do understand these sorts of sentences.

From game theory and economics

Economists use game theory to model the way people made decisions in the face of uncertainty. One imagines a set I of "players" and a set Ω of epistemically possible states of the world. Each state $\omega \in \Omega$ has associated with it some state $\Theta(\omega)$ of the world, which includes things like the payoff function of the game, and an object $t_i(\omega)$ to model the belief of each player $i \in I$ in that state ω.

The mathematician Aviad Heifetz has recently uncovered a hidden assumption in much of game theory, an assumption that each player knows the information structure of the other players. In trying to make this assumption an explicit part of the model leads to the conclusion that the states $\omega \in \Omega$ have an essentially circular nature, much as one might expect from the circularity of intentional states in general. Heifetz has used hypersets as a tool for modeling this aspect of game theory. We refer the reader to Heifetz (to appeara) and (to appearb) for details.

From mathematics

There are a number of areas of mathematics where one finds arguments that use circularity to some advantage. The interesting thing is that no paradoxes result: in fact, the circularity seems to be a powerful tool. We present two of these applications, from probability theory.

Suppose we flip a fair coin, and we count the flips beginning with the number 1. We expect to get heads at some point. We know that we have probability .5 of getting heads on the first flip. We also have probability .5 of getting heads on the second flip. The probability of heads happening *first* on the second flip is $(1 - .5)(.5) = .25$, since we would have to get tails on the first flip and heads on the second.

The question we would like to raise is: what is the probability that the first time we get heads it will be on an even-numbered flip? There are a number of ways to solve this, but we want to mention one that seems to appeal to circularity. Let x be the probability we are after. So $1 - x$ is the probability that the first heads will be on an odd-numbered flip.

We want to calculate x. Now let's say we flip a coin once. If it comes out heads, as it will do .5 of the time, then we have no chance to have a sequence of flips of the desired kind. But if it comes out tails, then there is still some chance that the first head will happen on an even-numbered flip. What is this chance? Why it is $1 - x$, since after we flip once, we are in the situation we began in, except that now we want to know about the probability of getting heads first on an odd-numbered flip. (This point is where the circularity lies.) So we have

$$x = (.5)(0) + (.5)(1 - x),$$

and we get $x = 1/3$. Incidentally, this shows that it is possible to use a fair coin to make a three-way random decision.

This example is from Newman (1982), as is our next one. It is a more sophisticated example of this kind of circularity. Suppose we have a *continuous roulette wheel*, labeled with the points t such that $0 \leq t < 1$. We spin the wheel, and we keep track of our sum. How many spins do we expect it to take before our sum is greater than 1?

The main trick in solving this is to attempt a harder problem. For $0 \leq t \leq 1$, let $E(t)$ be the number of spins which we expect it to take before our sum is greater than t. So $E(0) = 1$, and we want to find $E(1)$. We do this by finding a general formula for $E(t)$ for $0 < t \leq 1$.

Fix such a number t. When we spin the wheel once, we'll get some number x. If $x > t$, we're done on the first spin. If $x \leq t$, then we need to go on spinning. How many further spins would we need? We would expect to need $E(t - x)$. Overall, we need to take each possible x, find its probability $p(x)$, and then add up $p(x) \cdot 1$ for those $x > t$ and $p(x)E(t-x)$ for those $x \leq t$. Now $p(x) = 0$ in all cases, but to calculate such sums we really want integration. We get an equation defining E in terms of itself.

$$
\begin{aligned}
E(t) &= \int_t^1 1 \, dx + \int_0^t 1 + E(t - x) \, dx \\
&= 1 + \int_0^t E(t - x) \, dx \\
&= 1 + \int_0^t E(u) \, du
\end{aligned}
$$

(We have made the substitution $u = t - x$ in the last step.) By the Fundamental Theorem of Calculus, $E'(t) = E(t)$. Combined with the fact that $E(0) = 1$, we see that $E(t) = e^t$. So our original question is answered by saying that the expected number of spins is the classic constant e.

The point at which circularity enters in this discussion is when we defined E in terms of itself. As you can see, this is a powerful method. It does not depend on hypersets, but it does illustrate the utility of mathematical methods involving circularity.

5

Circularity and paradox

This is what I mean by a paradox: an apparently unacceptable conclusion derived by apparently acceptable reasoning from apparently acceptable premises. Appearances have to deceive, since the acceptable cannot lead by acceptable steps to the unacceptable.

R. M. Sainsbury, *Paradoxes*

It seems fitting to close this abbreviated tour of circular phenomena by discussing the phenomena which gave circularity its bad name: namely the semantical and logical paradoxes. We begin with the oldest and most famous.

5.1   The Liar Paradox

It is not uncommon to find announcements like the following on documents:

Notice: The copyright to this work is held by .... This work may be reproduced in whole or in part, provided this notice is included with the reproduced material.

While the intent of such a notice is clear, there is something obviously circular about it, in that it makes reference to itself. Similarly, it is not uncommon in airports to hear announcements similar to the following:

USAir Flight 205 will depart from gate H12 at six o'clock; this announcement will not be repeated.

This announcement too, while perfectly clear, has something circular about it. It refers to itself. In instruction booklets on preparing income tax statements, it is not uncommon to find paragraphs that refer to another. Such examples show that the relation "refers to" is circular, that is, that things sometimes refer to themselves, or to other things that refer back to them.

There seems to be nothing particularly problematic in general about sentences and other forms of communication that involve circular reference. But there are problematic cases, like the following:

1. This claim is not true.
2. Sentence (2) on page 56 of this book is not true.
3. What I am now saying is false.

The problem with these sentences arises once we ask whether or not they are true. It seems as if they are true if and only if they are not true. These are all forms of what is commonly known as the Liar Paradox.

Russell's reaction to the paradoxes was to blame them on the circularity. He formulated what he called the "vicious circle principle" which attempted to ban circularity from scientific discourse. Following Russell's lead, Tarski proposed to simply ban sentences of a language from talking about truth of sentences in that very language on the grounds that they somehow involve a vicious circle. On Tarski's account, to talk of the truth of sentences in one language, you need a different language, a "metalanguage." That language would have its own metalanguage, and so on, generating an infinite hierarchy of different truth predicates and different hierarchies. This means that English becomes an infinite hierarchy of languages.

For many years, Tarski's hierarchical approach was, by and large, the accepted wisdom on the semantical paradoxes. Things changed in 1975, though, with the publication of Saul Kripke's article "Outline of a theory of truth" (Kripke 1975). In the first half of this article, Kripke convincingly demonstrated that circularity of reference is much more common than had been supposed, and that whether or not something is paradoxical may well depend on non-linguistic, empirical facts. Kripke then went on to outline a theory of truth that allowed for circularity, but it is the first part of his article that concerns us here.

Consider the following pair of sentences from Kripke's paper:

1. Most of Nixon's statements about Watergate are false.
2. Everything Jones says about Watergate is true.

There is nothing intrinsically paradoxical about these statements. There are many plausible circumstances where both could be true. But Kripke points out that there are also circumstances where they are paradoxical. Suppose, for example, that Jones asserts only (1) about Watergate, while Nixon asserts (2) plus eight other claims about Watergate, half true, half false. What are we to say of (2)? Is it true, or false? It would seem that it is true if and only if it is not true.

As a result of Kripke's paper, there has been a great deal of subsequent work on these paradoxes. We return to the subject in Chapter 13, where we apply hypersets to pursue one line of thought on the Liar Paradox, as well as the paradoxes of denoting.

5.2 Paradoxes of denotation

Paradoxes involving truth arise when we allow a truth predicate in our language and make some seemingly reasonable assumptions about the way that the truth predicate works. Paradoxes involving denotation arise when we allow a denotation predicate in our language and make some parallel and seemingly reasonable assumptions about the way it works.

The most rudimentary examples of sets are those that are specifiable by enumeration, with terms like

(1)   the set whose elements are just 0 and π

This denotes the set $\{0, \pi\}$. Similarly the term

(2)   The set whose only member is the set denoted by (1)

denotes the set $\{\{0, \pi\}\}$. A set a is *reflexive* if $a \in a$. The simplest example of a reflexive set is $\Omega = \{\Omega\}$. It seems like it should be named by

(3)   The set whose only member is the set denoted by (3).

There are reasons to worry about this sort of circular use of the predicate "denotes." Consider the term

(4) $\begin{cases} 1 & \text{if (4) denotes 0} \\ 0 & \text{if (4) denotes 1} \\ 0 & \text{if (4) does not uniquely denote anything} \end{cases}$

It is very tricky to figure out what to say about (4). It seems that it must denote either 0, 1, or nothing at all, or more than one thing. But consideration of each case leads to a contradiction.

Let's look at an example like this in more detail, this time using "this" rather than the labels. And to simplify things, let's agree to use "denotes" so that a use of a term can denote at most one thing. Define the term t_0 as follows: Let 0, 1 and 2 be distinct numbers that have names $\ulcorner 0 \urcorner$, $\ulcorner 1 \urcorner$, and $\ulcorner 2 \urcorner$, and let t_0 be the term expressing:

$\begin{cases} 1 & \text{if this denotes 0} \\ 0 & \text{if this denotes 1} \end{cases}$

Imagine we were reading this expression out loud, simultaneously pointing at some name n as we said "this." For example, if we use "this" to denote $\ulcorner 0 \urcorner$, then t_0 denotes 1, and similarly, if we use "this" to denote $\ulcorner 1 \urcorner$ then t_0 denotes 0. If we use "this" to denote $\ulcorner 2 \urcorner$, or the term (1), then clearly t_0 does not denote anything at all. There is nothing paradoxical here. It just shows that some terms sometimes do not denote anything. Similarly, if we use "this" to denote t_0 itself, then t_0 cannot denote anything at all. It could only denote 0 or 1, but it cannot denote either of these, as we see by inspection.

But now consider the term t_1 expressing the following variant of t_0:

$$\begin{cases} 1 & \text{if this denotes 0} \\ 0 & \text{if this denotes 1} \\ 0 & \text{if this does not denote anything} \end{cases}$$

Again, if we use "this" to denote $\ulcorner 0 \urcorner$, then t_1 denotes 1; if we use "this" to denote $\ulcorner 1 \urcorner$ then t_1 denotes 0. If we let "this" denote $\ulcorner 2 \urcorner$, then again t_1 does not denote anything at all. What happens if we let "this" denote t_0? Well, we have seen that if "this" denotes t_0 then t_0 does not denote anything, so the final clause applies and insures that t_1 denotes 0.

So far, there has been nothing paradoxical. But what happens if we use "this" to denote t_1 itself? Again, it could only denote 0 or 1, but it cannot denote either of these, for the same reasons as before. So it would seem that t_1 does not denote anything. But then, by the third clause, it seems that t_1 denotes 0 after all.

We now have a paradox on our hands. Terms like t_0 and t_1 seem to make sense, and it seems that we can use them in sensible ways. If we use "this" to denote a name of 0, 1, or 2, or to t_0, then they have clear denotations, or no denotation at all. But if "this" denotes t_1, we seem to end up in paradox. What is going wrong? How are we to think about such terms?

5.3 The Hypergame Paradox

For our next paradox we discuss the Hypergame Paradox. This paradox is raised in Smullyan (1983). Zwicker (1987) discusses it in detail, points out its relation to other paradoxes, and suggests a resolution. We have some ideas about this paradox which we will present in Chapter 12. For now, we simply raise it.

The paradox concerns two-player games. Call the players I and II. In a game $\mathcal{G}$ players I and II take turns moving, with I going first. When the game starts, player I has a set of allowable first moves. Depending on what I plays, II has a set of allowable moves. And so on. If at any stage, the set of moves for one of the players is empty, then the game ends. The other player wins that play of the game. But some games can go on forever. Checkers, for example, has this property, unless you put an upper bound on the number of times a given position can occur. Let us call a game $\mathcal{G}$ "wellfounded" if no play of the game can go on forever.

Some games really consist of families of games. Dealer's Choice in poker, for example, is a game in which the first move by the dealer is the choice of a kind of poker to be played on a given hand. This suggests the following game, called the *Supergame*. Player *I* chooses a game. Then player *II* takes on the role of *I* in $\mathcal{G}$. Then they continue with $\mathcal{G}$.

Supergame is obviously not a wellfounded game, since *I* might choose a nonwellfounded game on the first move. So let's define *Hypergame* a bit differently, by restricting *I*'s first move to be the play of a *wellfounded* game. Hypergame itself is a wellfounded game, since once *I* has chosen a wellfounded game, any play in that game must terminate. But here comes the paradox. As we have just seen, Hypergame is a wellfounded game. Player *I* can therefore choose Hypergame itself as a first move. But then *II* could choose Hypergame as well. So the two players could continue playing forever, each playing Hypergame. This would result in a play of the game which does not terminate. So we seem to have shown that Hypergame both is and is not a wellfounded game.

This little paradox is intimately connected with hypersets. We will suggest a resolution in Chapter 12.

5.4 Russell's Paradox

Russell's Paradox has many forms. The one that infects set theory has to do with non-reflexive sets; that is, sets that are not elements of themselves. On Frege's analysis of the notion of set, every predicate has a set as its extension. Hence, there is a set R which consists of all and only non-reflexive sets:

$$R \;=\; \{x \mid x \text{ is non-reflexive}\}$$

But then we see that $R \in R$ iff R is non-reflexive, which holds iff $R \notin R$. Hence either assumption, that $R \in$ or $R \notin R$ leads to a contradiction.

Notice that we have not used any assumptions about the notion of set, other than that the displayed predicate determines a unique set. This point is made clearer by observing that the following is simply a theorem of first-order logic, independent of what E happens to mean:

$$\neg \exists x \forall y \, [yEx \leftrightarrow \neg(yEy)]$$

Exercise 5.1 Here are some more famous paradoxes involving circularity for you to ponder, if you enjoy such things.

1. Sentence 1 on page 59 cannot be known.
2. Sentence 2 on page 59 cannot be proven.
3. Let N be the least natural number that cannot be defined in English using fewer than twenty-three words.

4. If sentence 4 on page 60 is true, then *Vicious Circles* deserves a Pulitzer Prize.

What is paradoxical about these examples?

5.5 Lessons from the paradoxes

If we look back over these paradoxes, we see two commonalities. All of them somehow involve the interaction of circularity of some sort with negation. Bertrand Russell realized this and laid the blame squarely on circularity:

> This leads us to the following rule: "Whatever involves *all* of a collection must not be one of the collection."

> Bertrand Russell, 1908

This was a natural reaction, given Russell's views about mathematics being reducible to logic. This view led him to want to "define" everything in terms of logic. In particular, sets were not seen as existing, but as in need of definitions before they existed.

Russell's attitude to the paradoxes has dominated twentieth century logic, even among those who reject his logicism. It has resulted in the iterative conception of set, in Tarski's insistence on the hierarchy of language and meta-languages, and similar moves which replace circularity by hierarchies. However, Kripke has shown us in the case of the Liar Paradox that no simple-minded syntactic restriction will suffice. Circularity and seeming paradox can arise in other ways. The examples of circularity in computer science and philosophy have a similar moral.

The theory of hypersets gives more evidence that whatever is going on in the paradoxes, circularity itself is not the villain. To be convinced of this, you will need to read Parts II and III of this book, but we can get an initial reassurance by re-examining the Russell Paradox from the point of view of non-wellfounded sets.

Suppose we have some set b and form the Russell set using b as a universe. That is, let

$$R_b = \{c \in b \mid c \text{ is non-reflexive}\}$$

There is nothing paradoxical about R_b. The reasoning that seemed to give rise to paradox only tells us that $R_b \notin b$. In other words, the Russell construction gives us a way to take any set b whatsoever and generate a new set not in b. This has nothing to do with whether b is wellfounded.

Exercise 5.2 Suppose we have sets satisfying the following:

$$
\begin{aligned}
a &= \{0, 1, 2\} \\
b &= \{0, 1, b\} \\
c &= \{0, b\}
\end{aligned}
$$

(Here 0, 1 and 2 are treated as von Neumann ordinals.) Note that of these sets, only a is wellfounded. Compute the Russell sets R_a, R_b, and R_c. Verify by inspection in these cases the observation made above that for $d \in \{a, b, c\}$, $R_d \notin d$.

Tarski's Theorem Paradoxes are not only sources of frustration. They can also lead to great insight. The fact that $R_b \notin b$ is something we learned from Russell's paradox. Once we proved this fact, we may use it forevermore, whether or not we return to the paradox, the Frege conception of set, or anything else. There are a number of other, more famous, results that also can be regarded as lessons from the paradoxes. We want to describe two of these now.

Consider formulas in the language of set theory, such as

$$(\exists x_1)((x_1 \in x_3) \vee (2 = x_3)).$$

This formula has x_3 as its one free variable, and we sometimes want to make this clear by writing $\varphi(x_3)$. As in this example, we want to allow our formulas to contain sets inside. We also want to associate a set $\ulcorner \varphi(x_3) \urcorner$ with this formula. For example, we could agree that given an atomic formula like $v_3 \in 4$ we have $\ulcorner v_3 \in 4 \urcorner = \langle 0, \langle 0, 3 \rangle, \langle 1, 4 \rangle \rangle$. The outermost 0 tells us that we have an atomic formula for $\in$, the inner 0 tells us that 3 refers to v_3, and the 1 signals a constant. For something like $v_2 = v_3$ we could have $\ulcorner v_2 = v_3 \urcorner = \langle 1, \langle 0, 2 \rangle, \langle 0, 3 \rangle \rangle$.

Let $\mathcal{F}_0$ be the class of all formulas φ either of the form $x = y$ or $x \in y$. There is a formula $Sat(x, y, z)$ so that for all $\varphi \in \mathcal{F}_0$, and all sets a and b,

(5) $$Sat(\ulcorner \varphi \urcorner, a, b) \quad \leftrightarrow \quad \varphi(a, b).$$

We would define

$$Sat(c, a, b) \qquad \text{iff} \qquad (c = \langle 0, \langle 0, a \rangle, \langle 0, b \rangle \rangle \wedge a \in b)$$
$$\vee \quad (c = \langle 1, \langle 0, a \rangle, \langle 0, b \rangle \rangle \wedge a = b)$$

It is important to understand what (5); the details of the coding are much less important. The point is that Sat gives a *truth definition* for our (admittedly tiny) class $\mathcal{F}_0$.[1]

We could write Sat out more explicitly, by trading in the pairing notation for its logical definition. If we did that, we would get a formula expressed entirely in the formal language of set theory. It is not surprising that this can be done: after all, essentially every mathematical definition can be formalized in set theory and this is all we have done. The interesting question is whether we can do all of the coding in such a way that the formula Sat itself belongs to the class $\mathcal{F}_0$ that we started with. In the case at hand, $\mathcal{F}_0$ is simply too small for this to happen, since it has no formulas with three free variables. We might try to expand it to include a number of simple types of three variable formulas, say

---

[1] It is standard practice in model theory to speak of the truth of a formula *in a structure*. When we speak of truth in this section, we mean true *in the universe of all sets*.

Boolean combinations of the formulas we already have. This too would not work. However, the good news is that it is possible to expand $\mathcal{F}_0$ to a bigger class, say $\mathcal{F}$, then get truth-defining formulas for $\mathcal{F}$ inside of $\mathcal{F}$ itself. To make this point precise, we make the following definition.

Definition A set of formulas $\mathcal{F}$ is *self-sufficient* if $\mathcal{F}$ contains all atomic formulas (allowing constants for each set a), all negated atomic formulas, and is closed (up to logical equivalence) under $\wedge$, $\vee$, bounded quantifiers ($\exists x \in y$) and ($\forall x \in y$), unbounded existential quantification ($\exists x$); and if $\mathcal{F}$ satisfies the following conditions:

Satisfaction For every sequence $x_0, x_1, \ldots, x_n$ of $n + 1$ variables there is a formula $Sat_n(x_0, x_1, \ldots, x_n) \in \mathcal{F}$ such that for all $\varphi(v_1, \ldots, v_n) \in \mathcal{F}$ with free variables $v_1, \ldots, v_n$, the universal closure of the following is true:

$$Sat_n(\ulcorner \varphi \urcorner, x_1, \ldots, x_n) \quad \leftrightarrow \quad \varphi(x_1, \ldots, x_n).$$

S_n^m **Property** For natural numbers n, m, there is an operation S_n^m of $m + 1$ arguments definable in $\mathcal{F}$ such that for each formula $\varphi(x_1, \ldots, x_n)$ the universal closure of the following formula is true in V:

$$Sat_{m+n}(\ulcorner \varphi \urcorner, x_1, \ldots, x_m, y_1, \ldots, y_n) \leftrightarrow$$
$$Sat_m(S_n^m(\ulcorner \varphi \urcorner, x_1, \ldots, x_m), y_1, \ldots, y_n))$$

Example 5.1 The main examples of self-sufficient sets are the sets of Σ_n *formulas*, for $n \geq 1$. A formula is in Σ_k if it is equivalent to one written in the form $\exists y_1 \forall y_2 \exists y_3 \ldots y_n \psi$, where all the quantifiers in ψ are bounded quantifiers. If one wrote out the definition of Sat for the class $\mathcal{F}_0$ above, the definition would be Σ_1.

The point of a self-sufficient set is that it has enough resources to carry out truth definitions for itself. There is a strong analogy to the concepts from recursion theory, as we see in the following parallel to Theorem 3.2.

Theorem 5.1 (Recursion Theorem) *Let $\mathcal{F}$ be a self-sufficient set of formulas. Then for any $\varphi(x_0, x_1, \ldots, x_n) \in \mathcal{F}$ there is a formula $\psi(x_1, \ldots, x_n) \in \mathcal{F}$ such that the following holds:*

$$\psi(x_1, \ldots, x_n) \quad \leftrightarrow \quad \varphi(\ulcorner \psi \urcorner, x_1, \ldots, x_n).$$

Proof (Sketch for $n = 1$.) Given $\varphi(x_0, x_1)$, let $\theta(x_0, x_1)$ be the following $\mathcal{F}$-formula:

$$\exists z (S_1^1(x_0, x_0) = z \wedge \varphi(z, x_1))$$

Now let $\psi(x_1)$ be $\theta(\ulcorner\theta\urcorner, x_1)$ so that $S_1^1(\ulcorner\theta\urcorner, \ulcorner\theta\urcorner) = \ulcorner\psi\urcorner$. Then we have the following

$$\begin{aligned}
\varphi(\ulcorner\psi\urcorner, x_1) &\leftrightarrow \exists z(S_1^1(\ulcorner\theta\urcorner, \ulcorner\theta\urcorner) = z \wedge \varphi(z, x_1)) \\
&\leftrightarrow \theta(\ulcorner\theta\urcorner, x_1) \\
&\leftrightarrow \psi(x_1)
\end{aligned}$$

as desired. ⊣

With these preliminaries out of the way, we can given an example of an important lesson of the Liar Paradox.

Proposition 5.2 *Let $\mathcal{F}$ be a self-sufficient set of formulas, and let Sat_0 be the satisfaction formula for $\mathcal{F}$ sentences. Then the formula $\neg Sat_0$ is not equivalent to any formula in $\mathcal{F}$.*

Proof Assume toward a contradiction that $\neg Sat_0$ is in $\mathcal{F}$. Using the Recursion Theorem, pick a sentence φ equivalent to $\neg Sat_0(\ulcorner\varphi\urcorner)$. By the definition of Sat_0, we have

$$Sat_0(\ulcorner\varphi\urcorner) \leftrightarrow \varphi$$

but then

$$Sat_0(\ulcorner\varphi\urcorner) \leftrightarrow \neg Sat_0(\ulcorner\varphi\urcorner)$$

which is a contradiction. ⊣

Corollary 5.3 (Tarski) *No self-sufficient set of formulas is closed under negation. In particular, neither the set of all first-order formulas nor the set of all second-order formulas is self-sufficient.*

Fix a self-sufficient set $\mathcal{F}$. For each k, there *is* a formula $\neg Sat_0$ as in Proposition 5.2. It just won't belong to $\mathcal{F}$. This is exactly as with the Russell Paradox: the set R_b will be a set; it just won't belong to b. If $\mathcal{F}$ is the set of Σ_n formulas, then it is possible to show that $\neg Sat_0$ will belong to Σ_{k+1}. This shows that Σ_{k+1} formulas can express more than Σ_k ones. This result is called the *Arithmetical Hierarchy Theorem*. That's one lesson from the Liar Paradox. Here is another, Tarski's Undefinability Theorem for Truth:

Corollary 5.4 (Tarski) *Then there is no single formula $\varphi(v_1)$ of first-order logic such that for all sentences ψ, $\varphi(\ulcorner\psi\urcorner) \leftrightarrow \psi$.*

Proof If φ exists, then it must be a Σ_k formula for some k. The negation of a Σ_k formula is always a Σ_{k+1} formula, so $\neg\varphi$ will be Σ_{k+1}. But since the Σ_{k+1} formulas are a self-sufficient set, the proof of Proposition 5.2 (using φ in place of Sat_0) shows us that $\neg\varphi$ cannot be Σ_{k+1}. ⊣

This tells us one of the important results concerning the logical aspects of set theory: there is no single first-order formula that serves to define the truth of *all* sentences of first-order logic in the universe of sets.

In this book, besides presenting and advancing the theory of hypersets, we will be applying the theory to a number of circular phenomena, including the phenomena behind the paradoxes presented in this part of the book. In each case we will develop a mathematical model of the phenomenon in question, and show how the paradox turns into the proof of one or more theorems about that model. Such an analysis resolves the paradox in question, provided the model is a plausible model of the phenomena in question. As to whether our models satisfy this criteria, the reader will have to judge.

One thing is quite clear, though. Circularity is here to stay; it will not be banned by fiat. What we hope to convince you of in the rest of this book is that the theory of non-wellfounded sets gives us a beautiful, coherent, and powerful tool with which to study circular phenomena in all domains.

Historical Remarks

We discuss the semantical paradoxes further in Chapter 13, and the hypergame paradox in Chapter 12. The endnotes to those chapters also have some historical comments.

Theorem 5.1 is called a Recursion Theorem out of analogy to the parallel result in Chapter 3. The result here is patterned after Gödel's argument at the center of his proof of the Incompleteness Theorem for Peano Arithmetic. The method is also called *diagonalization*, reflecting the similarity to Cantor's "diagonal" proof of the uncountability of $\mathcal{P}(N)$. The basic idea has been used countless times in every area of computability and definability theory. Tarski's Theorem on the undefinability of truth, Tarski (1939) is one of the best known of such applications.

Part III

Basic Theory

6

The Solution Lemma

Putting the matter in a positive way, a set z can have as members only those sets which are formed before z. ... It is, of course, possible that there is a completely different analysis of the notion of a set, and this might lead to a different set of axioms. Up to the present, however, there has been no analysis of set essentially different from that given here which leads to a satisfactory set of axioms.

Joseph Shoenfield, 1977

There are various ways to present the Anti-Foundation Axiom (*AFA*). The first formulation, due to Forti and Honsell (1983), is that for every relational structures $\langle A, R \rangle$ there is a homomorphism onto a transitive set. Aczel (1988) states a form which is quite close to this, in terms of an operation of labeling the nodes of graphs with sets. We will take a different approach in this chapter, through what has become known as the Solution Lemma. In Chapter 10, we'll go back and examine the connection of this chapter's work to graphs.

In the previous chapters we have seen many examples of equations between sets whose solutions could be useful in modeling circular phenomena. To fix an example for discussion, suppose we want to model a device D which has one state d, and with the behavior which permits a single transition q which results in the same state d. (Think of a clock as in Section 3.2, but one which never breaks.) In our modeling, the set d should satisfy $d = \{\langle q, d \rangle\}$. To simplify the example even further, let's suppose we want a set d satisfying $d = \{q, d\}$. Why should we expect that there is a set d satisfying this equation? Or, put differently, how can we think about the universe of sets so that the existence of such a set becomes non-problematic?

It seems reasonable that we should be able to enlarge the usual universe *WF* of wellfounded sets so as to allow such a set. After all, why not just allow

$$\{q, \{q, \{q, \{q, \dots\}\}\}\}$$

as a set? The trick is to do this in a systematic way, so that the axioms of *ZFC⁻* (see page 28) remain true.

The basic idea is one that recurs throughout mathematics: faced with a universe which is somehow missing solutions to certain equations or other problems, enlarge the universe to one where those problems have solutions. Examples abound: the move from the natural numbers to the integers, or the integers to the rationals, or from the rationals to the reals, from the reals to the hyperreals or to the complex numbers, from the Euclidean plane to the projective plane, from a topological space to one of its various compactifications. All these represent instances of this process.

For inspiration, let's consider one of these constructions in more detail: the move from the integers to the rationals. The construction is outlined in the box on page 69.

In the same way, the universe of non-wellfounded sets is an enrichment of the ordinary universe of set theory. This time, the enrichment is needed to solve certain equations, like $x = \{q, x\}$. (More generally, we need to solve *systems* of equations. But for the moment, we ignore this.) We can carry out this enrichment formally by modeling these new sets as *equivalence classes* of equations. Why equivalence classes? Well, if

$$x = \{q, x\},$$

then the logical axiom of identity (which we take as a given, not something to discuss as we do the axioms of set theory) tells us that

$$x = \{q, \{q, x\}\},$$

so these two equations should have the same solutions.[1] Just as the same rational number satisfies the equations $2x = 3$ and $4y = 6$, so too the same hyperset should satisfy both the equations just displayed. And we want them to have the same solution as the equation $y = \{q, y\}$, since the particular indeterminate used to state the equation should not have anything to do with the solution.

The Anti-Foundation Axiom (in the form we present it) tells us that every system of equations of a certain form has a solution. But looking at the example $d = \{q, d\}$, we see there is still another problem to be solved. How

---

[1] Actually, what this argument shows is that every solution of the first should be a solution of the second. But we will later argue that we want such an equation to have a unique solution, so the two would have to have the same solution.

Outline of the Construction of the Rationals from the Integers

(0) We begin with the set of integers Z considered as an algebraic system $\langle Z, 0, 1, +, \times \rangle$. That is, we focus on certain elements and operations, and we are concerned about algebraic laws such as $x(y+z) = xy + xz$ which hold in the structure.

(1) Our extension of Z has a purpose: we would like to have a structure in which we can have unique solutions to all equations of the form $px = q$, where $p \neq 0$.

(2) Consider the set of all pairs $\langle a, b \rangle$ with $b \neq 0$. These correspond to equations $ax = b$ that we want to solve. Define $\langle a, b \rangle \equiv \langle c, d \rangle$ iff $ad = bc$. Check that this is an equivalence relation. Write a/b for the equivalence class of $\langle a, b \rangle$.

(3) Turn the set of equivalence classes into an algebraic system by declaring 0 to be $0/1$, 1 to be $1/1$, $(a/b) + (c/d) = ((ad + bc)/bd)$, and $(a/b)(c/d) = (ac/bd)$. Then check that these definitions do not depend on the particular choice of members of the equivalence classes used. Call the resulting structure Q.

(4) Check that Q satisfies all of the relevant algebraic laws.

(5) Note that Z is isomorphic to a part of Q; the association is $n \mapsto (n/1)$.

(6) Finally, we need to check that Q actually meets the purpose we mentioned in (1). Suppose $p = a/b$ and $q = c/d$ with $d \neq 0$. If we want x so that $px = q$, we take $x = cb/da$.

---

many solutions will an equation like $d = \{q, d\}$ have? One? Two? Infinitely many?

Suppose d_1 and d_2 are both solutions to this equation. Then, by expanding these sets indefinitely, we see that

$$d_i \quad = \quad \{q, \{q, \{q, \{q, \ldots\}\}\}\}$$

for $i = 1, 2$. Intuitively, there is nothing that can possibly distinguish these sets. They are the same all the way down. Any difference would violate the spirit, if not the letter, of the Axiom of Extensionality, the intent of which is to say that a set is completely determined by its members. As usually formulated, the Axiom of Extensionality does not allow us to conclude that $d_1 = d_2$ (see Exercise 2.4). But clearly the spirit of extensionality would have them be identical. Hence, we will assume:

Every system of equations has a *unique* solution.

This is the content of the Solution Lemma.

6.1 Modeling equations and their solutions

To turn the informal statement of the Solution Lemma into a precise axiom, we need to say just what we mean by a system of equations, and by a solution to a system of equations.

There are two problems to be faced. First, not just anything we might intuitively consider an equation *can* have a solution. For example, if $\mathcal{P}$ is the power set operation, then Cantor's Theorem tells us that there is no function from b onto $\mathcal{P}(b)$. (See Proposition 2.7 on page 26.) Consequently, there can be no set a satisfying the equation

$$a = \mathcal{P}(a).$$

If we were to try to add an ideal element to represent a solution to this equation, our notion of set would become incoherent. So just what equations *can* be solved in an enlargement of the universe of sets?

The second problem is that we want to state the axioms of set theory solely in terms of sets and things (like functions) that can be defined in terms of sets in ZFC^-. Thus, we do not want to have our axiom refer to actual equations.[2] Rather, we model the notions of system of equations, and solution to such a system, within set theory.

There are a number of equivalent ways to do this, and so a number of ways we can formulate our axiom.[3] And such is life that the easier it is to state and to prove to be consistent, the harder the axiom is to apply.

Example 6.1 Fix sets p and q. Suppose we need sets x, y, and z that satisfy the following conditions:

$$
(1) \qquad
\begin{aligned}
x &= \{x, y\} \\
y &= \{p, q, y, z\} \\
z &= \{p, x, y\}
\end{aligned}
$$

The desired sets are built "on top of" p and q. We call p and q *atoms* of our system of equations, and we write $A = \{p, q\}$.[4] The objects x, y, and z are *indeterminates* of the system. We'll write X for this set $\{x, y, z\}$ of indeterminates.

You should think of the indeterminates as urelements even though this is not necessary at this point. To see why, consider the analogous case of solving

---

[2]It would be possible to work out a theory which had as objects not just sets and urelements, but also equations. Our decision to take some things as primitive and others as defined is to some extent arbitrary.

[3]They are equivalent in that you can prove each from the other within ZFC^-.

[4]Warning: This terminology is potentially confusing, since many authors use the term *atom* and *urelement* synonymously. For us, the notion of an atom is relative to a system of equations, whereas the notion of urelement is primitive.

Page header

a system of linear equations such as

$$\begin{array}{rcl} x & = & y+z+3 \\ (2) \qquad y & = & 2x+w \\ z & = & x+y \end{array}$$

In formalizing this, it would seem to be a good idea to make sure that x, y, and z are *not* numbers. For if they were, then the question of whether they were a solution to (2) would be answerable outright, whereas the answer should be contingent on the values provided by some assignment of numbers to variables. So just as a way of keeping this straight, one should take x, y, and z in (1) to be non-sets, so that they themselves won't even be a candidate for a solution.

We therefore must say what it would mean to have sets which satisfy (1). But first we need something to model the equations themselves. For this, we use the function e with domain the set $X = \{x, y, z\}$ whose values are the right-hand sides of the equations in (1). So $e_x = \{x, y\}$, $e_y = \{p, q, y, z\}$, and $e_z = \{p, x, y\}$.

How should we model a solution s to (1)? As with e, s should be a function defined on X which gives, for each indeterminate $v \in X$, a set s_v. These various sets should satisfy the identities in (1), which is to say that $s_x = \{s_x, s_y\}$, $s_y = \{p, q, s_y, s_z\}$, and $s_z = \{p, s_x, s_y\}$. Note that the atoms p and q are not in the domain of s; this is because we are thinking of them as fixed objects in terms of which we want to define solutions to (1). A uniform statement of the three conditions on s is that

$$\begin{array}{rcl} s_v & = & \{s_w \ : \ w \in e_v \cap X\} \quad \cup \quad \{w \ : \ w \in e_v \cap A\} \\ & = & s[e_v \cap X] \qquad\qquad\quad \cup \qquad (e_v \cap A) \end{array}$$

for each $v \in X$.

We codify this in our first model of the notion of a system of equations and a solution to a system of equations.[5]

Definition

1. A *(flat) system of equations* is a tuple $\mathcal{E} = \langle X, A, e \rangle$ consisting of a set $X \subseteq \mathcal{U}$, a set A disjoint from X, and a function

$$e : X \to \mathcal{P}(X \cup A).$$

2. X is called the set of *indeterminates* of $\mathcal{E}$, and A is called the set of *atoms* of $\mathcal{E}$. For each $v \in X$, the set $b_v =_{df} e_v \cap X$ is called the set of indeterminates on which v immediately depends. Similarly, the set $c_v =_{df} e_v \cap A$ is called the set of atoms on which v immediately depends.

---

[5] We say "first" because a part of our later work will consist of refining, extending, and generalizing the notion of a system of equations. See especially Chapters 8 and 16.

3. A *solution* to $\mathcal{E}$ is a function s with domain X satisfying

$$s_x \;\; = \;\; \{s_y : y \in b_x\} \cup c_x,$$

for each $x \in X$.

One thing worth remembering is that a single equation may be a perfectly good system. (For that matter, it is possible to have $X = \emptyset$. Note that a flat system without indeterminates has the empty function as its solution.) Also, it is possible to have $e_x = \emptyset$ in any system.

Exercise 6.1 Consider Example 6.1 from page 70. Identify the sets b_v and c_v for $v \in X$. Show that our model of a solution to (1) gives us what we want.

6.2   The Solution Lemma formulation of *AFA*

Given these definitions, we can now state, in a precise way, the simplest of the many forms of the Anti-Foundation Axiom:

ANTI-FOUNDATION AXIOM: Every flat system $\mathcal{E}$ of equations has a unique solution s.

We sometimes call this form of the axiom the (Flat) Solution Lemma, out of deference to the existing literature, and also because we'll see several other forms of *AFA* in later chapters. We also define

ZFA: the axioms of our standard set theory, ZFC^- (see page 28), with *AFA* added.

Most of the results in this book are proved in ZFA. When we do not need either *FA* or *AFA* we usually point this out. (Of course, in a few places we discuss consequences of *FA*, to compare the situation with *AFA*.)

Once we have *AFA* at our disposal, we define the *solution set* of any flat system $\mathcal{E}$ of equations by:

$$solution\text{-}set(\mathcal{E}) \;\; = \;\; \{s_v : v \in X\} \;\; = \;\; s[X],$$

where s is the solution of $\mathcal{E}$. This is just the set of all sets of the form s_v, where s is the solution to $\mathcal{E}$, and v varies over the indeterminates of $\mathcal{E}$. Moreover, we define

$$V_{afa}[A] = \bigcup \{solution\text{-}set(\mathcal{E}) : \mathcal{E} \text{ a flat system with atoms } A\}.$$

That is, $V_{afa}[A]$ is the collection of all sets which are in the solution set of some system which uses A as its atoms.

Exercise 6.2 Use the Solution Lemma to show that there is one and only one set which is its own singleton. This set is called Ω. [Hint: To show that there is such a set, show how to view the equation $x = \{x\}$ as a flat system of equations. To show that there is only one such set, show that any set that is its own singleton would give rise to a solution to this system, so there can only be one.]

Proposition 6.1 *Let $A \subseteq \mathcal{U}$ and let $\mathcal{E} = \langle X, A, e \rangle$ be a flat system. Then solution-set($\mathcal{E}$) is transitive on sets: that is, if b and c are sets and $c \in b \in$ solution-set($\mathcal{E}$), then also $c \in$ solution-set($\mathcal{E}$).*

Proof Let $Z = $ solution-set($\mathcal{E}$). We show that Z is transitive on sets. Every element of one of the sets s_x is either an element of A, or itself is of the form s_y for some $y \in X$.   ⊣

Proposition 6.2 *For all $A \subseteq \mathcal{U}$, $V_{afa}[A] \subseteq V[A]$. That is, if $\mathcal{E} = \langle X, A, e \rangle$ is a flat system of equations whose atoms are urelements, and s is the solution of $\mathcal{E}$, then for all $x \in X$, support(s_x) $\subseteq A$.*

Proof By Proposition 6.1, $Z = $ solution-set($\mathcal{E}$) $\cup A$ is transitive. For all $x \in X$, $TC(s_x)$ is the smallest transitive set which includes s_x. Therefore, $TC(s_x) \subseteq Z$. None of the elements s_x of the solution set can possibly be urelements, since they all either have elements (when $e_x \neq \emptyset$) or are the empty set. So $TC(s_x) \cap \mathcal{U} = Z \cap \mathcal{U} = A$.   ⊣

This result immediately suggests the question as to whether $V_{afa}[A] = V[A]$, that is, whether every set in $V[A]$ can be obtained as a part of some solution to a system of flat equations. Later in the chapter we will show that the answer is affirmative.

Here is an example which shows why this particular "flat" form of the Solution Lemma is often awkward to use. It also suggests how the difficulty can be gotten around. We will later take the idea embodied in this example and use it to prove a much more general form of the Solution Lemma.

Example 6.2 Suppose we wanted a solution to the following equation: $x = \{\{x, q\}, p\}$. At first sight, we are out of luck. There is no way to model this equation as a flat system of equations in the sense used above, since we cannot write the right-hand side as a union of a set of indeterminates and a set of atoms. But we can get around this by looking at a different system of equations. Consider the following pair of equations:

$$x = \{y, p\}$$
$$y = \{x, q\}$$

Now each right-hand side can be written in the desired form. For example, the set of indeterminates on which x depends is just $\{y\}$, and the constant part of the equation for x is $\{p\}$. The set we want is one member of the solution set of this expanded system of equations.

Exercise 6.3 Use the Solution Lemma to show that there is a unique set

$$\{0, \{1, \{2, \{3, \dots\}\}\}\}.$$

[Hint: You will need an infinite system of equations. Take the atoms of your system to be the set N of natural numbers. The set in question will be one member of the solution set of your system of equations.]

6.3 An extension of the Flat Solution Lemma

If you read the previous two sections carefully, you might note that the requirement that the indeterminates of a flat system be urelements is not really needed. That is, although in many cases one would prefer to use non-sets as the indeterminates in a system, this is not a mathematical necessity. In this section, we want to discuss this a bit further.

Let's call a triple $\mathcal{E} = \langle X, A, e \rangle$ a *generalized flat system* if X and A are any two disjoint sets, and if $e : X \to \mathcal{P}(X \cup A)$ as before. We'll speak of the indeterminates and atoms of such a system, and we'll also define the notion of a solution s in the same way as before.

Theorem 6.3 *Every generalized flat system $\mathcal{E} = \langle X, A, e \rangle$ has a unique solution s. Moreover, there is a flat system $\mathcal{E}' = \langle Y, A, e' \rangle$ using the same set A of atoms such that solution-set($\mathcal{E}$) = solution-set($\mathcal{E}'$).*

Proof The idea here is more important than the result, and we'll see it in a number of places later in the book. We want to replace X by a set of urelements. Now we can't just use any old set of urelements, since we need to have a set disjoint from A. However, we have the Strong Axiom of Plenitude to work with. For $x \in X$, let $y_x = \mathsf{new}(x, A)$. Let

$$Y \quad = \quad \{y_x \mid x \in X\}.$$

Then $Y \subseteq \mathcal{U}$ and $Y \cap A = \emptyset$. Let $\mathcal{E}'$ be the flat system $\langle Y, A, e' \rangle$, where

$$e'(y_x) \quad = \quad \{y_z \mid z \in e_x \cap X\} \cup (e_x \cap A).$$

Then $\mathcal{E}'$ has a solution, say s'. We get a solution s to the original $\mathcal{E}$ by $s_x = s'(y_x)$. (You should check that this really is a solution to $\mathcal{E}$.) This proves the existence part of the result. For the uniqueness, any solution s to $\mathcal{E}$ gives a solution s' to $\mathcal{E}'$ (check this too!). So since s' is unique, so is s.

The final assertion of the theorem is a consequence of the proof. ⊣

With this result at hand, you might wonder why we bothered with urelements at all in the original definitions. The reason is that the Flat Solution Lemma will generalize in ways in which Theorem 6.3 will not. We don't want to discuss this now, but the point will be clearer as the book progresses. For now, we want to end this chapter with a consequence of Theorem 6.3 and some exercises. The theorem is not as important for us as the next definition and result, and we encourage you to focus on the Flat Solution Lemma in the sequel.

Definition Let a be a set, and consider the following generalized flat system $\mathcal{E} = \langle X, A, e \rangle$: $A = support(a)$, $X = TC(\{a\}) - A$, and for all $x \in X$, $e_x = x$. We call this system the *canonical flat system* for a.

We should check that this really is a generalized flat system. For this, note that each $x \in X$ is a set, and $x \subseteq X$ by transitivity.

Another important fact is that the identity function on X is the solution to the canonical system for a. To see this, note that for any $x \in X$,

$$x \;=\; (x \cap X) \,\cup\, (x \cap A),$$

since X is transitive and $support(x) \subseteq A$.

The following is a consequence of this observation.

Theorem 6.4 *For all $A \subseteq \mathcal{U}$, $V[A] = V_{afa}[A]$. That is, if a is a set and $support(a) \subseteq A$, then there is a flat system $\mathcal{E} = \langle X, A, e \rangle$ such that $a \in solution\text{-}set(\mathcal{E})$. In particular, $V[\mathcal{U}] = V_{afa}[\mathcal{U}]$.*

Proof We showed in Proposition 6.2 that $V_{afa}[A] \subseteq V[A]$, so we need only prove the converse. Let a be any arbitrary element of $V[A]$, that is, a set with $support(a) \subseteq A$. Let $\mathcal{E}_a$ be the canonical flat system for a. The solution to $\mathcal{E}_a$ is the identity on X. Since $a \in X$, $a = s_a$ belongs to the solution set of $\mathcal{E}_a$. By Theorem 6.3, there is a flat system $\mathcal{E}'$ using the same set A of atoms, with the same solution set as $\mathcal{E}$. Thus $a \in solution\text{-}set(\mathcal{E}')$. This proves that $V[A] \subseteq V_{afa}[A]$, as desired. $\dashv$

Exercise 6.4 (A continuation of Exercise 6.3 in light of Theorem 6.4.) Recall that in set theory one often defines the natural numbers to be certain sets (see Section 2.1). Using these definitions, solve Exercise 6.3 without using any atoms.

As we have seen, *AFA* insures that every system of equations has a solution. This makes it natural to ask what systems have solutions under *FA*. The question has an easy answer. Let $<$ be defined on the indeterminates of a flat system by $x < y$ iff $y \in e_x$. Call the system *wellfounded* if $<$ is wellfounded on X. A result known as the Mostowski Collapsing Lemma, provable in ZFC^-, shows that wellfounded systems always have a unique solution; indeed, this is really just a restatement of the Mostowski Collapsing Lemma. Under *FA*, these are the only systems that have solutions:

Proposition 6.5 *FA is equivalent to the assertion that only wellfounded flat system $\mathcal{E}$ have solutions.*

We will not prove this result, since it is not needed in what follows.

Historical Remarks

Many researchers have been intrigued by non-wellfounded sets, and there are many papers on the topic in the literature. Most of these papers discuss axioms other than *AFA*. The axiom *AFA* itself was first formulated by Forti and Honsell (1983). Aczel rediscovered the axiom in 1984, in a form which we'll see in Chapter 10. The history of the entire subject, as well as many of the different axioms, is discussed in detail in Aczel (1988).

7

Bisimulation

In discussing the standard extension of the integers to the rationals, we noted that one needs to consider equations of the form $ax = b$ under a certain equivalence relation, since many equations determine the same rational number. (See page 68.) So too, in constructing the hypersets as an extension of the more familiar wellfounded sets, we need to realize that different systems of equations can give rise to the same solutions. This leads us to one of the most important topics of this book, bisimulation.

7.1 Bisimilar systems of equations

To motivate the definitions that follow, let's start with an example.

Example 7.1 Consider the following two systems of equations. $\mathcal{E}_1$ is the system

$$x = \{x\},$$

while $\mathcal{E}_2$ is the system

$$x = \{y\}$$
$$y = \{x, z\}$$
$$z = \{x\}.$$

Clearly, these are quite different systems of equations. However, both systems have as their solution set the set Ω introduced in Exercise 6.2. $\Omega = \{\Omega\}$, and it is the only set which is its own singleton. Thus $s_x = \Omega$ satisfies the system $\mathcal{E}_1$. But notice that if we assign Ω to each of the indeterminates in $\mathcal{E}_2$, then we have a solution to that system. All you need besides the definition of Ω is the observation that $\{b, b\} = \{b\}$ for all b (and hence for Ω).

We can generalize this example as follows.

Exercise 7.1 Let $\mathcal{E} = \langle X, A, e \rangle$ be any system of equations such that $A = \emptyset$ but each $e_x \neq \emptyset$. Show that Ω is the solution set for this system.

Example 7.2 Here is a second kind of example. Consider the following two systems:

$$
\begin{aligned}
x &= \{y, z, w\} & \qquad x' &= \{y', z'\} \\
y &= \{p, w\} & \qquad y' &= \{p, z'\} \\
z &= \{w\} & \qquad z' &= \{z'\} \\
w &= \{z, w\} &
\end{aligned}
$$

Notice that p is an atom of both systems. Let s be the solution to the system on the left. The last two equations on the left suggest that $s_z = s_w$. After all, there is no reason why they *shouldn't* be the same. And if $s_z = s_w$, then we would get a solution s' to the system on the right by setting $s'(x') = s_x$, $s'(y') = s_y$, and $s'(z') = s_z$. Therefore s (or rather its restriction to $\{x', y', z'\}$) would seem to be a solution to the second system as well.

These sorts of examples show that we must come to grips with a fundamental question: Under what conditions will two systems of equations have the same solutions?

This question is relatively easy to answer. If we have two systems with the same solutions, then we can see that a special relationship must hold between the parts of the two systems. But then it is easy enough to see that this special relationship is also sufficient for systems to have the same solution sets. This relationship is explained in the next definition.

Definition Let $A \subseteq \mathcal{U}$, and let $\mathcal{E} = \langle X, A, e \rangle$ and $\mathcal{E}' = \langle X', A, e' \rangle$ be two generalized flat systems of equations which use A as their set of atoms.

1. An *A-bisimulation relation* between $\mathcal{E}$ and $\mathcal{E}'$ is a relation R on $X \times X'$ such that the following conditions hold:
   a. Suppose that xRx'. Then for every indeterminate $y \in e_x \cap X$ there is an indeterminate $y' \in e'_{x'} \cap X'$ such that yRy'.
   b. Suppose that xRx'. Then for every indeterminate $y' \in e'_{x'} \cap X'$ there is an indeterminate $y \in e_x \cap X$ such that yRy'.
   c. If xRx', then e_x and $e'_{x'}$ contain the same atoms. That is, $e_x \cap A = e'_{x'} \cap A$.
2. We say that the systems are *A-bisimilar*, and write $\mathcal{E} \equiv \mathcal{E}'$, if there is an A-bisimulation relation between them with the following two properties:
   a. For every $x \in X$ there is an $x' \in X'$ such that xRx'.
   b. For every $x' \in X'$ there is an $x \in X$ such that xRx'.

Frequently, we omit the "A" and just speak about bisimulation relations and bisimulation.

As a hint that we are on the right trail, let's look at our two examples.

Example 7.3 The two systems in Example 7.1 are bisimilar. To see this, let R relate x to each of x, y, z. That is,

$$R = \{\langle x, x \rangle, \quad \langle x, y \rangle, \quad \langle x, z \rangle\}.$$

Let's check that the conditions of the definition are met. First, consider $\langle x, x \rangle$. The first component of this pair is x. The only element of $e_1(x)$ is x itself. So we go to the second component of the pair, namely x as an indeterminate of $\mathcal{E}_2$. Fortunately, there is an element, y, of $e_2(x)$ so that $R(x, y)$. Going the other way, the only element of $e_2(x)$ is y, and as we now know, x is an element of $e_1(x)$ and $R(x, y)$. This takes care of $\langle x, x \rangle$. We still have to consider the other two pairs in R. These verifications are similar so we leave them to you.

The systems in Example 7.2 are also bisimilar. Here the bisimulation relation is the following:

(1) $$\{\langle x, x' \rangle, \quad \langle y, y' \rangle, \quad \langle z, z' \rangle, \quad \langle w, z' \rangle\}.$$

We suggest you check that this is indeed a bisimulation relation.

The following result shows us that this is the special relationship we are after.

Theorem 7.1 *Let $\mathcal{E}$ and $\mathcal{E}'$ be flat systems over the same set $A \subseteq \mathcal{U}$. $\mathcal{E}$ and $\mathcal{E}'$ have the same solution sets if and only if they are bisimilar.*

Proof First, assume that $\mathcal{E}$ and $\mathcal{E}'$ have the same solution sets. Let s and s' be the solutions to the two systems. Define a relation R on $X \times X'$ by:

$$x R x' \quad \text{iff} \quad s(x) = s'(x').$$

It follows from the fact that the systems have the same solution sets that R is an A-bisimulation. We verify some of the conditions. Suppose first that $x \in X$, so that $s_x \in solution\text{-}set(\mathcal{E})$. Then there is some $x' \in X'$ such that $s_x = s'(x')$. (Again we write $s'(x')$ rather than $s'_{x'}$ for readability.) Hence $x R x'$. The converse is similar. Next, suppose that $x R x'$ and $y \in e_x \cap X$. Then since $s_y \in s_x = s'(x')$, we must have some $y' \in e'(x')$ such that $s_y = s'(y')$. Thus $y R y'$. Once again, the converse is similar. Finally, if $s_x = s'(x')$, then these sets must have the same urelements. The set of urelements in s_x is $e_x \cap A$, since each s_y is a set. The same holds for $s'(x')$. Therefore $e_x \cap A = e'_{x'} \cap A$. This concludes the verification that R is an A-bisimulation.

To prove the converse, suppose we have an A-bisimulation R between $\mathcal{E}$ and $\mathcal{E}'$. Let us suppose we could prove that $s_x = s'(x')$ if $x R x'$. This would allow us to prove that the solution sets of the two given systems are the same. For suppose $a \in solution\text{-}set(\mathcal{E})$. Then $a = s_x$ for some $x \in X$. By condition (2a) in the definition of bisimulation, there is some x' such that $x R x'$. Hence $s_x = s'(x')$ by our assumed result. But then $a = s_x = s'(x')$ would belong to *solution-set*($\mathcal{E}'$). This proves that *solution-set*($\mathcal{E}$) $\subseteq$ *solution-set*($\mathcal{E}'$). The converse is proved similarly.

Thus it remains only to prove that $s_x = s'(x')$ whenever xRx'. To do this, we construct a new generalized flat system $\mathcal{E}^*$ over the same $A \subseteq \mathcal{U}$. The set X^* of indeterminates of $\mathcal{E}^*$ will be the set of pairs $\langle x, x' \rangle$ such that xRx'.

Before exhibiting the system of equations for these new indeterminates, we first illustrate what happens in Example 7.2. Call the two systems $\mathcal{E}_1$ and $\mathcal{E}_2$. A bisimulation R between $\mathcal{E}_1$ and $\mathcal{E}_2$ was exhibited in (1). In this case, the equations of $\mathcal{E}^*$ would be

$$\begin{aligned}
\langle x, x' \rangle &= \{\langle y, y' \rangle, \langle z, z' \rangle, \langle w, z' \rangle\} \\
\langle y, y' \rangle &= \{p, \langle w, z' \rangle\} \\
\langle z, z' \rangle &= \{\langle w, z' \rangle\} \\
\langle w, z' \rangle &= \{\langle z, z' \rangle, \langle w, z' \rangle\}
\end{aligned}$$

The general formula is that for all $\langle u, u' \rangle \in X^*$,

$$e^*_{\langle u, u' \rangle} = \{\langle v, v' \rangle \in X^* \mid v \in e_u \text{ and } v' \in e'_{u'}\} \cup (A \cap e_u)$$

This defines our system $\mathcal{E}^*$. Here are two candidates for solutions of this system, functions s^1 and s^2 defined on X^* by

$$s^1_{\langle u, u' \rangle} = s_u \qquad s^2_{\langle u, u' \rangle} = s'_{u'}$$

Using the fact that R is an A-bisimulation, we show that these are in fact both solutions of $\mathcal{E}^*$. This is an important point, so we spell out some of the details. We'll show that s^1 is a solution since the proof for s^2 is very similar. Suppose $\langle u, u' \rangle$ is one of our new indeterminates. We must show that

(2) $\qquad s^1_{\langle u, u' \rangle} = \{s^1_{\langle v, v' \rangle} \mid \langle v, v' \rangle \in e^*_{\langle u, u' \rangle}\} \cup (A \cap e^*_{\langle u, u' \rangle})$.

To do this, take some element $b \in s^1_{\langle u, u' \rangle} = s_u$. Since s is a solution of $\mathcal{E}$, b is either of the form s_w for some $w \in X \cap e_1(u)$, or an urelement $z \in A \cap e^1_u$. In the first case, there is some $w' \in X' \cap e'_v$ such that wRw'. This means that $\langle w, w' \rangle \in X^*$. This is the key step in the proof because now we see that $b = s_w = s^1_{\langle w, w' \rangle}$ belongs to the set on the right-hand side of (2). And in the second case, $z \in A \cap e^*_{\langle u, u' \rangle}$ by the definition of $e^*_{\langle u, u' \rangle}$.

This completes half of the verification of (2). The other half is simpler but we prove it for the sake of completeness, and to stress the importance of this proof for what is to come. Suppose we have an element $s^1_{\langle v, v' \rangle}$ of the right hand side of (2), where $v \in e_u$ and $v' \in e'_{u'}$. We need to prove that $s^1_{\langle v, v' \rangle} \in s^1_{\langle u, u' \rangle}$. But $s^1_{\langle v, v' \rangle} = s_v$ and $s_u = s^1_{\langle u, u' \rangle}$. And $s_v \in s_u$ since s is a solution of $\mathcal{E}$. Further, suppose we have $z \in A \cap e^*_{\langle u, u' \rangle}$. Then $z \in (A \cap e_u)$ by definition of $e^*_{\langle u, u' \rangle}$. So $z \in s^1_u = s^1_{\langle u, u' \rangle}$.

STRONG EXTENSIONALITY OF SETS / 81

We have now shown that s^1 and s^2 are both solutions to $\mathcal{E}^*$. By the uniqueness part of *AFA*, $s^1 = s^2$. This means that for all $\langle u, v \rangle \in R$, $s_u = s_v$. This is what we needed to prove. ⊣

Corollary 7.2 *The relation of bisimulation on generalized flat systems over A is an equivalence relation; that is, it is reflexive, symmetric, and transitive.*

Proof The relation of having the same solution set is an equivalence relation on systems of equations. Based on the Theorem, this is now seen to be equivalent to the bisimulation relation. ⊣

Exercise 7.2 Our proof of this corollary uses the Solution Lemma, since the theorem used it. Give a direct proof which does not use the Solution Lemma.

The fact that bisimilar systems of equations give rise to the same solution sets can sometimes be a nasty shock when one is using *AFA* to model various non-wellfounded phenomena. Sometimes the natural way of writing down equations will assign bisimilar systems to objects which one might take to be distinct. But then these objects get modeled by the same set. What this means, in practice, is that we have to be careful to code any important distinction into our model explicitly. We will see an example of this in Section 7.3.

7.2 Strong extensionality of sets

As we mentioned earlier, the ordinary Axiom of Extensionality does not always help us to decide whether two sets a and b are distinct or not. For example, if $a = \{b\}$ and $b = \{a\}$, then all the usual formulation of Extensionality tells is that $a = b$ iff $b = a$. However, the results of the previous section show us that $a = b = \Omega$, because they are both solutions to the system of equations $x = \{y\}$ and $y = \{x\}$.

We would like to be able to have a criterion for set identity which does not require us to turn to systems of equations. Fortunately, the work that we've done on bisimulation gives us such a criterion.

Definition A *bisimulation relation on sets* is a binary relation R on sets which satisfies the following condition: if aRb then

(1) for every set $c \in a$ there exists a set $d \in b$ such that cRd.

(2) for every set $d \in b$ there exists a set $c \in a$ such that cRd.

(3) $a \cap \mathcal{U} = b \cap \mathcal{U}$.

We say that sets a and b are *bisimilar* if there is some bisimulation relation R on sets such that aRb.

Example 7.4 Suppose that p is an urelement and let $a = \{p, a\}$ and $b = \{p, \{p, b\}\}$. Here is a bisimulation R such that aRb:

$$R = \{\langle a, b \rangle, \langle a, \{p, b\} \rangle\}.$$

We stress that under our definition, R is a relation on sets; we do not put the pair $\langle p, p \rangle$ into R.

The following simple consequence of Theorem 7.1 gives us what we need to determine when sets are identical.

Theorem 7.3 (Strong Extensionality) *Let I be the identity relation on sets. Then I is the largest bisimulation relation on sets. That is,*

1. *I is a bisimulation relation on sets.*
2. *If R is a bisimulation relation on sets, then R is a subrelation of the identity relation. That is, if aRb then $a = b$.*

Proof Part (1) follows easily from the definition of a bisimulation. For part (2), assume that aRb. Recall the canonical generalized flat system $\mathcal{E} = \langle X, A, e \rangle$ for a from Theorem 6.4. We took $A = support(a)$, $X = TC(\{a\}) - A$, and $e_x = x$ for all $x \in X$. The identity is the solution to $\mathcal{E}$. Similarly, we get a canonical system $\mathcal{E}' = \langle X', A', e' \rangle$ for b.

We check at this point that $A = A'$. Suppose that $p \in support(a)$, say by having $p \in a' \in \mathcal{U}$ for some $a' \in TC(\{a\})$. Then there is a finite sequence

$$a = a_0 \ni a_1 \ni \cdots \ni a_n = a'.$$

By applying condition (1) in the definition of bisimulation on sets n times, we see that there is some $b' \in TC(\{b\})$ such that $a'Rb'$. But then $p \in support(b)$. This argument proves that $A \subseteq A'$, and a similar one proves the converse inclusion.

Let R^* be the restriction of R to $X \times X'$. We'll use the assumption that R is an A-bisimulation on sets to show that R^* is an A-bisimulation between $\mathcal{E}$ and $\mathcal{E}'$. We'll only verify half of the conditions, since the other halves are shown the same way. Let

$$Y = \{x \in X \mid \text{for some } x' \in X', xR^*x'\}.$$

We must show that $Y = X$, for this means that every element of X is related to some element of X'. Y contains a, and also if $x \in Y$ and $y \in x$, then $y \in Y$. (This uses the transitivity of X'.) So Y is a transitive set containing a. Since $TC(\{a\})$ is the smallest such set, $X = TC(\{a\}) \subseteq Y$.

Next, suppose that xR^*y and also that $x' \in e_x \cap X$. This just means that $x' \in x \cap X$. In particular, x' is a set. So as R is an A-bisimulation on sets, there is some set $y' \in y$ such that $x'Ry'$. By transitivity of X', $y' \in X'$. Thus $y' \in e'_y \cap X'$, and $x'R^*y'$.

Finally, we check the condition on urelements. Let xR^*y, so $x\cap\mathcal{U} = y\cap\mathcal{U}$. But $x\cap\mathcal{U}$ is a subset of $support(a) = A$, so $x\cap\mathcal{U} \subseteq A$. Similarly, $y\cap\mathcal{U} \subseteq A$. So

$$e_x\cap A \quad = \quad x\cap A \quad = \quad x\cap\mathcal{U}$$

and similarly for e'_y. Thus $e_x\cap A = e'_y\cap A$.

This proves that R^* is a bisimulation on flat systems over A. Since the identity maps are the solutions to $\mathcal{E}$ and $\mathcal{E}'$, Theorem 7.1 tells us that

$$
\begin{aligned}
a \quad &= \quad s_a \\
&= \quad s_b \\
&= \quad b
\end{aligned}
$$

This completes the proof. ⊣

Exercise 7.3 Does Theorem 7.3 hold in the universe of wellfounded sets?

Exercise 7.4 In our work from this section, a bisimulation relation could be a set or a proper class. Suppose that there is a bisimulation relation R between a and b such that aRb. Show that R may be taken to be a set.

7.3 Applications of bisimulation

We gather in this section some applications of bisimulation.

Stream bisimulations

For any set A, the *streams* over A, A^∞, is the largest collection such that $A^\infty = A \times A^\infty$. We will show in Theorem 14.1 that there always is a non-empty set A^∞ with this property. You should take this on faith for now. (That is, none of the points in this section actually require there to be any streams, and the only important fact we need about A^∞ is that $A^\infty = A \times A^\infty$.)

In modeling the informal notion of stream, we are concerned to get the right identity conditions. So we must ask under what conditions would we expect streams s and t to be distinct? Or, to put it the other way around, under what conditions would we want to say that $s = t$? Clearly, if $s = t$ then s and t have the same first element and have the same second element. Conversely, if s and t have the same first element and have the same second element then $s = t$. Thus we might try to characterize this by saying that $s = t$ iff $1^{st}(s_1) = 1^{st}(s_2)$ and $2^{nd}(s_1) = 2^{nd}(s_2)$. While true, this is not very helpful in certain cases. Suppose, for example, that s, t each have 1 as a first element, and each has itself as its second element. Our criterion, applied in this case, results in a circularity: it just says $s = t$ iff $s = t$.

What we want, of course, is that distinct streams must eventually result in distinct behavior. That is, if $s \neq t$ then one of the following holds:

$$1^{st}(s) \neq 1^{st}(t),$$

$$1^{st}(2^{nd}(s)) \neq 1^{st}(2^{nd}(t)),$$
$$1^{st}(2^{nd}(2^{nd}(s))) \neq 1^{st}(2^{nd}(2^{nd}(t))),$$

etc. Or, to put it the other way around, if $1^{st}(s) = 1^{st}(t)$, $1^{st}(2^{nd}(s)) = 1^{st}(2^{nd}(t))$, $1^{st}(2^{nd}(2^{nd}(s))) = 1^{st}(2^{nd}(2^{nd}(t)))$, ..., then $s = t$. In analogy with the case of sets, we make the following definition.

Definition Let A be a set, and consider A^∞, the streams over A.

A *stream bisimulation* is a binary relation R on A^∞ such that if sRt, then

$$1^{st}(s) = 1^{st}(t) \quad \text{and} \quad 2^{nd}(s) \quad R \quad 2^{nd}(t).$$

Proposition 7.4 *Let s and t be streams from a set A, and suppose that there is a stream bisimulation R on A^∞ such that sRt. Then $s = t$.*

Proof We extend R to the following relation R^* by

$$
\begin{array}{rl}
R & \cup \\
& \cup \quad \{\langle b, b \rangle \mid b \in TC(A) \text{ is a set}\} \\
& \cup \quad \{\langle \{a\}, \{a\} \rangle \mid a \in A\} \\
& \cup \quad \{\langle \{a, s\}, \{a, t\} \rangle \mid a \in A, sRt\}
\end{array}
$$

Then R^* is a bisimulation on sets: the extra pairs are added to R because each stream $\langle a, s \rangle$ is actually the set $\{\{a\}, \{a, s\}\}$. ⊣

On the successor operation

In this section let us write $s(a) = a \cup \{a\}$ since this is the usual successor function when applied to ordinals. The following result is easily shown for wellfounded sets. It also holds, though, on the non-wellfounded sets, which is perhaps a bit more surprising.

Proposition 7.5 (MOC Workshop) *The successor function is injective. That is, if $s(a) = s(b)$ then $a = b$.*

Proof Suppose towards a contradiction that $a \cup \{a\} = b \cup \{b\}$, but $a \neq b$. We first claim that $a \in b$ and $b \in a$. To see this, note that if $a \notin b$ then $a \notin b \cup \{b\}$. But then $a \notin a \cup \{a\}$, and this is a contradiction. Similarly, $b \in a$.

Next, we observe that

$$a - \{a, b\} \quad = \quad b - \{a, b\}.$$

That is, a and b have the same elements, except possibly for a and b themselves. To prove the inclusion from left to right: let $u \in a - \{a, b\}$. Then $u \in a \cup \{a\} = b \cup \{b\}$. Now u is not equal to a or to b, so it's a member of $(b \cup \{b\}) - \{a, b\} = b - \{a, b\}$. The other inclusion is similar.

Now we'll construct a bisimulation R on sets which relates a and b, and so the conclusion $a = b$ follows by strong extensionality. Define R by:

$$uRv \quad \text{iff} \quad \text{either} \quad u = v$$
$$\text{or } (u = a \text{ and } v = b)$$
$$\text{or } (u = b \text{ and } v = a)$$

We show that R is a bisimulation on sets. Suppose that u and v are sets and uRv. We don't have to worry about urelements. We check half of the bisimulation condition: that for every $u' \in u$ there is some $v' \in v$ such that $u'Rv'$. (By symmetry, the other half holds as well.) So let $u' \in u$. By definition, we have three cases:

(i) $u = v$.

(ii) $u = a$ and $v = b$.

(iii) $u = b$ and $v = a$.

In case (i), take $v' = u'$. Obviously $u'Rv'$.

Here is how to act in case (ii). If $u' = a$ or $u' = b$, take $v' = a$. (This is possible since, as we know, $b \in a$.) Either way, $u'Rv'$. Otherwise, $u' \in a - \{a, b\} = b - \{a, b\}$, so we can take $v' = u'$.

Case (iii) is similar to case (ii).

Thus R is a bisimulation so $a = b$ after all. ⊣

As a consequence of this result we obtain the following strange characterization of reflexive sets involving the successor function.

Proposition 7.6 *For all sets a, the following are equivalent:*

1. $s(a) \in a$
2. $s(a) = a$
3. a *is reflexive.*

Proof It's enough to prove (1) ⇒ (2), since easily (2) ⇔ (3) and (2) ⇒ (1). So suppose that $s(a) \in a$. Let $b = s(a)$. Then we have $s(b) = s(a) \cup \{s(a)\}$. But by assumption, $s(a)$ is a member of $a \subseteq s(a)$, so $s(b) = s(a)$. Now, by Proposition 7.5, it follows that $a = b$. So $a = s(a)$. ⊣

The Difference Lemma

We present here a simple but perhaps surprising result that will also be used much later in the book, in Chapter 20. Recall that a set a is *reflexive* if $a \in a$. Using the Solution Lemma, it is easy to construct many examples of reflexive sets. For example, given any set b, we can obtain a reflexive set $a = \{a, b\}$. The following shows a limit on the kinds of reflexive sets we can construct.

Proposition 7.7 (Difference Lemma) *Let V be a transitive set and let $b = V - \{V\}$. Then $b \notin V$.*

Proof If V is not reflexive, then $b = V$ so $b \notin V$ follows from the fact that V is not reflexive.[1] The basic idea is that if $b \in V$ then there would be no way to tell the difference between b and V, which is wrong since $V \in V$ and $V \notin b$. So suppose towards a contradiction that $b \in V$. Hence $b \in (V - \{V\}) = b$. Hence b is also reflexive.

We claim that the following relation R is a bisimulation:

$$R \;=\; \{\langle V, b \rangle\} \cup \{\langle a, a \rangle \mid a \in b\}$$

To check this, we only need to consider the pair $\langle V, b \rangle$. We must first show that for all $a \in V$ there is some $c \in b$ such that $\langle a, c \rangle \in R$. Well, if $a = V$, then we take $c = b$. If $a \neq V$, then $a \in b$ and we take $c = a$.

Second, we must show that for all $a \in b$ there is some $c \in V$ such that $\langle a, c \rangle \in R$. But since V is transitive, we can take $c = a$. ⊣

Exercise 7.5 We say that a set b is *closed under singletons* if $x \in b$ implies $\{x\} \in b$; b is *closed under differences* if for all sets $c, d \in b$, $c - d \in b$.

1. Find a transitive, reflexive set closed under singletons.
2. Find a transitive, reflexive set closed under differences.
3. Show that there is no transitive, reflexive set which is closed under both singletons and differences.

Exercise 7.6 Find a reflexive, transitive set V with transitive element c such that $(V - \{c\}) \in V$.

A characterization of the ordinals

It is a standard result that the ordinals are exactly those wellfounded sets that are transitive and linearly ordered by $\in$. We next have a parallel result characterizing the ordinals among all the hypersets, presented as an exercise.

Exercise 7.7 Prove that a is an ordinal iff a is a transitive set, linearly ordered by $\in$, and if no member of a is reflexive. [Hint: use the fact that if a is a wellfounded transitive set which is linearly ordered by $\in$, then a is an ordinal. You also will need to use some standard facts about ordinals.]

An example of the misuse of hypersets

Suppose that Jones likes to keep track of birds which live near his seaside home. Each time he sees a bird, he makes note of some feature which sets it apart from all the birds which he has ever seen. When a gull with a cracked beak lands on his porch, he can find no feature that sets it apart from a certain gull with a cracked beak three weeks ago and described in his notes as having

---

[1] Our original statement of this result assumed that V was reflexive. Gerry Wojnar pointed out that this is not needed.

a cracked beak. So he decides it is the same gull. Is this belief wellfounded? Probably not: there is no reason to suppose that any feature will be found on just one bird.

This is just one of a great number situations in which some model (features) has lead someone to inadvertently identify two objects being modeled (birds), when they might in fact be distinct. This is a pervasive problem in mathematical modeling, something to be guarded against.

An example of this using hypersets has to do with the Liar sentence "This sentence is not true." We won't discuss the matter at length until Chapter 13, but we want to point out where trouble can occur. Suppose we consider (3) and (4):

(3)   Sentence (3) is not true.

(4)   Sentence (3) is not true.

These sentences are versions of what are known as the Liar (which we discussed in Chapter 5) and the Strengthened Liar, respectively. As noted in Chapter 5, some people have argued that (3) can be neither true nor false, since assuming it is either leads to a contradiction. But if (3) can be neither true nor false, then (3) is not true so it seems that (4) must be true. (This, together with the observation that the two sentences seem to be making the same claim, is sometimes taken as a refutation of the position that the Liar is neither true nor false. For a discussion of a hole in the argument, see page 189.)

Suppose that Jones, in pondering these puzzling sentences one stormy night, decides to give mathematical model of their content as sets s_3 and s_4 in the following way:

$$s_3 = \langle \neg, Tr, s_3 \rangle$$
$$s_4 = \langle \neg, Tr, s_3 \rangle$$

where $\neg$ and Tr are urelements. The idea would be that $\neg$ stands for negation, and Tr for the truth predicate. This seems to reflect their logical structure pretty well. But notice that with this definition, s_3 and s_4 are bisimilar and hence identical! In this case, Jones might jump to the conclusion that the two sentences have the same content. In doing so, though, he might be making the same mistake all over again. If (3) and (4) are true under different conditions, for example, then the fact that s_3 and s_4 are identical is evidence of the inadequacy of his modeling scheme. The fact that s_3 and s_4 are identical shouldn't be regarded as a discovery about the meaning of the sentences (3) and (4). In Chapter 13, we will discuss an approach to modeling the semantics of these sentences that allows them to come out with different truth values.

7.4   Computing bisimulation

In this section we restrict attention to *finite* flat systems $\mathcal{E}_1 = \langle X_1, A_1, e_1 \rangle$ and $\mathcal{E}_2 = \langle X_2, A_2, e_2 \rangle$ of equations in order to ask the question: How difficult is it to tell if two such systems are bisimilar? One might think that the only way is to examine all possible relations R between X_1 and X_2, and then check each to see if it is a bisimulation. This would mean that the notion might not be practically computable: if X_1 has n elements and X_2 has m, then there are 2^{mn} possible relations to check.

As it happens, there is a much more efficient procedure which can be used to tell whether systems are bisimilar or not. We will present an algorithm for *pointed flat systems*, but the same general idea will work for sets, and also for pointed graphs (see Chapter 10). (Concerning sets, note that by Theorem 7.3, bisimilar sets are equal. However, this uses *AFA*. But the method of this section would give an algorithm to tell whether two finite sets are bisimilar, even in the absence of *AFA*.) It is also straightforward to extend the idea to flat systems which do not come with distinguished indeterminates.

Definition A *pointed flat system* is a tuple $\mathcal{E} = \langle X, A, e, x \rangle$ where $\langle X, A, e \rangle$ is a flat system and $x \in X$. A *bisimulation* of pointed flat systems $\mathcal{E}_1$ and $\mathcal{E}_2$ is a bisimulation R in our earlier sense which is also required to have $R(x_1, x_2)$.

The basic idea is that a pointed flat system not only describes a set to us, but also an element of particular interest. Two pointed flat systems are bisimilar if and only if the elements of interest are equal.

We now present an algorithm to tell whether two pointed systems

$$\mathcal{E}_1 = \langle X_1, A_1, e_1, x_1 \rangle \text{ and } \mathcal{E}_2 = \langle X_2, A_2, e_2, x_2 \rangle.$$

are bisimilar or not. To do this, build a relation $R \subseteq X_1 \times X_2$ by carrying out the following steps:

(Step 1) Let R_1 be all pairs $\langle u, v \rangle$ from $X_1 \times X_2$ such that either e_u and e_v differ on some atom, or one is empty and the other is not.

(Step $n + 1$) Given R_n, let R_{n+1} be R_n together with all pairs $\langle u, v \rangle$ such that either

    (a) there is some $u' \in e_u \cap X$ such that for all $v' \in e_v \cap X$, $R_n(u', v')$; or

    (b) there is some $v' \in e_v \cap X$ such that for all $u' \in e_u \cap X$, $R_n(u', v')$; or

Do this until $R_n = R_{n+1}$. Call this relation R, and let $\overline{R}$ be the complement of R. If $\langle x_1, x_2 \rangle \in \overline{R}$, then there is a bisimulation relating x_1 and x_2; it is $\overline{R}$. If $\langle x_1, x_2 \rangle \notin \overline{R}$, then there is no such bisimulation.

Here is how we analyze the algorithm: The process of building R can only go on for $m_1 m_2$ steps, where m_1 and m_2 are the numbers of elements in X_1 and X_2, respectively. The reason is that each step must add at least one pair to R, and the number of possible pairs is $m_1 m_2$. Each step involves looking at

$m_1 m_2$ pairs and deciding whether or not each is added to the relation. So the total number of steps is at most $m_1^2 m_2^2$.

We claim that $\overline{R}$, the complement of R, is the maximum bisimulation relation between the pointed systems $(\mathcal{E}_1)_{x_1}$ and $(\mathcal{E}_2)_{x_2}$. An easy induction on n shows that if $R_n(u, v)$, then there can be no bisimulation between $(\mathcal{E}_1)_u$ and $(\mathcal{E}_2)_v$. In addition, if $R_n = R_{n+1} = R$, then for all $\langle u, v \rangle \in \overline{R}$, $\overline{R}$ is a bisimulation between $(\mathcal{E}_1)_u$ and $(\mathcal{E}_2)_v$.

Historical Remarks

The concept of bisimulation was discovered several times in different places. A version for modal logic was proposed by van Benthem in 1976 and one for processes by Park in 1981. Theorem 7.1 and 7.3 are due to (1988). The Difference Lemma is new here, as are Propositions 7.5 and 7.6. Exercise 7.7 was suggested by G. A. Antonelli. The algorithm for computing bisimulation seems to be a folklore result.

8

Substitution

The Flat Solution Lemma is conceptually very straightforward. It is awkward to use in practice, however, since it forces us to write out everything in terms of sets of atoms and indeterminates. We cannot use familiar constructions involving natural numbers or ordered pairs, for example.

Example 8.1 Suppose we want to find a set $c = \langle p, c \rangle$ which is the ordered pair of p with c itself. Since $\langle p, c \rangle = \{\{p\}, \{p, c\}\}$, such a set will be non-wellfounded. However, to construct it, we will have to replace the simple equation $x = \langle p, x \rangle$ with a flat system of equations, using p as an atom. This reflects the way ordered pairs are usually modeled:

$$
\begin{aligned}
x &= \{y, z\} \\
y &= \{p\} \\
z &= \{p, x\}.
\end{aligned}
$$

It would obviously be more convenient to be able write $x = \langle p, x \rangle$ and have a form of the Solution Lemma that applied directly.

Example 8.2 Suppose that we have a set $Act = \{a, b\}$, and that we want a set S containing two elements s_1 and s_2 satisfying the following equations:

$$
\begin{aligned}
s_1 &= \{\langle a, s_1 \rangle, \langle b, s_2 \rangle\} \\
s_2 &= \{\langle a, s_1 \rangle, \langle a, s_2 \rangle\}
\end{aligned}
$$

We might want such an example if we were trying to build canonical labeled transition systems (see page 38). Once again, we could "flatten" these equations into a bigger system and then use the Solution Lemma. But the work here

91

becomes considerable, and so we look for a simplification. We want to prove a theorem which carries out this messy rewriting of systems once and for all.

8.1 General systems of equations

The first step in generalizing the Solution Lemma is to generalize the notion of a system of equations.

Definition A *(general) system of equations* is a tuple $\mathcal{E} = \langle X, A, e \rangle$ consisting of a set $X \subseteq \mathcal{U}$, a set $A \subseteq \mathcal{U}$ disjoint from X, and a function $e : X \to V_{afa}[X \cup A]$.

The point of having e take values in $V_{afa}[X \cup A]$ is twofold. First, it allows any set on the right hand side of the equation whose support is included in $X \cup A$. Second, it prohibits bare urelements from appearing on the right hand side, since urelements are not in $V_{afa}[X \cup A]$. The reason for this prohibition is to avoid systems like $x = x$; such a system would not have a unique solution. Besides, we could not hope to solve a system like $x = a$ (where $a \in A$) in which the right hand side of one of the equations was an atom, since our solutions take sets as values.

Example 8.3 Let p be any urelement, and let $y, z \in \mathcal{U}$. Define a general system $\mathcal{E}$ by taking $X = \{y, z\}$, $A = \{p\}$, $e_y = \{\Omega, p, \{z\}\}$, and $e_z = \langle y, z, z \rangle$. In Section 8.3, we'll define what it means to be a solution to a general system. The solution of this $\mathcal{E}$ will be a function s defined on $\{y, z\}$, such that

$$
\begin{aligned}
s_y &= \{\Omega, p, \{s_z\}\} \\
s_z &= \langle s_y, s_z, s_z \rangle.
\end{aligned}
$$

Notice that A is the support of the solution.

The restriction in the definition of a general system of equations that X be a set of urelements is crucial; we'll see why in Example 8.5, page 95. This is in contrast to the situation with flat systems. Some of our results would hold if we allowed the indeterminates to be arbitrary sets, but the formulation above seems to make for the most elegant overall theory.

8.2 Substitution

The main problem in generalizing the Solution Lemma comes in defining what it means for a function s be a solution to a general system of equations. Intuitively, we want s to be a solution if it assigns to each $v \in X$ a set s_v in such a way that s_v is the result of substituting, for each occurrence of an indeterminate x anywhere in the transitive closure of e_v, the value s_x. But first we must prove that this process of substitution is well defined. This may seem like a rather technical problem, but it is a problem nonetheless. And its solution is a good illustration in the use of *AFA*.

To show that the process of substitution mentioned above can be justified within *AFA*, we want to show that there is a substitution operation $sub(s, b)$, read "the result of substituting s_x for x in b."

Definition A *substitution* is a function s whose domain is a set of urelements. A *substitution operation* is an operation *sub* whose domain consists of a class of pairs $\langle s, b \rangle$ where s is a substitution and $b \in \mathcal{U} \cup V_{afa}[\mathcal{U}]$, such that the following conditions are met.

(1) If $x \in dom\ s$, then $sub(s, x) = s_x$.
(2) If $x \in \mathcal{U} - dom\ s$, then $sub(s, x) = x$.
(3) For all sets b, $sub(s, b) = \{sub(s, p) \mid p \in b\}$.

Usually, *sub* is taken to be the unique largest substitution operation, as shown to exist in the main theorem of this section, and we write $b[s]$ for $sub(s, b)$. Thus $[s]$ represents the operation taking each set or urelement b to $b[s] = sub(s, b)$.[1]

In words, substitution takes a given s and uses it directly on all urelements in its domain. For all sets b, it works "recursively," by substituting into the elements of b and gathering up the results. But as is typical with hypersets, this is not ordinary recursion since there is no "base case."

If we were dealing only with wellfounded sets b, there would be no difficulty in proving that *sub* exists and is unique. The proof would be a simple instance of the principle of definition by recursion on wellfounded sets. However, we are not restricted to wellfounded sets, so we must justify the definition differently than we would justify a definition by transfinite recursion. To remind ourselves of the analogy, though, we call the points (1)–(3) above the *corecursion conditions for sub*.

Before we turn to the proof that *sub* exists, let's look at an example which illustrates some problems to be overcome in proving this result, and the main ideas for the proof.

Example 8.4 Let $A = \{x, y, z\}$ be a set of three distinct urelements and let

$$b \;=\; \{x, y, \Omega, \{b, x, z\}\}$$

Let's suppose we want to substitute 3 for x, x for y, leaving z alone. That is,

---

[1] This makes our use of square brackets potentially ambiguous; $f[X]$ means the image of X under f when f is a function and $X \subseteq dom\ f$, whereas $b[s]$ means $sub(s, b)$ when s is a function with domain a subset of $\mathcal{U}$. In practice this ambiguity will not cause any confusion as we will always make sure it is clear from context which is intended. The square bracket notation for substitution is very convenient since it gives a convenient way to refer to the total operation $[s]$ on $\mathcal{U} \cup V_{afa}[\mathcal{U}]$. Incidentally, in writing $b[s]$ we depart from our practice of writing functions on the left of their arguments. This means that when you see terms like $f(b)[s]$ or $f(b[s])$ you need to be sure you see how the expression should be read.

we are considering an s with domain $\{x, y\}$ defined by $s(x) = 3, s(y) = x$ and calculating $b[s]$.

First, let's ask what the answer should be. A first guess might be the set $\{3, x, \Omega, \{b, 3, z\}\}$. A second's thought shows that this is not correct, since it does not satisfy the third corecursion condition. We have not reached deeply enough inside and done the substitution to b when it is considered as a element of an element of b. What we need in order to verify the corecursion conditions is that $b[s]$ would satisfy the equation

(1) $$b[s] \quad = \quad \{3, x, \Omega, \{b[s], 3, z\}\}$$

By now you should be able to figure out how to construct such a set $b[s]$ without much difficulty. We are going to do it in a very explicit and systematic way, one that suggests how to prove that *sub* exists.

We first form the canonical (generalized) flat system $\mathcal{E}$ for $b \cup \{3\}$ (cf. page 75.) The indeterminates of this system are the elements of

$$\begin{aligned} X &= (TC(\{b\}) \cup TC(\{3\})) - \mathcal{U} \\ &= \{b, \{b, x, z\}, \Omega, 3, 2, 1, 0\}, \end{aligned}$$

the set of atoms of the system is the same set $A = \{x, y, z\}$, and the equations are shown below:

$$\begin{aligned} e_b &= \{x, y, \Omega, \{b, x, z\}\} \\ e_{\{b,x,z\}} &= \{b, x, z\} \\ e_\Omega &= \{\Omega\} \\ e_3 &= \{0, 1, 2\} \\ e_2 &= \{0, 1\} \\ e_1 &= \{0\} \\ e_0 &= 0 \end{aligned}$$

Of course, this e is the identity on X. But writing it as we have should clarify things. As we know, this same identity map on X is the solution to $\mathcal{E}$. For what follows it is important to remember that the set $\{b, x, z\}$ in e_b is an indeterminate of this system of equations.

We now consider the system $\mathcal{E}' = \langle X, A, e' \rangle$ with the same indeterminates and atoms but where all occurrences of x on the right-hand sides are replaced by 3 and all occurrences of y on the right-hand sides are replaced by x. In other words:

$$
\begin{aligned}
e'_b &= \{3, x, \Omega, \{b, x, z\}\} \\
e'_{\{b,x,z\}} &= \{b, 3, z\} \\
e'_\Omega &= \{\Omega\} \\
e'_3 &= \{0, 1, 2\} \\
e'_2 &= \{0, 1\} \\
e'_1 &= \{0\} \\
e'_0 &= 0
\end{aligned}
$$

Notice that the indeterminate $\{b, x, z\}$ in e_b is not altered in going to e'_b, only the top-level occurrences of x and y get replaced by their values 3 and x respectively.

The identity map is not a solution to this system $\mathcal{E}'$, but the system does have a solution, call it sol, by *AFA*. We take $b[s]$ to be sol_b. Also, let $c = sol_{\{b,x,z\}}$ so that $c = \{b[s], 3, z\}$. Then

$$b[s] \;=\; \{3, x, \Omega, c\} \;=\; \{3, x, \Omega, \{b[s], 3, z\}\}.$$

So $b[s]$ does indeed satisfy equation (1).

Example 8.5 We now present the promised example showing why we require the domain X of a general system to be a set of urelements. Suppose $x = \emptyset$, $y = \{\emptyset\}$, and $X = \{x, y\}$. Let s be defined on X so that $s(x) = y$ and $s(y) = y$. Consider what happens if we try to calculate $sub(s, \{\emptyset\})$. On the one hand, this is $sub(s, y)$, so we should get $x = \emptyset$. On the other, $sub(s, \{x\})$ looks like it should be $\{y\}$. It is to prevent this kind of problem that we require $X \subseteq \mathcal{U}$.

Theorem 8.1 (Existence and Uniqueness of sub) *There is a unique operation $sub(s, b)$ which obeys the corecursion conditions for substitution and which is defined for all pairs $\langle s, b \rangle$ such that $dom\, s \subseteq \mathcal{U}$ and $b \in \mathcal{U} \cup V_{afa}[\mathcal{U}]$.*

Proof We define "$sub(s, b) = c$" if s is a function with domain a set of urelements and one of the following conditions hold:

(α) $b \in dom\, s$ and $c = s_b$

(β) $b \in \mathcal{U} - dom\, s$ and $c = b$

(γ) $b \in V_{afa}[\mathcal{U}]$, and the following holds.
  Let $X = (TC(\{b\} \cup rng(s))) - \mathcal{U}$, and let $A = (TC(\{b\} \cup rng(s))) \cap \mathcal{U}$.
  Let $\mathcal{E}' = \langle X, A, e' \rangle$ be the generalized flat system given by

$$e'_z \;=\; \{s_x \mid x \in z \cap dom\, s\} \cup \{x \mid x \in z \cap (A - dom\, s)\} \cup (z \cap X)$$

  Let sol be the solution of $\mathcal{E}'$. Then $c = sol(b)$.

The complicated definition in (γ) is intended to parallel the work we did in

Example 8.4. For example, $e'_{\{b,x,z\}}$ naturally splits into three sets: $\{3\} = \{s_b\}$; the set $\{z\}$ of other atoms in $\{b, 3, z\}$; and $\{b\} = \{b, x, z\} \cap X$.

To see that the overall definition of *sub* works, we need to see that *sub* has the intended domain and that it satisfies the corecursion equations. It is clear that it has the right domain, so we only need verify the corecursion equations. (1) and (2) are clear. To verify (3), let us write $\mathcal{E}_b$ and sol_b for the system of equations and its solution. For another set b', $sub(s, b')$ will be determined from another system $\mathcal{E}_{b'}$ via its solution $sol_{b'}$. Notice, though, that if $b' \in b - \mathcal{U}$, then $\mathcal{E}_{b'}$ is a subsystem of $\mathcal{E}_b$. Hence

$$sol_{b'}(b') \;=\; sol_b(b') \;\in\; sol_b(b).$$

(This also uses the fact that b' belongs to e'_b of $\mathcal{E}'$.) So

$$
\begin{aligned}
sub(s, b) \;=\;& sol_b(b) \\
=\;& \{s_x \mid x \in b \cap dom\, s\} \\
& \cup \{x \mid x \in b \cap (A - dom\, s)\} \\
& \cup \{sub(s, b') \mid b' \in b - A\}.
\end{aligned}
$$

For $x \in b \cap dom\, s$, $sub(s, x) = s_x$. And for $x \in b \cap (A - dom\, s) \subseteq \mathcal{U} - dom\, s$, $sub(s, x) = x$. Therefore,

$$sub(s, b) \;=\; \{sub(s, p) \mid p \in b\}.$$

This is just what corecursion (3) requires.

Turning to the uniqueness assertion, suppose that sub' also obeyed the corecursion conditions and was defined on all pairs $\langle s, b \rangle$. Fix s. Let R be the following relation on sets and urelements:

$$R \;=\; \{\langle sub(s, b), sub'(s, b) \rangle \mid b \in \mathcal{U} \cup V_{afa}[\mathcal{U}]\}.$$

We claim that the restriction of R to sets is a bisimulation on sets. We take a pair $\langle sub(s, b), sub'(s, b) \rangle$ and verify the conditions.

Suppose that $z \in \mathcal{U} \cap sub(s, b)$. Then since b is a set, z is of the form $sub(s, x)$ for some $x \in b \cap \mathcal{U}$. And $z = sub(s, x) = s_x = sub'(s, x)$, so z belongs to $sub'(s, b)$ as well. The other direction for urelements is similar.

Turning to sets, suppose that c is a set and belongs to $sub(s, b)$. Then if $b \in \mathcal{U}$, we again have $c \in s_b = sub'(s, b)$. On the other hand, if b is a set, then c is of the form $sub(s, p)$ for some $p \in b$. And $\langle sub(s, p), sub'(s, p) \rangle$ is a pair of sets in R.

Thus for all sets b, $sub(s, b) = sub'(s, b)$. For urelements, this assertion follows immediately from the definition. ⊣

Exercise 8.1 Prove that there is a unique operation Sk on sets with the property that

$$Sk(a) = \{Sk(b) \mid b \in a \text{ is a set}\}.$$

This operation throws out urelements of a set and gives the remaining "skeleton." [Hint: modify the method of Theorem 8.1.]

Exercise 8.2 Call an operation on urelements and sets *substitution-like* if for all sets a,
$$F(a) \;=\; \{F(p) \mid p \in a\}.$$
Prove that if F and G are substitution-like, then so is $F \circ G$.

Exercise 8.3 Let A be a set or class of urelements, let F and G be substitution-like, and suppose that for all $x \in A$, $F(x) = G(x)$. Prove that for all sets $a \in V_{afa}[A]$, $F(a) = G(a)$.

8.3 The general form of the Solution Lemma

Having convinced ourselves that substitution makes sense, we are now in a position to define what it means for s to be a solution of a general system of equations.

Definition Let $\mathcal{E} = \langle X, A, e \rangle$ be a general system of equations. A *solution* to $\mathcal{E}$ is a function s with domain X such that for each $x \in X$,

$$s_x \;=\; e_x[s].$$

To see that this is what we want, have another look at Example 8.3. We had a general system $\mathcal{E}$ there, and we also said that a certain function s was the solution to $\mathcal{E}$. You should go back and make sure that this s really is a solution to $\mathcal{E}$, by using the definition of substitution.

Exercise 8.4 Let p and q be urelements, and let $\mathcal{E}$ be the flat system with $X = \{x, y, z\}$, $A = \{p, q\}$, and equations given by:

$$
\begin{aligned}
x &= \{p, x, y\} \\
y &= \{q, x, z\} \\
z &= \{y\}
\end{aligned}
$$

Let s be the solution to $\mathcal{E}$. Show that s takes values in $V_{afa}[A]$. Show that s is also a solution to the following system $\mathcal{E}'$:

$$
\begin{aligned}
x &= \{p, \{p, x, y\}, \{q, x, z\}\} \\
y &= \{q, \{p, x, y\}, \{y\}\} \\
z &= \{\{q, x, z\}\}
\end{aligned}
$$

Our next result will be used many times throughout the rest of the book.

Theorem 8.2 (The General Solution Lemma) *Every general system of equations $\mathcal{E}$ has a unique solution s. Moreover, the solution set of $\mathcal{E}$ is a subset of $V_{afa}[A]$, where A is the set of atoms of $\mathcal{E}$.*

Proof The existence and uniqueness of s are immediate consequences of the next result and the Flat Solution Lemma. The solution set is included in $V_{afa}[A]$ by the next result and Proposition 6.2. $\dashv$

Lemma 8.3 *Let $\mathcal{E} = \langle X, A, e \rangle$ be a general system of equations. There is a flat system $\mathcal{E}^\flat = \langle Y, A, e' \rangle$ with $X \subseteq Y$ such that the following hold:*

1. *If s is a solution of $\mathcal{E}$, then s extends to a solution s' of $\mathcal{E}^\flat$. (Note that s' is necessarily unique, being a solution of $\mathcal{E}^\flat$.)*
2. *If s' is a solution of $\mathcal{E}^\flat$ and $s = s' \upharpoonright X$ (the restriction of s' to X), then s is a solution to $\mathcal{E}$.*

In particular, there is a one-to-one correspondence between solutions of $\mathcal{E}$ and $\mathcal{E}^\flat$.

Proof We have already seen the idea several times. Let

$$Y \;=\; \left(X \cup \bigcup_{x \in X} TC(e_x) \right) - A.$$

For future reference, note that

$$Y \cup A \;=\; (Y - X) \cup X \cup A.$$

For $x \in X$, let $e'_x = e_x$. Note that each such e'_x is a subset of $Y \cup A$. Each $y \in Y - X$ is also subset of $Y \cup A$, and we let $e'_y = y$. This gives us a flat system $\mathcal{E}^\flat$.

We verify the assertions concerning solutions. Let s be a solution to $\mathcal{E}$. Define a map s' on Y by

(2) $$s'_y \;=\; y[s].$$

This s' extends the function s on X, since for $x \in X$, $s_x = x[s]$. We check that s' as in (2) is a solution to $\mathcal{E}^\flat$.

$A \cap X = \emptyset$, so for all $a \in A$, $a[s] = a = a[s']$. The definition of substitution and the transitivity of Y imply that for all $y \in Y - X$,

$$
\begin{aligned}
s'_y \;&=\; y[s] \\
&=\; \{z[s] \mid z \in y\} \cup (y \cap A) \\
&=\; \{s'_z \mid z \in y\} \cup (y \cap A) \\
&=\; e'_y[s']
\end{aligned}
$$

For $x \in X$, $s'_x = s_x = e_x[s]$. Therefore

$$
\begin{aligned}
s'_x \;&=\; \{z[s] \mid z \in e_x\} \\
&=\; \{z[s] \mid z \in e_x \cap (Y - X)\} \cup \{s_z \mid z \in e_x \cap X\} \cup (e_x \cap A) \\
&=\; \{s'_z \mid z \in e_x \cap (Y - X)\} \cup \{s'_z \mid z \in e'_x \cap X\} \cup (e'_x \cap A) \\
&=\; e'_x[s']
\end{aligned}
$$

This proves that s' is a solution of $\mathcal{E}^\flat$.

For the converse, suppose s' is a solution of $\mathcal{E}^{\flat}$. Let $s = s' \upharpoonright X$. We need to see that s is a solution to the original system $\mathcal{E}$.

Before we get to that, we claim that for all $y \in Y$, $s'_y = y[s]$. To see this, let

$$R = \{\langle s'_y, y[s] \rangle \mid y \in Y\}.$$

We check that R is a bisimulation relation on sets. Suppose that $R(s'_y, y[s])$. Let $z \in s'_y \cap \mathcal{U}$. Then by the definition of the solution of a flat system, $z \in e_y \cap A = y \cap A$. Since $X \cap Z = \emptyset$, $z = z[s] \in y[s]$. Going the other way, suppose that $z \in y[s] \cap \mathcal{U}$. Now since s takes sets as its values, $p[s] \in \mathcal{U}$ is only possible when $p \notin dom\,(s)$. So since $z = z[s]$, we see that $z \notin dom\,(s)$. It follows that $z \in A$. Therefore $z \in s'_y$.

Further, suppose that c is a set and $c \in s'_y$. Then c must be of the form s'_v for some $v \in Y$. But $R(c, v[s])$, and so we've shown this part of the bisimulation condition. For the final part of the verification, suppose that c is a set belonging to $y[s]$. Then by corecursion condition (3) for substitution, c is of the form $v[s]$ for some $v \in y$. Since v is a set and Y is transitive, $v \in Y$. And again, $R(s'_v, v[s])$.

Having shown that $s'_y = y[s]$ for all $y \in Y$, we return to a proof that s is a solution to $\mathcal{E}$. Since $s_x = s'_x = e'_x[s'] = e_x[s']$, we see that for all $x \in X$,

$$
\begin{aligned}
s_x &= \{s'_z \mid z \in e_x - A\} \cup (e_x \cap A) \\
&= \{s'_z \mid z \in e_x \cap (Y - X)\} \cup \{s'_z \mid z \in e_x \cap X\} \cup (e_x \cap A) \\
&= \{z[s] \mid z \in e_x \cap (Y - X)\} \cup \{z[s] \mid z \in e_x \cap (X \cup A)\} \\
&= e_x[s]
\end{aligned}
$$

This concludes the verification. ⊣

Exercise 8.5 Following the proof above, compute $\mathcal{E}^{\flat}$ for the following system $\mathcal{E}$:

$$
\begin{aligned}
x &= \{\{\{y\}, \emptyset\}, x\} \\
y &= \{\{y\}, p\}
\end{aligned}
$$

From now on, we will work with the general form of the Solution Lemma.

Theorem 8.4 *AFA is equivalent (in ZFC$^-$) to the assertion that every general system of equations has a unique solution.*

Proof Recall that the original formulation of *AFA* is the Flat Solution Lemma on page 72. We have proved Theorem 8.2 assuming ZFC$^-$ and this original formulation.

Going the other way, let $\mathcal{E} = \langle X, A, e \rangle$ be a flat system. The original definition in Chapter 6 permits A to be any set whatsoever. Let $B = support(A)$, so that $B \subseteq \mathcal{U}$. If need be, we can replace X by a new set of urelements which is disjoint from B; the new system will have a unique solution if and only if the old one has a unique solution. Then each e_x belongs to $\mathcal{P}(A \cup X) \subseteq V_{afa}[X \cup B]$.

So we may regard e as a map $e : X \to V_{afa}[X \cup B]$. The upshot is that $\mathcal{E}' = \langle X, B, e \rangle$ is a general system. Let s be its solution (in the sense of this chapter). Since X is disjoint from the support of A, we see that $a[s] = a$ for all $a \in A$. It follows from this that s is a solution of $\mathcal{E}$ (in the sense of Chapter 6). Conversely, if s is any such solution, then s is a solution to $\mathcal{E}'$. ⊣

We close this section with a surprising result: assuming *AFA* for general systems with one indeterminate, we can prove *AFA* for general systems with two indeterminates. (It is possible to extend this result to get *AFA* for all finite systems, but not for arbitrary systems.) We won't use this result, but we present it in case you want a challenge. Also, the ideas in the solution are related to those of the next section.

Exercise 8.6 Working in set theory without *AFA*, assume that every general system $\mathcal{E} = \langle X, A, e \rangle$ *with X a singleton* has a unique solution. Prove that every general system $\mathcal{E}$ with X a set of two elements has a unique solution. [Hint: to get started, let r be the solution of the system $x = e_x$. The next system to solve is *not $y = e_y$*.]

8.4   The algebra of substitutions

This section contains some abstract material related to the material in this chapter but that we will not need until we return to theoretical material in Part V of the book. Feel free to omit it until you need it.

By a *substitution* we mean a function s defined on a subset of $\mathcal{U}$. Every s extends to a function $[s]$ which is defined on the whole universe $V_{afa}[\mathcal{U}] \cup \mathcal{U}$ of sets and urelements. What we want to do here is to explore some of the algebra of substitutions.

Let s and t be substitutions. If $s_x \in \mathcal{U}$ for all $x \in \mathcal{U}$, then we can define a composition $t \circ s$. This map will then be another substitution. However, there is no reason why s should be so restricted. And indeed, the substitutions that *AFA* gives us take sets for some of their values. So the relevant concept of composition will have to be a bit different than $t \circ s$.

Definition Let s and t be substitutions. Then $t \star s$ is the substitution whose domain is $dom(s)$, and such that for $x \in dom(s)$,

$$(t \star s)_x = s_x[t].$$

(To see that this is on the right track, note that if s happens to map its domain $dom(s)$ into $dom(t)$, then $t \star s = t \circ s$.) This defines $t \star s$ as a substitution; of course, we extend it to an operation $[t \star s]$ defined on all sets and urelements. We leave the main properties of this operation as exercises. You may omit them until you need the results, but be sure to read Theorem 8.5 below.

Exercise 8.7 Suppose that $dom(t) \subseteq dom(s)$. Prove that for all sets a, $a[t \star s] = (a[s])[t]$. That is, $[t \star s] = [t] \circ [s]$. [Hint: use Exercise 8.2.]

Exercise 8.8 Prove the following associativity property of $\star$:

if $dom(u) \subseteq dom(t)$, then $u \star (t \star s) = (u \star t) \star s$.

Also, find a left-identity element i for $\star$, that is, some i such that $i \star s = s$ for all s. Prove that i is unique.

A substitution s is called *proper* if $s_x \in V_{afa}[\mathcal{U}]$ whenever $s_x \neq x$. That is, s takes sets as its values whenever it is not the identity. Proper substitutions exclude cases which are sometimes interesting, where s shuffles urelements around. On the other hand, solutions of general systems of equations are proper substitutions. Also, if s and t are proper, so is $s \star t$.

Now we come to the main point of this section. This is the observation that *a proper substitution is basically the same as a general system of equations!*[2] This is startling, because it shows that two of the main concepts of the book, that of general system of equations and that of solution, are modeled by the same thing.

Theorem 8.5 *AFA is equivalent to the assertion that for every proper e there is a unique proper s such that $s = s \star e$.*

Proof By Theorem 8.4, *AFA* is equivalent to the assertion that every general system of equations has a unique solution.

Assume that every general system of equations has a solution and let e be any proper substitution. Then $\mathcal{E} = \langle X, A, e \rangle$ is a general system of equations, where $X = dom(e)$, and $A = support(e) - X$. Let s be the solution to $\mathcal{E}$. The definition of solution implies that $s = s \star e$.

Conversely, assume the hypothesis on proper substitutions. Let $\mathcal{E} = \langle X, e \rangle$ be a general system of equations. So $X \subseteq \mathcal{U}$. Then let s be such that $s = s \star e$. This implies that $dom(s) = X$, and that for all $x \in X$, $s_x = e_x[s]$. So X is a solution to $\mathcal{E}$.   ⊣

Exercise 8.9 This exercise is an abstract form of Exercise 8.4.

1. Let e be the proper substitution given in Exercise 8.4. What is $e \star e$?
2. Prove as a general fact that e and $e \star e$ have the same solutions.

There is a second way to define a composition on substitutions. We'll say that the *codomain* of a substitution s is $support(dom(s)[s])$. In other words, the codomain of s consists of all the urelements that occur in objects of the form s_x, as x ranges over the domain of x. We'll write $cod(s)$ for this codomain. Now we define $t \cdot s$ *only when* $cod(s) = dom(t)$; in this case, $t \cdot s$

---

[2]We say "basically" because a general system is strictly speaking a tuple, and substitutions are just functions. What we mean is that there is such a natural correspondence between the formal definitions that we'll slide over the small difference and pretend that the concepts are identical.

will be the substitution with domain $dom(s)$ and given again by the formula $(t \cdot s)_x = s_x[t]$.

The main advantage of this operation over $\star$ is that it is automatically associative: if $u \cdot (t \cdot s)$ is defined, then $u \cdot (t \cdot s) = (u \cdot t) \cdot s$. On the other hand, Theorem 8.5 would have to change. *AFA* is equivalent to the assertion that for every proper e there is a unique proper s such that $s = s' \cdot e$, where

$$s' \quad = \quad s \cup \{\langle z, z \rangle \mid z \in cod(e) - dom(e)\}.$$

Because of this complication, we have decided to use $\star$ as our composition operation in later parts of this book.

Historical Remarks

As far as we can tell, the work on flattening and the equivalence of the Flat Solution Lemma with the assertion that every general system has a unique solution is new.

The material in Section 8.4 is in some ways related to the *algebraic theories* introduced by Lawvere as a category-theoretic framework for studying substitution. One branch of this work studies the connection of substitution to fixed-point operations. A reference for this subject is the book Bloom and Ésik (1993). That book is also relevant to our work in that it studies algebraic properties of the operation of solving a system of equations. Exercise 8.6 is an adaptation of a result from this literature, originally from Bloom et al. (1977).

9

Building a model of *ZFA*

We have asserted more than once that the universe of hypersets is richer than
the universe of wellfounded sets, richer in that you can think of the former as
an enlargement of the latter obtained by "throwing in" solutions to systems of
equations. This chapter makes this claim precise and proves it. As a corollary
of this result, we obtain what is called a relative consistency proof, a proof that
if *ZFC* is consistent, then so is *ZFA*. That is, if one could prove a contradiction
in *ZFA*, then you could also prove a contradiction in *ZFC*. Since intuitions and
years of experience have convinced most mathematicians that *ZFC* is indeed
consistent, this is a persuasive argument that *ZFA* is consistent. This chapter
contains a proof of these results. The rest of the book is intended to make the
case for the usefulness of *ZFA*. This chapter can be skipped without loss of
continuity, though we think you should understand the gist of the construction,
at least as summarized in the box on page 104.

Our proof will proceed as follows. We'll start with a model M of ZFC^-
and show how to extend M to a universe M_{afa} where all flat systems of
equations have solutions, and where the axioms of ZFC^- still hold. In order
that M_{afa} be an extension of M, we'll assume that M is strongly extensional;
that is, that sets which are bisimilar in M are identical. Every model of *ZFC* is
strongly extensional, and every model of ZFC^- contains within it a model of
ZFC. Hence this section shows how to take any model M of ZFC^-, strongly
extensional or not, and get a model M_{afa} of *ZFA*. It is only to insure that M is
embedded in M_{afa} that we require M to be strongly extensional.

The idea of the construction is reasonably straightforward, but there are many details to check. For comparison with the extension of the integers to the rationals, discussed earlier, we give a parallel outline.

Outline of the Construction of M_{afa} from M

(0) We begin with a strongly extensional model M of *ZFC⁻*. We focus on the membership relation and the urelements, and we are concerned about the axioms of set theory.

(1) Our extension of M has a *purpose*: we would like to have a structure in which we have unique solutions to all flat systems.

(2) Consider the class of all pointed flat systems $e = \langle \mathcal{E}, x \rangle$ which belong to M. Define

$$\langle \mathcal{E}, x \rangle \quad \equiv \quad \langle \mathcal{E}', x' \rangle$$

iff there is a bisimulation R of $\mathcal{E}_x$ and $\mathcal{E}'_{x'}$ such that xRx'. Check that this is an equivalence relation. Write [e] for the equivalence class of e.

(3) Turn the set of equivalence classes into a structure for set theory by $[\langle \mathcal{E}, x \rangle] \in [\langle \mathcal{E}', y \rangle]$ iff there is an element $z \in e'_y \cap X'$ such that $\langle \mathcal{E}, x \rangle \equiv \langle \mathcal{E}', z \rangle$. Check that this does not depend on the choice of elements of the equivalence classes. For the urelements of M_{afa}, we use the same urelements as in M.

(4) Check that M_{afa} satisfies all of the axioms of *ZFC⁻*.

(5) Find an embedding $a \mapsto \overline{a}$ of M into M_{afa}. The embedding must be one-to-one, and it must have the property that $a \in b$ in M iff $\overline{a} \in \overline{b}$ in M_{afa}. Roughly speaking, the association takes a to the canonical system for a.

(6) Finally, we check that M_{afa} actually meets the purpose we mentioned in (1). That is, that the Solution Lemma holds in M_{afa}.

9.1 The model

Given a flat system $\mathcal{E} = \langle X, A, e \rangle$ and an indeterminate x of $\mathcal{E}$, we want to winnow all the indeterminates that are irrelevant to the value assigned to x by the solution. To do this, let X_x be the smallest subset of X containing x such that for all y, if $y \in X_x$ then $b_y \subseteq X_x$. (Here and in the next sentence we are using the notation b_y and c_y from the definition of a flat system of equations.) Let A_x be the set of atoms which occur in any e_y, as y ranges over X_x. So

$A_x = \bigcup\{c_y \mid y \in X_x\}$. Now let $\mathcal{E}_x$ be the system $\langle X_x, A_x, e_x \rangle$, where e_x is the restriction of e to X_x. The point of this construction is captured by the following simple exercise.

Exercise 9.1 Let $\mathcal{E}$ be a system of equations, x be an indeterminate, and let s be the solution to the system. Let s' be the solution to the system $\mathcal{E}_x$. Show that s' is the restriction of s to X_x. In particular, it follows that $s'_x = s_x$.

Our model is constructed from pairs $\langle \mathcal{E}, x \rangle$ where $\mathcal{E} = \langle X, A, e \rangle$ is a system of equations over A (and which belongs to M), $x \in X$, $X_x = X$, $A_x = A$, and $A \subseteq \mathcal{U}$. The requirements that $X_x = X$ and $A_x = A$ just insure that everything in $\mathcal{E}$ actually pertains to x and hence to the set s_x. We require that $A \subseteq \mathcal{U}$ to simplify details to come.[1] We call pairs like $\langle \mathcal{E}, x \rangle$ *pointed flat systems (of equations)*. In what follows we'll sometimes denote pointed systems by sans-serif letters like e and e'. Moreover, we let $\mathcal{PS}$ be the class of all pointed flat systems or urelements. Finally, we let p and q range over $\mathcal{PS}$. Define

$$\langle \mathcal{E}, x \rangle \quad \equiv \quad \langle \mathcal{E}', x' \rangle$$

iff $A = A'$ and there is an A-bisimulation R of $\mathcal{E}_x$ and $\mathcal{E}'_{x'}$ such that $x \, R \, x'$.

The sets in our model M_{afa} are essentially the equivalence classes of pointed flat systems with respect to $\equiv$. We say "essentially" here, because this is the idea. However, we have a problem because these equivalence classes are proper classes, and we would like the elements of M_{afa} to be sets. Here is a way around this difficulty. Exercise 2.12 shows that the Strong Axiom of Plenitude gives us a proper class $C \subseteq \mathcal{U}$ and a proper class $<$ of pairs of elements of C which is a wellorder of C. Each $x \in C$ gives an ordinal α_x. For each $X \subseteq C$, we then get an ordinal $\alpha_X = \sup_{x \in X} \alpha_x$. Now we say that e $= \langle \langle X, A, e \rangle, x \rangle$ is *standard* if $X \subseteq C$, and if $\langle Y, A, e' \rangle \equiv \mathcal{E}$, then $\alpha_X \le \alpha_Y$. For each pointed system $\mathcal{F}$, there is an equivalent standard pointed system $\mathcal{F}'$; furthermore, there is only a set of such $\mathcal{F}'$. So we define

$$[e] \quad = \quad \{f \mid f \text{ is standard and } f \equiv e\}.$$

In order to define membership in M_{afa}, we first define a relation E on pointed systems of M as follows:

$$\langle \mathcal{E}, x \rangle \, E \, \langle \mathcal{E}', y \rangle \quad \text{iff} \quad \text{there is some } z \in e'_y \cap X' \text{ such that } \langle \mathcal{E}, x \rangle \equiv \langle \mathcal{E}'_z, z \rangle$$

This means that we look at $\mathcal{E}' = \langle X', A, e' \rangle$ and more specifically at $e'_y \cap X'$, the indeterminates in e'_y. If one those indeterminates, say z, has the property that $\langle \mathcal{E}, x \rangle \equiv \langle \mathcal{E}'_z, z \rangle$, then we put $\langle \mathcal{E}, x \rangle; E \, \langle \mathcal{E}', y \rangle$. It need not be the case that $\langle \mathcal{E}, x \rangle$ and $\langle \mathcal{E}', y \rangle$ have the same atoms in order that these pointed systems

---

[1] For reference later in the chapter, recall that our flat systems of Chapter 6 were allowed to have any set A whatsoever. By Theorem 8.4, *AFA* is equivalent to the assertion that all flat systems $\mathcal{E} = \langle X, A, e \rangle$ with $A \subseteq \mathcal{U}$ have unique solutions.

be related by E. (But if they are related, then for whichever z witnesses the definition, $\mathcal{E}_x$ and $\mathcal{E}'_z$ will have the same sets of atoms.)

Here is the idea behind this definition: Suppose that $\langle \mathcal{E}, x\rangle$ and $\langle \mathcal{E}', y\rangle$ belong to M and are related by E. Let s and s' be the solutions of $\mathcal{E}$ and $\mathcal{E}'$, respectively. Then the sets s_x and s'_z are equal, since $\equiv$ is a bisimulation. But as $s'_z \in s'_y$, this means that $s_x = s'_z \in s_y$.

The key is that the pointed systems *belong to* M but *name sets* in M_{afa}. So the membership relation in M_{afa} should be defined in terms of systems in M. This is where E comes in. We define membership between sets in M_{afa} as follows:

(1) $\qquad [\langle \mathcal{E}, x\rangle] \in [\langle \mathcal{E}', y\rangle]$ in M_{afa} iff $\langle \mathcal{E}, x\rangle\, E\, \langle \mathcal{E}', y\rangle$ in M.

We also want to extend $\equiv$ and E to relate urelements to pointed systems. If $x \in \mathcal{U}$ and $p \in \mathcal{PS}$, then we define $x \equiv p$ iff $x = p$. If p is an urelement of M, we say that

$$p\, E\, \langle \mathcal{E}, x\rangle \quad \text{iff} \quad p \in c_x.$$

Turning to M_{afa}, we say that

(2) $\qquad p \in [\langle \mathcal{E}, x\rangle]$ in M_{afa} iff $p \in e_x$.

in the model M_{afa} which we are building.

At this point, we need to check that our definition of membership in M_{afa} is independent of the choice of representatives of equivalence classes. This is the content of parts (3) and (4) of the following collection of useful results concerning $\equiv$ and E.

Lemma 9.1 *Concerning the relations $\equiv$ and E on $\mathcal{PS}$:*

1. $\equiv$ *is an equivalence relation.*
2. *If $y \in e_x \cap X$, then $\langle \mathcal{E}_y, y\rangle\, E\, \langle \mathcal{E}, x\rangle$.*
3. *If $e_1 \equiv e_2$, $f_1 \equiv f_2$, and $e_1\, E\, f_1$, then also $e_2\, E\, f_2$.*
4. *If $e \equiv f$ and $p\, E\, e$, then $p\, E\, f$.*
5. $e \equiv f$ *iff the following conditions hold:*
    (a) For every $e'\, E\, e$ there is some $f'\, E\, f$ such that $e' \equiv f'$.
    (b) For every $f'\, E\, f$ there is some $e'\, E\, e$ such that $e' \equiv f'$.
    (c) For all urelements p, $p\, E\, e$ iff $p\, E\, f$.

Proof Part (1) was essentially done in Exercise 7.2. (That result did not mention pointed systems, but taking account of the pointedness is easy.)

Part (2) is a simple consequence of our definitions.

Part (3) follows from transitivity of $\equiv$ and the definition of E.

Part (4) also follows easily from the definitions.

For part (5), write $\langle \mathcal{E}, x\rangle$ for e and $\langle \mathcal{E}', y\rangle$ for f. Suppose first that $\langle \mathcal{E}, x\rangle \equiv \langle \mathcal{E}', y\rangle$. We verify (a), since (b) is similar. So assume that $\langle \mathcal{F}, u\rangle\, E\, \langle \mathcal{E}, x\rangle$.

This means that there is some $z \in e_x \cap X$ such that $\langle \mathcal{F}, u \rangle \equiv \langle \mathcal{E}_z, z \rangle$. And since $\langle \mathcal{E}, x \rangle \equiv \langle \mathcal{E}', y \rangle$, there is some $v \in e'_y \cap X'$ such that $\langle \mathcal{E}_z, z \rangle \equiv \langle \mathcal{E}'_v, v \rangle$. By transitivity of $\equiv$, $\langle \mathcal{F}, u \rangle \equiv \langle \mathcal{E}'_v, v \rangle$. And by part (2), $\langle \mathcal{E}'_v, v \rangle \; E \; \langle \mathcal{E}', y \rangle$. Condition (c) is immediate.

Conversely, assume (a) – (c). Consider the relation R^* on $X \times X'$ given by $u R^* v$ iff $\langle \mathcal{E}_u, u \rangle \equiv \langle \mathcal{E}'_v, v \rangle$. Then using (a) – (c) and part (2), R^* is a bisimulation between $\mathcal{E}$ and $\mathcal{E}'$, and $x R^* y$. ⊣

To summarize: we take the sets in M_{afa} to be the set of equivalence classes of pointed systems of M, and the urelements of M_{afa} to be those of M. The membership relation in M_{afa} is defined by (1) and (2).

9.2 Bisimulation systems

At this point, we have a strongly extensional model $M \models ZFC^-$ and the definition of a structure M_{afa} based on M. We have not yet verified that M_{afa} has any nice properties; we'll turn to this in Sections 9.3 and 9.4 below. In this section, we want to look a bit further at pointed systems.

First, we need a general construction which takes a set of pointed systems or urelements and gives another pointed system.

Proposition 9.2 *Let $S \subseteq \mathcal{PS}$. There is a pointed system S^+ such that for all $\mathsf{p} \in \mathcal{PS}$, $\mathsf{p} \; E \; S^+$ iff for some $\mathsf{q} \in S$, $\mathsf{p} \equiv \mathsf{q}$. Moreover, if e is any pointed system with this property, then $\mathsf{e} \equiv S^+$.*

Proof Let $S_0 = S - \mathcal{U}$. By the Axiom of Plenitude, we may assume that the sets of indeterminates used in the systems of S_0 are pairwise disjoint. The pointed system S^+ we want has as its set of indeterminates the unions of the sets of indeterminates from elements of S, together with a new indeterminate, say z. We'll call this big set X. The atoms of S^+ are the unions of the atoms the systems in S_0, together with all the urelements in $S \cap \mathcal{U}$. The equations of S^+ are the union of the equations of the $\mathcal{E} \in S_0$, plus one more:

$$z \;\; = \;\; \{x_{\mathcal{E}} : \mathcal{E} \in S_0\} \cup (S \cap \mathcal{U}).$$

Here $x_{\mathcal{E}}$ is the distinguished indeterminate of the pointed system $\mathcal{E}$.

The uniqueness assertion is by part (5) of Lemma 9.1. ⊣

Exercise 9.2 Consider the flat system $\mathcal{E} = \langle \{x, y\}, \{a, b\}, e \rangle$, where e is given by

$$
\begin{aligned}
x &= \{a, y\} \\
y &= \{b, x\}
\end{aligned}
$$

Write e_x for the pointed system $\langle \mathcal{E}, x \rangle$, and e_y for $\langle \mathcal{E}, y \rangle$. Exhibit the bisimulations which show that

$$
\begin{aligned}
\mathsf{e}_x &\equiv \{a, \mathsf{e}_y\}^+ \\
\mathsf{e}_y &\equiv \{b, \mathsf{e}_x\}^+
\end{aligned}
$$

Now we want to take this exercise and turn the result around. Instead of taking a flat system and getting some bisimulation equations, we want to take a "system of bisimulation equations" and solve it. Here are the relevant definitions, paralleling what we did in Section 6.1:

Definition

1. A *flat bisimulation system* is a tuple $\mathcal{E} = \langle X, A, e \rangle$ consisting of a set $X \subseteq \mathcal{U}$, a set $A \subseteq \mathcal{U}$ disjoint from X, and a function
    $$e : X \to \mathcal{P}(X \cup A).$$

2. X is called the set of *indeterminates* of $\mathcal{E}$, and A is called the set of *atoms* of $\mathcal{E}$. For each $v \in X$, let $b_v =_{\mathrm{df}} e_v \cap X$ and let $c_v =_{\mathrm{df}} e_v \cap A$.

3. A *bisimulation solution* to $\mathcal{E}$ is a function s with domain X, giving a pointed system for each $x \in X$ and satisfying each equation in the sense of bisimulation:
    $$s_x \quad \equiv \quad (\{s_y : y \in b_x\} \cup c_x)^+.$$

Let $\mathcal{E}$ be any flat system of equations. Working in set theory without *AFA* we cannot hope to solve $\mathcal{E}$. However, $\mathcal{E}$ is also a flat bisimulation system, and we can ask whether it has a (unique) solution in the sense of the definition above. Luckily, this is the case.

Theorem 9.3 (*ZFC⁻*) *Let* $\mathcal{E} = \langle X, A, e \rangle$ *be a flat system of bisimulation equations. Then* $\mathcal{E}$ *has a bisimulation solution* s. *Moreover,* s *is unique in the following sense: if* t *is any bisimulation solution to* $\mathcal{E}$, *then for all* $x \in X$, $t_x \equiv s_x$.

Proof The bisimulation solution is given by $s_x = \mathcal{E}_x$. To check that this works, we must show that for all $x \in X$,
$$\mathcal{E}_x \quad \equiv \quad (\{\mathcal{E}_y : y \in b_x\} \cup c_x)^+.$$

This verification is a generalization of Exercise 9.2.

The uniqueness is more involved. We'll illustrate the ideas by working an example. Suppose we have the following bisimulation system $\mathcal{E}$:

$$
\begin{aligned}
x &= \{p, y, z\} \\
(3) \qquad y &= \{q, x\} \\
z &= \emptyset
\end{aligned}
$$

A bisimulation solution would give us three pointed systems, say $\mathcal{F}$, $\mathcal{G}$, and $\mathcal{H}$ so that

$$
\begin{aligned}
\mathcal{F} &\equiv \{p, \mathcal{G}, \mathcal{H}\}^+ \\
\mathcal{G} &\equiv \{q, \mathcal{F}\}^+ \\
\mathcal{H} &\equiv \emptyset^+
\end{aligned}
$$

Let's write $\mathcal{F} = \langle\langle X, A, e\rangle, x'\rangle$, $\mathcal{G} = \langle\langle Y, B, f\rangle, y'\rangle$, and $\mathcal{H} = \langle\langle Z, C, g\rangle, z'\rangle$. Actually, it is not hard at this point to check that $A = B = \{p, q\}$, and $C = \emptyset$.

Note that the prime letters are intended to correspond to the variables used in the original system $\mathcal{E}$ above. We may assume that the sets X, Y, and Z are pairwise disjoint. Having these variables disjoint, we can put all the pointed systems together into one big system. Let's call this system s. Also, $\mathsf{s}_{x'} = \mathcal{F}$, $\mathsf{s}_{y'} = \mathcal{G}$, and $\mathsf{s}_{z'} = \mathcal{H}$.

Let T be the following relation:

$$T = \{\langle u, v\rangle \in (X \cup Y \cup Z) \times \{x, y, z\} \mid \mathsf{s}_u \equiv \mathsf{s}_{v'}\}.$$

The notation here is a bit "hairy" because of the primes. For example, $\langle u, x\rangle$ in T means that $\mathsf{s}_u \equiv \mathsf{s}_{x'}$. For another example, note that T contains the pairs $\langle x', x\rangle$, $\langle y', y\rangle$, and $\langle x', x\rangle$. As you can see, the reason that we are playing this trick with the primes is that we really want relations between the variables in $X \cup Y \cup Z$ and the set $\{x, y, z\}$ of atoms of the original $\mathcal{E}$ from (3).

Let $T_x = T \cap (X \times \{x, y, z\})$, and let T_y and T_z be defined similarly. We claim that T_x is a bisimulation between s_x and $\mathcal{E}_x$, and similarly for y and z.

Suppose that $\langle u, x\rangle$, say belongs to T_x. This means that $\mathsf{s}_u \equiv \mathsf{s}_{x'}$. Since $\mathsf{s}_{x'} \equiv \{p, \mathsf{s}_{y'}, \mathsf{s}_{z'}\}^+$, we see that $e_u \cap \mathcal{U} = \{p\}$, and that every element $v \in e_u$ which is not an urelement is bisimilar to either $\mathsf{s}_{y'}$ or $\mathsf{s}_{z'}$.

Going to the unprimed letters, we see that for every $v \in e_u$ which is not an urelement, either $\langle v, y\rangle$ or $\langle v, z\rangle$ belongs to T_x. The converse assertion holds also. This proves that T_x has the back-and-forth property.

To check that T_x preserves the atoms in the right way, note that the set of atoms of $\mathsf{s}_{x'}$ is the same as the atoms of $\mathcal{E}_x$, namely $\{p, q\}$; similar assertions hold for the other two relations. This means that if $\langle u, v\rangle \in T$, the atoms of s_u are the same as those of $\mathcal{E}_v$.

T_x is a bisimulation because in addition, every element of X is related to some element of $\{x, y, z\}$. This is an easy consequence of the fact that our pointed systems are all accessible; every variable gets used. ⊣

Theorem 9.3 is the most important step in proving the consistency of ZFA. It says, informally, that in set theory *without* AFA we can already solve bisimulation equations, and the solutions are "nearly unique". The remainder of the model construction passes to a quotient in order to get actual solutions (and unique ones).

9.3 Verifying *ZFC⁻*

We sketch some of the proofs that the axioms of ZFC^-, as well as the Solution Lemma, do hold in M_{afa}. To verify that some axiom φ holds in M_{afa}, we translate φ into a statement φ^{tr} about pointed systems in M. Then we prove φ^{tr} in ZFC^-.

Here is how the translation φ^{tr} goes: Write φ out using just the symbols $\in$ and $=$. Then replace $\in$ by E and $=$ by $\equiv$. Also, replace a quantifier "for all sets x" by a quantifier over pointed systems: "for all pointed systems e." Do the same for existential quantifiers over sets.

For example, the Axiom of Extensionality (see p. 28) says:

EXTENSIONALITY $(\forall a)(\forall b)[(\forall p)(p \in a \leftrightarrow p \in b) \rightarrow a = b]$.
Translating, we get

EXTENSIONALITY^{tr} $(\forall \mathsf{e})(\forall \mathsf{f})[(\forall \mathsf{p})(\mathsf{p}\, E\, \mathsf{e} \leftrightarrow \mathsf{p}\, E\, \mathsf{f}) \rightarrow \mathsf{e} \equiv \mathsf{f}]$.
We've changed the set variables a and b to sans-serif letters e and f to remind ourselves that we are dealing with pointed systems. In translation, variables like p and q range over arbitrary elements of $\mathcal{PS}$; these may be urelements or pointed systems.

Why does this translation procedure work? Well, M_{afa} is the set of equivalence classes of pointed systems (and urelements). The formal verification of an axiom in M_{afa} requires us to work with equivalence classes. We first must show that the truth or falsity of any formula in the language of set theory is independent of the choice of representatives of the equivalence classes. This is not hard to show, by induction. Furthermore, for any formula

$$M_{afa} \models \varphi([\mathsf{e}_1], \dots, [\mathsf{e}_n]) \quad \text{iff} \quad M \models \varphi^{tr}(\mathsf{e}_1, \dots, \mathsf{e}_n).$$

So given an axiom φ of *ZFA*, we'll consider its translation φ^{tr}. We'll prove φ^{tr} in ZFC$^-$. It follows that M will satisfy φ^{tr}, and so M_{afa} will satisfy φ. So from now on, we'll usually forget about M and just prove the translations of the axioms.

Return now to the translation of EXTENSIONALITY^{tr}. This follows from Lemma 9.1: If e and f have the same E-members, then reflexivity of $\equiv$ and part (5) of the Lemma tell us that $\mathsf{e} \equiv \mathsf{f}$.

Turning to the Pairing Axiom, note that the translation PAIRING^t is $(\forall \mathsf{e})(\forall \mathsf{f})(\exists \mathsf{g})[\mathsf{e}\, E\, \mathsf{g} \wedge \mathsf{f}\, E\, \mathsf{g}]$. This follows from Theorem 9.3.

Here is how we verify the Union Axiom. Let $\mathsf{e} = \langle \langle X, A, e \rangle, x \rangle$ be a pointed system. Let

$$S \quad = \quad \{y \in X \mid (\exists z \in e_x)\, y \in e_z\}.$$

Let $\mathsf{f} = \{\mathsf{e}_y \mid y \in S\}^+$. Then $(\mathsf{f} = \bigcup \mathsf{e})^{tr}$.

Exercise 9.3 Prove POWERSET^{tr}.

The key to the verification of the Collection and Separation Schemes is that the M_{afa} is "essentially definable" as a class inside of M. We omit these details.

For the Axiom of Choice, it is convenient to first build up some more machinery. We'll do this in the next section, and return to AC in Exercise 9.5.

To check the Axiom of Infinity, we need a result of independent interest, namely M_{afa} is an extension of M. Translating to sets, this means that the class of sets is nicely embedded in the class of pointed systems. To see this, let a be a set. By Theorem 6.4, there is a flat system $\mathcal{E}$ such that $a \in$ *solution-set*$(\mathcal{E})$. In fact, $\mathcal{E}$ from that theorem is a flat system isomorphic to the canonical system for a. For each set a, fix such a flat system $\mathcal{E}_a$. Then let $\bar{a}$ be the pointed system $\langle \mathcal{E}_a, a \rangle$. For all sets $b \in a$, $\bar{b} \ E \ \bar{a}$. Conversely, if f $E \ \bar{a}$, then for some $b \in a$, f $\equiv \bar{b}$.

Going back to the Axiom of Infinity, let ω be the natural numbers (see page 13). The property of ω in the axiom is that

$$\emptyset \in \omega \land (\forall b)[b \in \omega \to (\exists c \in \omega)\, c = b \cup \{b\}].$$

So $\bar{\emptyset} \ E \ \bar{\omega}$, and every e $E \ \bar{\omega}$ is $\equiv$ to $\bar{n}$ for some n.

Let e be a pointed system such that e $E \ \bar{\omega}$. Then for some $n \in \omega$, e $\equiv \bar{n}$. Now $\bar{n}$ has the property that for all pointed systems f, f $E \ \overline{n+1}$ iff either f $\equiv \bar{n}$ or f $E \ \bar{n}$. It follows that $(\overline{n+1} = \bar{n} \cup \{\bar{n}\})^{tr}$. This implies that $(\overline{n+1} = \text{e} \cup \{\text{e}\})^{tr}$.

This argument proves that $\bar{\omega}$ satisfies the translation of the Axiom of Infinity:

INFINITY^{tr} $(\exists a)[\emptyset \in a \land (\forall b)[b \in a \to (\exists c \in a)\, c = b \cup \{b\}]]^{tr}$.

We pause for a moment to mention a property of the mapping $a \mapsto \bar{a}$. The following relation S is a bisimulation on sets: aSb iff $\bar{a} \equiv \bar{b}$. We want to take this back to our models M and M_{afa}. Since we have assumed that the original M is strongly extensional, we see that $a \mapsto [\bar{a}]$ is an injection of M into M_{afa}. Moreover, it is easy to check at this point that $a \in b$ in M iff $[\bar{a}] \in [\bar{b}]$ in M_{afa}. The existence of this embedding on M inside of M_{afa} was one of our overall goals for this construction.

Exercise 9.4 Recall that our basic set theory ZFC$^-$ includes the Strong Axiom of Plenitude (see page 23).[2] For this, we also have a binary function symbol new in our language. In doing the work of this section, we would need some function, say n, so that we could translate new(a, b) by $n(a, b)$. The function n should take pairs consisting of pointed systems and urelements, and it should return an urelement.

Come up with a definition of n, and then verify φ^{tr}, where φ is the Strong Axiom of Plenitude.

9.4 Verifying *AFA*

Now we turn to a discussion of *AFA* in M_{afa}; we'll show that M_{afa} satisfies the Flat Solution Lemma. Recall that this axiom says that every flat system has a solution. We stated this axiom in terms of the defined notions of unordered pair,

---

[2]It is also possible to assume just the ordinary Axiom of Plenitude and then prove its translation.

ordered pair, function, and domain. These would have to be written out fully before translating the *AFA* into a form that we could work with. We won't give the official statements or the translations, but we will indicate how everything works.

We need a few definitions. Given $a, b \in \mathcal{PS}$, we have a pointed system $\{a, b\}^+$ as described in Proposition 9.2. This pointed system has a and b as its E-elements. Now we define $\langle a, b \rangle^+ = \{\{a\}^+, \{a, b\}^+\}^+$. This would be a pointed system which acts like the ordered pair of a and b. We extend this operation to triples by defining $\langle a, b, c \rangle^+ = \langle a, \langle b, c \rangle^+ \rangle^+$.

If $f : \mathcal{PS} \rightharpoonup \mathcal{PS}$, then f^+ is the pointed system given by

$$f^+ \quad = \quad \{\langle a, f(a) \rangle^+ \mid a \in dom(f)\}^+.$$

Note that we are overloading the $+$ notation, using it to mean different things for pairs, triples, and functions.

We'll need one more piece of notation, and then some lemmas on the whole machinery. If $\mathcal{E} = \langle X, A, e \rangle$ is a flat system in M, then $\mathcal{E}^+$ is the pointed system of M given by

$$\mathcal{E}^+ \quad = \quad \langle X^+, A^+, \{\langle x, (e_x)^+ \rangle^+ \mid x \in X\} \rangle^+.$$

Note that here, the $+$ on X, A, and each e_x is the operation for sets from Proposition 9.2. The $+$ on the pairs and triples are the ones described earlier in this section.

Lemma 9.4 *Let* e, f, g, *and* h *be pointed systems.*[3]

1. $(\text{``}g = \{e, f\}\text{''})^{tr}$ *iff* $g \equiv \{e, f\}^+$.
2. $(\text{``}g = \langle e, f \rangle\text{''})^{tr}$ *iff* $g \equiv \langle e, f \rangle^+$.
3. $(\text{``}g = \langle e_1, e_2, e_3 \rangle\text{''})^{tr}$ *iff* $g \equiv \langle e_1, e_2, e_3 \rangle^+$.
4. $(\text{``}e \subseteq \mathcal{U}\text{''})^{tr}$ *iff for some* $A \subseteq \mathcal{U}$, $e \equiv A^+$.

Proof We sketch the proofs. In all of the parts, the right-to-left direction is easy, and we omit it.

For part (1), note that "$g = \{e, f\}$" really means $e \in g$, $f \in g$, and every $h \in g$ is equal to either e or f. Writing this out and translating it, we get $e\,E\,g$, $f\,E\,g$, and every $h\,E\,g$ is $\equiv$ to either e or f. Assuming this, we use Theorem 9.3 to see that $g \equiv \{e, f\}^+$. (That is, we consider the bisimulation system $x = \{e, f\}$.)

For part (2), suppose that $(\text{``}g = \langle e, f \rangle\text{''})^{tr}$. This means that there are $h_1$ such that $(\text{``}h_1 = \{e, f\}\text{''})^{tr}$, $(\text{``}h_2 = \{f, f\}\text{''})^{tr}$, and $(\text{``}g = \{h_1, h_2\}\text{''})^{tr}$. By (1),

---

[3]The quotes around the assertions in this lemma are used because the translation operation is defined on formulas of the language of set theory, not on expressions of natural language. In the cases at hand, we prefer to write the ordinary language expressions instead of rendering them by formulas. So to be more readable, we put quotes around an informal expression to refer to any formula which formalizes it.

we have $h_1 \equiv \{e, f\}^+$, $h_2 \equiv \{f, f\}^+$, and $g \equiv \{h_1, h_2\}^+$. We can put these bisimulations together to see that $g \equiv \langle e, f \rangle^+$.

Part (3) is similar to part (2).

In the last part, assume that ("$e \subseteq \mathcal{U}$")^{tr}. This means that for all $p \in \mathcal{PS}$, if $p \, E \, e$, then $p \in \mathcal{U}$. Write e as $\langle X, A, e, x \rangle$. We claim that $e(x) \cap X = \emptyset$. For if $y \in e(x) \cap X$, then $e_y E e$. This would contradict the assumption that ("$e \subseteq \mathcal{U}$")^{tr}. The fact that $e(x) \cap X = \emptyset$ implies that that $X = X_x = \{x\}$. By definition of $\mathcal{PS}$, A must be a set of urelements. So $e \equiv A^+$. ⊣

Lemma 9.5 *Let* e, f, g, *and* h *be pointed systems.*

1. ("f *is a function*")^{tr} *iff there is a function* $f : \mathcal{PS} \rightharpoonup \mathcal{PS}$ *such that*
   a. *If* $f(p) \equiv f(q)$, *then* $p \equiv q$.
   b. $f^+ \equiv f$.
2. *Suppose that* $f : \mathcal{PS} \rightharpoonup \mathcal{PS}$ *and* $f^+ \equiv f$. *Then* ("f(g) = h")^{tr} *iff there is some* g$'$ $\equiv$ g *and* h$'$ $\equiv$ h *such that* $f(g') = h'$.
3. *Let* $S \subseteq \mathcal{PS}$, *and let* $f : S \to \mathcal{PS}$ *and* $g : S \to \mathcal{PS}$. *Then* $f(p) \equiv g(p)$ *for all* $p \in S$ *iff* $f^+ = g^+$.
4. ("e *is a flat system whose atoms are urelements*")^{tr} *iff there is a flat bisimulation system* $\mathcal{E}$ *such that* $e \equiv \mathcal{E}^+$.
5. ("e *is a flat system and* s *is a solution to* e")^{tr} *iff there is a flat bisimulation system* $\mathcal{E}$ *such that* $e \equiv \mathcal{E}^+$ *and also a bisimulation solution* s *of* $\mathcal{E}$ *such that* $s^+ \equiv$ s.

Proof We continue to give some of the details of the left-to-right directions.

For (1), let $f = \langle \langle X, A, e \rangle, x \rangle$. For each $y \in e_x$,

$$(\text{"f}_y \text{ is an ordered pair"})^{tr}.$$

For each $y \in c$ there is a pair $\langle p_y, q_y \rangle$ of pointed systems such that $\langle p_y, b_y \rangle^+ \equiv f_y$; this is by Lemma 9.4.2. By Collection, let S be a set containing a pair $\langle p_y, q_y \rangle$ for each $y \in e_x$. By the Axiom of Choice, we can find a subset $f \subseteq S$ such that

1. For every $\langle p, q \rangle \in S$ there is some $\langle p', q' \rangle \in f$ such that $p \equiv p'$.
2. If $\langle p, q \rangle$ and $\langle p', q' \rangle$ are distinct elements of f, then $p \not\equiv p'$.

Condition (2) insures that f is a function and that (1a) holds. We claim that $f^+ \equiv f$. It is sufficient to show that for each $y \in e_y$, there is some $\langle p, q \rangle \in f$ such that $\langle p, q \rangle^+ \equiv f_y$. We do know that there is some $\langle p_y, q_y \rangle \in S$ such that $\langle p_y, q_y \rangle^+ \equiv f_y$. By condition (1), there is some $\langle p, q \rangle \in f$ such that $p \equiv p_y$. For this $\langle p, q \rangle$, there is some $z \in e_x$ such that $\langle p, q \rangle^+ \equiv f_z$. Then

$$(\text{"f is a function, } \langle p_y, q_y \rangle^+, p_y \equiv p, \text{ and } \langle p, q \rangle^+ \in f\text{"})^{tr}.$$

So $q_y \equiv q$. Therefore $\langle p, q \rangle^+ \equiv \langle p_y, q_y \rangle^+ \equiv f_y$. This concludes the proof of part (1).

In (2), we see that ($``\langle g, h \rangle \in f"$)^{tr}. So as $f^+ \equiv f$, there is a pair $\langle g', h' \rangle \in f$ such that $\langle g, h \rangle \equiv \langle g', h' \rangle$. By Lemma 9.4(2), $g \equiv g'$ and $h \equiv h'$.

Next, we verify the extensionality assertion of (3). Let $e \ E \ f^+$. Then there is some $p \in S$ such that $e \equiv \langle p, f(p) \rangle^+$. Since $f(p) \equiv g(p)$, $e \equiv \langle p, g(p) \rangle^+ \ E \ g^+$. The same argument works in reverse, of course, and we conclude that f^+ and g^+ have the same E-members. Thus $f^+ \equiv g^+$.

Turning to (4), we know that there are e_1, e_2, and e_3 such that $e = \langle e_1, e_2, e_3 \rangle^+$. Moreover, there are $X, A \subseteq \mathcal{U}$, and e_1 such that $e_1 \equiv X^+$ and $e_2 \equiv A^+$. By part (1), there is a function e such that $e^+ \equiv e_3$. Now for each $x \in X$, ($``e_3(x) \subseteq e_1 \cup e_2"$)^{tr}. So for each x there is some $b_x \subseteq X \cup A$ such that $e_3(x) \equiv (b_x)^+$. Let e be the function defined on X by $e_x = b_x$. So $\mathcal{E} = \langle X, A, e \rangle$ is a flat system of equations, Then

$$\{ \langle x, (e_x)^+ \rangle^+ \mid x \in X \} \quad \equiv \quad e_3.$$

and therefore

$$\mathcal{E}^+ \quad = \quad \langle X^+, A^+, e^+ \rangle^+ \quad \equiv \quad \langle e_1, e_2, e_3 \rangle^+ \quad = \quad e.$$

Part (5) is the goal of this lemma, and the main result leading up to Theorem 9.6. It relates the concept of a solution of a flat system to that of a bisimulation solution. By part (4), we can find $\mathcal{E} = \langle X, A, e \rangle$ such that $\mathcal{E}^+ \equiv e$. By part (1), we can find a function s defined on X so that $s^+ \equiv s$. For all $x \in X$,

$$(``s_x \quad = \quad \{ s_y : y \in b_x \} \cup c_x ")^{tr}$$

This means that for all $p \in \mathcal{PS}$, $p \ E \ s_x$ iff either for some $y \in b_x$, $s_x \equiv b_x$, or $p \in c_x$. By Proposition 9.2,

$$s_x \quad \equiv \quad (\{ s_y : y \in b_x \} \cup c_x)^+.$$

By part (2), for each $x \in X$, $s_x \equiv s_x$. So for all $x \in X$,

$$s_x \quad \equiv \quad (\{ s_y : y \in b_x \} \cup c_x)^+.$$

This shows that s is a bisimulation solution of $\mathcal{E}$. $\qquad\qquad \dashv$

Theorem 9.6 ANTI-FOUNDATION AXIOM^{tr}: *For all pointed systems* e, *if* ($``e$ *is a flat system*"$)^{tr}$, *then exists* s *such that* ($``s$ *is a solution to* e"$)^{tr}$, *and for all* f, *if* ($``f$ *is a solution to* e"$)^{tr}$, *then* $s \equiv f$.

Proof We know ZFA^{tr}, and by Theorem 8.4 we only need to consider flat systems whose atoms are urelements. Assume that

($``e$ is a flat system whose atoms are urelements"$)^{tr}$.

Then by Lemma 9.5(4), let $\mathcal{E}$ be a flat bisimulation system equivalent to it; let X be the set of indeterminates of $\mathcal{E}$. By Theorem 9.3, let s be a bisimulation solution to $\mathcal{E}$. By Lemma 9.5(5), ($``s^+$ is a solution to $\mathcal{E}$"$)^{tr}$. Moreover, if ($``f$ is a solution"$)^{tr}$, then there is a bisimulation solution t to $\mathcal{E}$ such that $t^+ \equiv f$.

But then for all $x \in X$, $t_x \equiv s_x$, by the uniqueness assertion of Theorem 9.3. By Lemma 9.5(3), $t^+ \equiv s^+$. Since $\mathsf{f} \equiv t^+$, we are done. $\dashv$

This concludes our proof of the translation of the Solution Lemma. We close this section with the verification of the Axiom of Choice, and then two exercises on the model construction.

Exercise 9.5 Use the Axiom of Choice to prove (Axiom of Choice)^{tr}. You may use any form of the Axiom of Choice you wish.

Exercise 9.6 Suppose that M happened to satisfy the Solution Lemma. Prove that M and M_{afa} are isomorphic. (As a consequence of this, we can see that if we take any model M, and then extend it to M_{afa}, and then extend again to get $(M_{afa})_{afa}$, we would not get anything new in the second extension.)

Exercise 9.7 (Only for those who have seen consistency arguments in set theory.) If ϕ is a formula in the language of set theory, let ϕ^{wf} be the formula obtained by replacing quantifiers over sets by quantifiers over wellfounded sets.

1. Prove that if $b \in M_{afa}$ is wellfounded, then $b = [\bar{a}]$ for some a which is wellfounded in M.

2. Show that if *ZFA* $\vdash \phi^{wf}$, then *ZFC*$^-$ $\vdash \phi^{wf}$.

3. Prove that if *ZFA* could prove the Continuum Hypothesis (*CH*), then so could *ZFC*$^-$. (And we know that if *ZFC*$^-$ is consistent, then it cannot prove *CH*. The point of this exercise is that *AFA* will not help us to solve questions like *CH* which deal completely with wellfounded sets.)

Historical Remarks

The consistency of *ZFA* first appears in Forti and Honsell (1983). Aczel (1988) discusses it along with consistency results for other set theories admitting non-wellfounded sets.

Part IV

Elementary Applications

10

Graphs

This chapter discusses a graph-theoretic version of the Anti-Foundation Axiom. This is the version of the axiom presented by Aczel. It is an elegant and insightful alternative to our presentation, and gives one a different way to think about hypersets, namely, as things that can be "pictured" by graphs. We are going to start with a version that does not deal with urelements, just for historical reasons, since that is the way Aczel formulated things. In the second section we will redo things to handle urelements.

10.1 Graphs and the sets they picture

A *graph* is just a set G of objects (i.e., sets or urelements) called "nodes," together with a binary relation $E \subseteq G \times G$. If $\langle a, b \rangle \in E$ we say there is an *edge* from a to b. Also, E is usually understood, so we usually write $a \to_G b$ (or even just $a \to b$) instead of $a E b$. Formally speaking, our graphs will be ordered pairs of the form $\langle G, \to_G \rangle$.

Example 10.1 Consider, for example, the graph shown on the left of Figure 2. As you can see, there are two nodes (u and v) and the edge relation only has the one pair $u \to v$. Let's call the node set G_1, so $G_1 = \{u, v\}$. Formally, the

FIGURE 2 Graphs G_1 and G_2

graph as a whole is

$$\langle \{u, v\}, \{\langle u, v \rangle\} \rangle.$$

Remark At this point we need to remind you about, or introduce you to, an important convention in mathematics. We often want a name for a graph, so that we can refer to it later. In the case of the graph we are discussing, it seems natural to use G_1 for the name of the graph as well as the name of the node set. We indicate this by writing $G_1 = \langle G_1, E \rangle$. This use of the equals sign is quite unlike other uses. What it really means is something like: "Let's agree to use the symbol G_1 for both the graph and its node set, depending on the context."

Some books would use a different font for the two uses. For example, they might reserve sans-serif letters like G for graphs and then write: "A graph is a pair $G = \langle G, E \rangle$ such that ...." This usage is clearer. But even in this case, there is an unwritten convention that if one finds G in one sentence and G in the next, then the second symbol refers to the node set of the graph denoted by the first symbol.

The next definition presents concepts which are central to our work in this chapter.

Definition A *decoration* of a graph G is a function $d : G \to V_{afa}$ satisfying the condition:

(1) $$d(a) \;=\; \{d(b) \mid a \to_G b\}.$$

If $b = \{d(a) \mid a \in G\}$, then we call G a *picture* of b.

Here is an example of a decoration of our graph G_1: d is the function defined on $\{u, v\}$ such that $d(v) = \emptyset$, and $d(u) = \{\emptyset\}$. Let's verify that d really works in (1). There are two equations to verify: one for u and one for v. We first start with v. Note that the right hand side of (1) would be $\{d(b) \mid u \to b\}$. No matter what d is, this set is empty. So having $d(v) = \emptyset$ conforms to the definition of decoration. Next, we look at u. Now the right hand side of (1) would be $\{d(b) \mid u \to b\}$. This is just $\{d(v)\}$. Since $d(v) = \emptyset$, our set is $\{\emptyset\}$. This shows that the second required equation really does hold. In terms of pictures, it means that G_1 is a picture of $\{\emptyset, \{\emptyset\}\}$, which is just the von Neumann ordinal 2.

Example 10.2 Consider the graph on the right side of Figure 2. This time, the node set is $\{w, x\}$. Let's call this set G_2; by our convention we also use G_2 to refer to the whole graph. You might want to write down the explicit form of G_2, as we did with G_1. The following function defined on G_2 is a decoration of G_2: $d(w) = \{\emptyset\}$, $d(x) = \emptyset$. (Check this!) Like G_1, G_2 is a picture of $2 = \{\emptyset, \{\emptyset\}\}$.

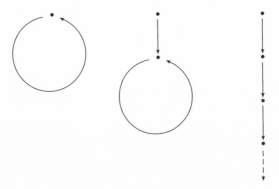

FIGURE 3 Three graphs which picture Ω.

Example 10.3 Consider a graph with three nodes, x, y and z, where y and z are childless children of x. What set does this picture? Well, if d is a decoration then $d(y) = d(z) = \emptyset$ and $d(x) = \{d(y), d(z)\} = \{\emptyset, \emptyset\} = \{\emptyset\}$. So this graph also pictures $2 = \{\emptyset, \{\emptyset\}\}$.

Example 10.4 The prototypical example graph for hyperset theory is the "loopy" graph $G = \{g\}$ with $g \to g$. This graph is shown in Figure 3; it is the graph on the left. Let d decorate this graph. Then $d(g)$, the set of decorations of children of g, is just $\{d(g)\}$. That is, $d(g) = \{d(g)\}$. Since the equation $x = \{x\}$ has Ω as its solution, we see that $d(g) = \Omega$.

There are many other graphs which picture Ω, including the two others shown in Figure 3.

If $a \to_G b$, we call b a *child* of a in G. (However, note that in contrast to the everyday usage, a node can be a child or descendant of itself.) Thus the defining condition on a decoration is that the decoration of a node is the set of decorations of its children.

Exercise 10.1 Draw two pictures of 3, one with three nodes, one with four.

Be sure you understand the examples above before reading further.

Example 10.5 Our next example is the graph G_3 shown in Figure 4. The nodes are six sets u, v, w, x, y, and z, and there are seven edges. This time, it is not so easy to read off properties of a decoration. Clearly $d(w) = \emptyset$, but that seems to be all that we can say. However, decoration function does give

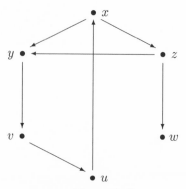

FIGURE 4 The graph G_3

us some equations that d must satisfy:

(2)
$$
\begin{aligned}
d(u) &= \{d(x)\} \\
d(v) &= \{d(u)\} \\
d(w) &= \emptyset \\
d(x) &= \{d(y), d(z)\} \\
d(y) &= \{d(v)\} \\
d(z) &= \{d(w), d(y)\}
\end{aligned}
$$

It should be noted that at this point, we have not presented any reason to think that there is such a decoration d. Further, even if we somehow come up with a decoration d, we have not given any reason why it would be unique.

Do you think that you have seen this last example, or something closely related to it, before? If you solved the exercises of Chapter 2, then you did. But it will take us a while to explain just how.

Example 10.6 Now we present a graph G_4 closely related to G_3. Let $\{34, 76, 0, 3, 6, 83\}$ be the set of nodes of G_4, and let

$$\{\langle 34, 76\rangle,\ \langle 76, 0\rangle,\ \langle 76, 6\rangle,\ \langle 0, 3\rangle,\ \langle 0, 6\rangle,\ \langle 6, 83\rangle,\ \langle 83, 34\rangle\}$$

be its edge relation.

The way in which G_3 and G_4 are related is that u corresponds to 34, v to 83, etc. In fact, the set

(3) $\qquad f \;=\; \{\langle u, 34\rangle, \langle v, 83\rangle, \langle w, 3\rangle, \langle x, 76\rangle, \langle y, 6\rangle, \langle z, 0\rangle\}$

is a function from G_3 to G_4. Further, it is one-to-one and onto. It also has the property that for all a and b in G_3, $a \to b$ in G_3 if and only if $f(a) \to f(b)$ in G_4. In general, we call such a function an *isomorphism* of the graphs involved.

Now that we have G_4, we might try to see what properties a decoration d' of it should satisfy. We encourage you to write out the equations for d' which correspond to (2). In addition, you might try to think about the relation between a decoration d of G_3 and a decoration d' of G_4.

Recall that a set a is *pure* if $support(a) = \emptyset$. At this point, we present a definition which gives us a graph for each pure set.

Definition Let a be any pure set, and let G_a be the graph whose node set is $TC(\{a\})$ and whose edge relation is

$$b \to c \quad \text{iff} \quad c \in b.$$

Note that the edge relation is the *converse* of the membership relation on G_a.

More generally, a graph G is *canonical* if its node set is a pure transitive set and its edge relation is the converse of the membership relation on G.

The terminology of canonical graphs as a parallel to ideas seen in Chapters 2 and 3. The idea is that in a canonical graph, the children of a node a are exactly the elements of a. The idea is that a decoration gives the "essential set-theoretic structure" of the graph, forgetting about the objects that happened to serve as nodes.

Example 10.7 This example is related to G_1 and G_2 from above. Consider the von Neumann ordinals 0 and 1. These are the sets $0 = \emptyset$ and $1 = \{0\} = \{\emptyset\}$. These are pure sets, and $TC(\{1\}) = \{0, 1\}$. So when $a = 1$, G_a is the following graph:

$$G_a \quad = \quad \langle \{0, 1\}, \{\langle 0, 1 \rangle\} \rangle.$$

You might check at this point that the following function d is a decoration of G_a: $d(0) = 0$, $d(1) = 1$. This is no accident, as the next result shows.

Proposition 10.1 *If G is a canonical graph then the identity function on G is a decoration of G.*

Proof Let d be the identity function on G. The condition that we need to check is that for all $b \in G$,

$$d(b) \quad = \quad \{d(c) \mid c \to_G b\}.$$

Since d is the identity function on G, we must check that for all $b \in G$,

$$(4) \qquad\qquad b \quad = \quad \{c \mid c \to b\}$$

So fix some $b \in G$. Since G is transitive and pure, G contains all elements of b. Indeed, for all sets c, c is a child of b in G if and only iff $c \in b$. So the set on the right side of (4) is $\{c \mid c \in b\}$. Now (4) holds by the Axiom of Extensionality. Since b is arbitrary, we are done. ⊣

Example 10.8 Have a look back at Exercise 2.3 on page 15 and at its solution at the end of the book. We considered two sets x and y satisfying $x = \{y, \{y, \emptyset\}\}$ and $y = \{\{\{x\}\}\}$. In the exercise we computed the transitive closures of those sets and got x, y and four other sets which we name as follows

(5) $u = \{x\}, \quad v = \{\{x\}\}, \quad w = \emptyset, \quad z = \{y, \emptyset\}$

The sets above are all pure, so we can consider G_x. It would have as its node set $\{u, v, w, x, y, z\}$. The edge relation would have seven pairs. Now have a look back at Figure 4. When we discussed that example, we just said that u, v, ..., z were just six sets. It made no difference what sets they are (provided they are different). If we take them to be the sets described above, then Figure 4 depicts G_u.

Also have a second look at our conditions on a possible decoration d of G_a, as written out in (2). We know now from Proposition 10.1 that the identity is a decoration of G_u. You might want to check that the identity really satisfied all the conditions in (2), with u, v, etc. as chosen above.

Exercise 10.2 In terms of the preceding material, give a decoration of G_4.

Exercise 10.3 Prove that if the decoration of a graph G is the identity function on G, then G must be canonical. (This is the converse to Proposition 10.1.)

Proposition 10.1 gives us lots of examples of graphs which have decorations. The natural questions at this point are: exactly which graphs have decorations? Are the decorations unique? What relation on graphs corresponds to having the same, or related, decorations?

In order to attack these related problems, we first reformulate all of our work on decorating graphs in terms of solving sets of equations.

Proposition 10.2 *In ZFC⁻ (standard set theory, but without the Foundation or Anti-Foundation Axioms), the following are equivalent:*

1. *Every flat system of equations without atoms $\mathcal{E} = \langle X, \emptyset, e \rangle$, where $e : X \to \mathcal{P}X$, has a unique solution.*
2. *Every graph G has a unique decoration.*

Proof $(1) \Rightarrow (2)$. Let $G = \langle G, \to \rangle$ be a graph. Let X be a set of urelements in correspondence with G, and write x_a for the urelement corresponding to a. Let e be given by

$$e_{x_a} = \{x_b \mid a \to_G b\}.$$

By tracing through the definitions, we see that a solution s to $\mathcal{E}_G$ gives a decoration d of G in the following way: $d(a) = s(x_a)$. Conversely, if d is a decoration of G, then we get a solution s of $\mathcal{E}_G$ by setting $s(x_a) = d(a)$. The association of decorations and solutions is a bijection. Our assumption in (1) tells us that $\mathcal{E}_G$ has a unique solution. It follows that G has a unique decoration.

(2) $\Rightarrow$ (1).   Going the other way, we show how to take a flat system $\mathcal{E}$ without atoms, and to associate a graph $G_{\mathcal{E}}$ with it in such a way that solutions to $\mathcal{E}$ are the same as decorations of G. The nodes of G are the indeterminates of $\mathcal{E}$, and we put $x \to y$ in G iff $y \in e_x$ in $\mathcal{E}$. The rest of the proof is as in converse direction above. ⊣

Corollary 10.3 *Assuming AFA, every graph has a unique decoration.*

Aczel originally formulated *AFA* as the following statement: every graph has a unique decoration. We have chosen to emphasize the formulation via systems of equations because it is much easier to use in the applications, and because of the striking parallel with other extension processes in mathematics. But since decorations of graphs and solutions of systems correspond, the essential mathematical content of the two approaches is the same.

10.2   Labeled graphs

Proposition 10.2 establishes an important connection between the concept of a system of equations and that of a graph. However, there is a slight mismatch in the result. All of the work earlier in this book allows systems of equations to make use of urelements, and yet these have so far been missing from our discussion of graphs. We want to generalize the notions of graph and decoration to get a perfect match.

There are two ways to achieve this match-up, both by allowing graphs that have labeled nodes. In one approach, you allow only childless nodes to be labeled; such a node can be labeled by a single urelement. This is natural if you are thinking of the graph as a complete picture of the set with the edge relation as the (converse of the) membership relation. This was the approach taken in Barwise and Etchemendy (1987).

The other way to achieve the desired match is to think of the edge relation as depicting the restriction of the membership to *sets* alone. But then you need to allow any node in a graph to be labeled by a set of urelements. Either of these approaches gives a natural graph-theoretic reformulation of *AFA*. The first is in some ways most natural if you want to have a graph-theoretic conception of non-wellfounded set. The second approach, however, turns out to be more flexible and more useful, due to its connections with modal logic (see Chapter 11), and so is the one we pursue here.

Definition Let $A \subseteq \mathcal{U}$. A *(labeled) graph* $G = \langle G, \to, l \rangle$ *over* A is a triple such that $\langle G, \to \rangle$ is a graph, and $l : G \to \mathcal{P}(A)$. A *decoration* of a graph G over A is a function $d : G \to V[A]$ such that for all $g \in G$

$$d(g) \quad = \quad \{d(h) \mid g \to h\} \cup l(g).$$

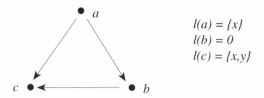

$$l(a) = \{x\}$$
$$l(b) = 0$$
$$l(c) = \{x,y\}$$

FIGURE 5 A labeled graph

One must be careful when drawing labeled graphs, because it is easy to confuse the labeling function l with the names of the nodes which we normally put into a picture. For example, Figure 5 shows a labeled graph G. Its nodes are a, b, and c, and its labeling function l is given by $l(a) = \{x\}$, $l(b) = \emptyset$, and $l(c) = \{x, y\}$. The point is that a, b, and c are not the labels even though may they look like them.

In this example, a decoration d would give sets $d(a)$, $d(b)$ and $d(c)$ satisfying the following equations:

$$
\begin{aligned}
d(a) &= \{d(b), d(c), x\} \\
d(b) &= \{d(c)\} \\
d(c) &= \{x, y\}
\end{aligned}
$$

Now we have a result which parallels Proposition 10.2.

Theorem 10.4 *Let $A \subseteq \mathcal{U}$. In ZFC⁻ (set theory without the Foundation or Anti-Foundation Axioms), the following are equivalent:*

1. *Every flat system of equations*

$$\mathcal{E} \quad = \quad \langle X, A, e \rangle, \quad \text{where } e : A \to \mathcal{P}(X \cup A),$$

   has a unique solution.

2. *Every graph G over A has a unique decoration.*

Proof The proof here is essentially the same as that of Proposition 10.2: the earlier constructions extend to the labeled case. Given $\mathcal{E}$, we construct $G_\mathcal{E}$ as before, and we label it by $l(b) = e_b \cap A$. And given $G = \langle G, \to, l \rangle$, we now take $\mathcal{E}_G$ to be $\langle G, e \rangle$, where

$$e_g \quad = \quad \{h \mid g \to h\} \cup l(g).$$

The only small point to be aware of is that a decoration of a graph G over A must map the nodes of G into elements of $V_{afa}[A]$; this is because a solution of a system of equations over A must do the same thing. ⊣

Representing binary relations in terms of membership

Theorem 10.4 allows us to give an easy proof of a nice little result. It shows that if we are working with hypersets, then every binary relation is isomorphic to the membership relation on some set.

Let $\langle A, R \rangle$ and $\langle B, S \rangle$ each consist of a set together with a binary relation on that set. Recall that an *isomorphism* from $\langle A, R \rangle$ to $\langle B, S \rangle$ is a bijective function f from A onto B such that for all $x, y \in A$, $x \ R \ y$ iff $f(x) \ S \ f(y)$.

Proposition 10.5 *For every binary relation $\langle A, R \rangle$ there is a set B such that $\langle A, R \rangle$ is isomorphic to $\langle B, \ni_B \rangle$, where $\ni_B$ is the converse of the membership relation on B.*

Proof By the Axiom of Plenitude, we can assume that A is a set of urelements. We can think of A as the set of nodes of a labeled graph G. The edge relation in G is given by $n \to m$ iff nRm. (That is, R is the edge relation on G.) Label each node n with $\{n\}$. Now take the decoration d of this graph and let B be the range of d. Notice that for each $n \in A$, n is the only urelement in $d(n)$. Consequently, d is one-to-one. Also, for each node n, $d(n)$ is a set, hence not an urelement.

We check that nRm iff $d(n) \ni d(m)$. The definition of a decoration insures that if nRm, then $d(n) \ni d(m)$. Going the other way, suppose $d(n) \ni d(m)$. Since d is a decoration,

$$d(n) \;\; = \;\; \{d(m') \mid nRm'\} \cup \{n\}$$

Being a set, $d(m)$ cannot be n so it must be of the form $d(m')$ for some m' such that nRm'. But as d is one-to-one, $m = m'$. Thus nRm. Therefore d is an isomorphism. ⊣

Corollary 10.6 *For every binary relation $\langle A, R \rangle$ there is a set B such that $\langle A, R \rangle$ is isomorphic to $\langle B, \in_B \rangle$, where $\in_B$ is the membership relation on B.*

Proof Apply Proposition 10.5 to the structure $\langle A, R^{-1} \rangle$. ⊣

Exercise 10.4 Give a proof of Proposition 10.5 using the Solution Lemma, without the detour through graphs.

Exercise 10.5 Proposition 10.5 is false in **ZFC**. How would one have to modify it to obtain a true result? That is, what kinds of binary relations are representable as restrictions of $\in$ in **ZFC**?

Exercise 10.6 The proof of Proposition 10.5 made crucial use of urelements. However, the conclusion does not depend on them. Use the result to conclude that for every binary relation $\langle A, R \rangle$ there is a *pure* set B such that $\langle A, R \rangle$ is isomorphic to $\langle B, \ni_B \rangle$, where $\ni_B$ is the converse of the membership relation restricted to B. [Note: B need not be transitive. Also, you will need a fact from set theory which we have not covered in this book.]

Exercise 10.7 (Only for those who have read Chapter 9 or otherwise know about consistency proofs in set theory.) Let AFA_0 be the statement that all *unlabeled* graphs have unique decorations. Assuming that ZFA is consistent, prove that $ZFC^- + AFA_0 + \neg AFA$ is also consistent. In other words, if we work in set theory with urelements, then AFA for flat systems without indeterminates is weaker than the Flat Solution Lemma.

10.3 Bisimilar graphs

Theorem 10.4 suggests that sets and graphs are closely related. Indeed, there are ways to go back and forth. Before making the needed definition, we state the following simple observation:

Lemma 10.7 *If G is any labeled graph and d is its decoration, then $d[G]$ is a set of sets that is* transitive on sets; *that is, if $a \in b \in d[G]$ and a is a set (not an urelement) then $a \in d[G]$.*

Proof If $b \in d[G]$ then $b = d(g)$ for some node g of G. Let $a \in b = d(g)$. Recall that $d(g) = \{d(h) \mid g \to h\}$. So for some $h \in G$, $a = d(h)$. Hence $a \in d[G]$. ⊣

Definition A *canonical labeled graph* is a graph $G = \langle G, \to, l \rangle$ where G is a set of sets that is transitive on sets, $\to$ is defined by

$$a \to b \quad \text{iff} \quad b \text{ is a set and } b \in a$$

and $l(a) = a \cap \mathcal{U}$.

Notice that if G is a canonical labeled graph, then its decoration d is the identity: $d(b) = b$.

Given any set a, we know that there is a canonical system of equations for a, one which uses $support(a)$ as atoms. We can turn this system into a canonical labeled graph G_a over $support(a)$ as follows: The nodes of G_a are the sets belonging to $TC(\{a\})$. And we put an edge from b to c in this graph if $c \in b$. The labeling function l is defined by: $l(b) = b \cap support(a)$.

Assuming AFA, we have also a way of going from graphs to sets, by decoration. This is not quite correct, since decorating a graph doesn't really give us a single set at all. That is, given a graph G and its decoration d, the image set $d[G] = \{d(g) \mid g \in G\}$ is a set of sets. But it is not possible to get every set as the image set of some graph. For example, consider a singleton $\{x\} \subseteq \mathcal{U}$. If $\{x\} = d[G]$, then G would have to be non-empty (since $d[\text{empty graph}] = \emptyset$). Let $g \in G$. So $d(g) = x$. But this is impossible, since a decoration function maps nodes to *sets*.

On the other hand, for every set a, there *is* a graph G and a node $g \in G$ so that $d(g) = a$. (The canonical graph for a will do.) To make the match between sets and graphs tighter, we introduce *accessible pointed graphs*, or *apg's*. These are labeled graphs with a distinguished node $\top$ called the *top point* of G, from which every node is reachable.[1] So formally, an apg is a tuple $\langle G, \to, l, \top \rangle$ with $\top \in G$, and with the property that every point is reachable by a finite path starting from this $\top$. Now given an apg G, we have a set $d(\top_G)$. We say that G *pictures* this set $d(\top_G)$. To save on notation, we'll write $pict(G)$ for $d(\top_G)$.

In this way, we have operations $G \mapsto pict(G)$ taking apgs to sets, and $a \mapsto G_a$ taking sets back to apg's. We can see that $pict(G_a) = a$, since the identity map decorates the canonical graph of any set. On the other hand, it is not in general true that $G_{pict(G)} = G$. For example, there are infinitely many graphs which picture $\emptyset$ (in fact, there are a proper class of them), but $G_\emptyset$ is the one-point graph.

Suppose we are given two labeled graphs G and G' over $\mathcal{U}$, and let d and d' be their decorations. Under what conditions on the graphs is it the case that $d[G] = d[G']$? This is very much like the question as to when two systems of equations have the same solution set, and has a very similar answer.

Definition Let G and G' be labeled graphs over $\mathcal{U}$. A *bisimulation relation* between G and G' is a relation $R \subseteq G \times G'$ satisfying the following conditions:

(1) If gRg' and $g \to_G h$ then there is h' with hRh' and $g' \to_{G'} h'$.

(1') If gRg' and $g' \to_{G'} h'$ then there is h with hRh' and $g \to_G h$.

(2) If gRg' then $l(g) = l'(g')$.

The bisimulation R is said to be a *total bisimulation* of G and G' if, in addition,

(3) For every $g \in G$ there is $g' \in G'$ such that gRg'.

(3') For every $g' \in G'$ there is $g \in G$ such that gRg'.

The labeled graphs G and G' are said to be *bisimilar* if there is a total bisimulation from G to G'.

Proposition 10.8 *Let g and g' be nodes in labeled graphs G and G' over $\mathcal{U}$. There is a bisimulation relation R between G and G' such that gRg' iff $d(g) = d'(g')$, where d, d' are the decorations of G and G' respectively. Hence, G and G' are bisimilar iff $d[G] = d'[G']$.*

Proof Assume R is a bisimulation between G and G'. Define a relation R' on sets by $aR'b$ iff there are nodes g, g' of G, G' respectively such that gRg', $a = d(g)$ and $b = d'(g')$. It is routine to check that R' is a bisimulation on sets, so the result follows from Theorem 7.3. To prove the converse, define gRg'

---

[1] These apg's are the graph-theoretic version of our pointed systems of equations. Also, the $\top$ node is the analogue for graphs of the initial state of an automaton.

iff $d(g) = d'(g')$. It is easy to check that R is a bisimulation on the labeled graphs. ⊣

Corollary 10.9 *Every labeled graph is bisimilar to a unique canonical labeled graph.*

Our results in this chapter show that labeled graphs and systems of equations are the same except for a cosmetic difference. However, cosmetic differences can be important. The ability to draw pictures makes the graphs easier to work with, especially for small examples. On the other hand, the notion of a system of equations will generalize better, as we'll see in Chapter 16.

11

Modal logic

Modal logic developed early in the twentieth century as a way to try to formalize intuitions about necessity and possibility. For many years, though, systems of modal logic proliferated without much of a way to judge between them, and the subject was not held in very high repute in certain quarters. In the late 1950s and 60s, though, Kanger, Kripke and others developed the so-called possible worlds framework for modal logic. This framework has had enormous influence, not just on the logic of necessity and possibility, but in other areas as well. In particular, the ideas in this approach have been applied to develop formalisms for describing many other kinds of structures and processes in computer science, giving the subject applications that would have probably surprised the subject's founders and early detractors alike.

In this chapter we want to show how modal logic might have developed if hypersets had been around at the time that the possible worlds framework was being developed. We will take *canonical* Kripke structures to be subsets $W \subseteq V_{afa}[A]$ which are *transitive on sets* (that is, if $a \in b \in W$ and a is a set, then $a \in W$). So the worlds of such Kripke structures are just sets a in $V_{afa}[A]$. The sets b in a are the worlds accessible from a. The atomic propositions in a are the atomic propositions true in a.

We think of modal sentences as ways of describing the internal structure of such worlds. We will see that this approach is in some ways equivalent to the standard presentation, but that new and interesting questions arise. It also has the advantage of leading to generalizations of such notions as labeled transition systems and feature structures.

11.1 An introduction to modal logic

Modal logic has three basic ingredients: the syntax and semantics of the sentences of various modal languages, proof systems for such languages, and relations between the semantics and the proof systems.

The syntax of modal logic

In this section we introduce the basic modal language. Let $A \subseteq \mathcal{U}$. The set $\mathcal{L}(A)$ of *finitary modal sentences over A* is the smallest set containing the elements $p \in A$, a single primitive sentence T, and which is closed in the following ways:

If $\varphi \in \mathcal{L}(A)$, so are $\neg\varphi$ and $\Diamond\varphi$.

If $\varphi, \psi \in \mathcal{L}(A)$, then $\varphi \wedge \psi \in \mathcal{L}(A)$.

We introduce F, $\vee$, $\rightarrow$, $\leftrightarrow$, and $\square$ as defined symbols as follows:

F	is	$\neg\mathsf{T}$
$\varphi \vee \psi$	is	$\neg(\neg\varphi \wedge \neg\psi)$
$\varphi \rightarrow \psi$	is	$\neg(\varphi \wedge \neg\psi)$
$\varphi \leftrightarrow \psi$	is	$(\varphi \rightarrow \psi) \wedge (\psi \rightarrow \varphi)$
$\square\varphi$	is	$\neg\Diamond\neg\varphi$

We also define an infinitary extension $\mathcal{L}_\infty(A)$ of $\mathcal{L}(A)$ by the following inductive definition:

If $\varphi \in \mathcal{L}_\infty(A)$, so are $\neg\varphi$ and $\Diamond\varphi$.

If $\varphi, \psi \in \mathcal{L}_\infty(A)$, then $\varphi \wedge \psi \in \mathcal{L}_\infty(A)$.

If $\Phi \subseteq \mathcal{L}_\infty(A)$ is a set, then $(\bigwedge \Phi) \in \mathcal{L}_\infty(A)$.

We have included binary conjunction as well as conjunction over sets only so as to have $\mathcal{L}(A) \subseteq \mathcal{L}_\infty(A)$. We could have done things differently, by treating binary conjunction as a conjunction of a set of size at most two. We could then have defined T to be $\bigwedge \emptyset$. We define $\bigvee \Phi$ to be $\neg \bigwedge \Phi^\neg$, where $\Phi^\neg$ is the set $\{\neg\varphi \mid \varphi \in \Phi\}$.

Semantics

We now define the *truth relation* $a \models \varphi$ between sets and sentences of $\mathcal{L}_\infty(A)$. To understand this definition, one should think of sets as "worlds" in some big collection of such worlds, and then think of $a \models \Diamond\varphi$ as meaning that "φ is true in some world accessible to (i.e., in) a."

Definition Let A be a set of urelements. The relation $a \models \varphi$, read φ is true in (or holds in) a, is defined inductively for sentences $\varphi \in \mathcal{L}_\infty(A)$ as follows:

$$
\begin{array}{ll}
a \models p & \text{if } p \in a, \text{ for all } p \in A \\
a \models \mathsf{T} & \text{for all } a \\
a \models \neg\varphi & \text{iff } a \not\models \varphi \\
a \models \Diamond\varphi & \text{iff for some set } b \in a,\, b \models \varphi \\
a \models \varphi \wedge \psi & \text{iff } a \models \varphi \text{ and } a \models \psi \\
a \models \bigwedge \Phi & \text{iff } a \models \varphi \text{ for all } \varphi \in \Phi
\end{array}
$$

The modal sentences that hold in a can be thought of as describing a. They can tell us whether a contains some particular urelement (or not), or whether it contains an element which has some other property, and so on.

Exercise 11.1 Show that $a \models \neg\neg\varphi$ iff $a \models \varphi$.

The following shows that the defined symbols work as one would expect. The proof is a routine induction on sentences.

Lemma 11.1 *For all sentences φ and ψ, and for all sets of sentences Φ:*

$$
\begin{array}{ll}
a \models \mathsf{F} & \textit{never holds} \\
a \models \Box\varphi & \textit{iff for every set } b \in a,\, b \models \varphi \\
a \models \varphi \vee \psi & \textit{iff } a \models \varphi \textit{ or } a \models \psi \\
a \models \varphi \rightarrow \psi & \textit{iff } a \not\models \varphi \textit{ or } a \models \psi \\
a \models \varphi \leftrightarrow \psi & \textit{iff } (a \models \varphi \textit{ if and only if } a \models \psi) \\
a \models \bigvee \Phi & \textit{iff } a \models \varphi \textit{ for some } \varphi \in \Phi
\end{array}
$$

Example 11.1 Consider $A = \{p, q\}$ and two sets a and b so that

$$
\begin{array}{rcl}
a & = & \{p, b\} \\
b & = & \{q, a\}
\end{array}
$$

Think of a and b as modeling two states of affairs or worlds, each accessible from the other but not from itself, a where the proposition p holds, b where q holds. According to the above definition, $a \models p$. Thus $b \models \Diamond p$ and $a \models \Diamond\Diamond p$. However, $a \models \Box\neg p$ since the only set which belongs to a (i.e., the only world accessible from a) is b, and $b \models \neg p$.

With this logic comes a natural consequence relation and notion of validity.

Definition A sentence $\varphi \in \mathcal{L}_\infty(A)$ is said to be *valid*, written $\models \varphi$, if $a \models \varphi$ for every set $a \in V_{afa}[A]$. If $T \subseteq \mathcal{L}_\infty(A)$ then φ is said to be *consequence of* T, written $T \models \varphi$, if every $a \in V_{afa}[A]$ such that $a \models \psi$ for all $\psi \in T$ also has the property that $a \models \varphi$.

We relativize the semantical notions of consequence to a collection $W \subseteq V_{afa}[A]$ that is transitive on sets, by restricting the quantifiers to range over W in the above definition. In this case we say W-*valid* and W-*consequence*. We write $\models^W \varphi$ and $T \models^W \varphi$ respectively.

Exercise 11.2 Let W be transitive on sets. Prove that if $\models^W \varphi$, then $\models^W \Box\varphi$.

Exercise 11.3 Let W be transitive on sets. Call W *modally normal* if for all theories T and all sentences φ, if $T \models^W \varphi$, then $T \models^W \Box\varphi$. Is every class W that is transitive on sets modally normal? If not, can you characterize those that are?

At this point we know how to interpret the modal languages $\mathcal{L}(A)$ and $\mathcal{L}_\infty(A)$ on arbitrary sets. This means that we know how to spell out the conditions under which $a \models \varphi$, whenever a is a set. However, we would like to broaden our discussion to interpret $\mathcal{L}(A)$ on other structures.

Definition Let $A \subseteq \mathcal{U}$. A *Kripke structure over A* is just a labeled graph G over A. If G is a canonical labeled graph, then we call it a *canonical Kripke structure.*

The reason graphs over sets of urelements are called "Kripke structures" has to do with the use of such structures in giving semantics for modal languages, an approach pioneered by Saul Kripke in the late 1950's. When used in this setting, the nodes of the graph are called *possible worlds*, the edge relation is called the *accessibility relation* on possible worlds, the elements of A are thought of as *atomic propositions*, and the labeling function l is called an *assignment*. The intuitive idea is that $l(w)$ should be the set of atomic propositions true in the world $w \in G$. Kripke then interpreted a statement like "it is possible that p" holding at a world w if the statement p holds at some world w' accessible from w, that is, at some child of w, thinking of the structure as a graph.

In terms of Kripke structures, we have defined what it means for $a \models \varphi$ when a is a world in some canonical Kripke structure. But what about other Kripke structures?

Definition Let G be any labeled graph over A, i.e., any Kripke structure over A. Let d_G be its decoration. For any node w of G, define $w \models_G \varphi$ iff $d_G(w) \models \varphi$. We say that φ is *valid* in G, and write $\models^G \varphi$ iff $w \models_G \varphi$ for all $w \in G$.

Our definition of $w \models_G \varphi$ differs from the definition which is usually presented in modal logic. However, the next result shows that it is equivalent to the usual definition.

Proposition 11.2 *Let G be a Kripke structure. The following hold for all $w \in G$ and all sentences φ of $\mathcal{L}(A)$.*

$w \models_G p$	*iff*	$p \in l(w)$ *(for all $p \in A$)*
$w \models_G \top$		*for all w*
$w \models_G \neg\varphi$	*iff*	$w \not\models_G \varphi$
$w \models_G \Diamond\varphi$	*iff*	*there is a $w' \in G$ such that $w \to w'$ and $w' \models_G \varphi$*
$w \models_G \varphi \wedge \psi$	*iff*	$w \models_G \varphi$ *and* $w \models_G \psi$
$w \models_G \bigwedge \Phi$	*iff*	$w \models_G \varphi$ *for all* $\varphi \in \Phi$

Proof The proof is a routine induction on sentences. ⊣

We say that φ is valid in all Kripke structures if for all Kripke structures G, $\models^G \varphi$. We write $T \models_G \varphi$ to mean that for all $n \in G$, if $n \models_G \psi$ for all $\psi \in T$, then also $n \models_G \varphi$, and write $T \models_{Kr} \varphi$ if $T \models_G \varphi$ for all G. Our definition of the semantics for Kripke structures via decorations makes the following result obvious:

Proposition 11.3 *Let R be a bisimulation relation between Kripke structures G and G' over A. If nRn' then for all $\varphi \in \mathcal{L}_\infty(A)$, $n \models_G \varphi$ iff $n' \models_{G'} \varphi$. Hence if R is total, then φ is valid in G iff it is valid in G'.*

Proof By Proposition 10.8, we know that if nRn', then $d_G(n) = d_{G'}(n')$, so this follows immediately from the definition of satisfaction in Kripke structures. ⊣

We will prove a converse to this result later. For now, let's observe the following corollary.

Corollary 11.4 *Let $\varphi \in \mathcal{L}_\infty(A)$.*

1. *Let G be a Kripke structure and let W be the range of G's decoration. Then $\models_G \varphi$ iff $\models_W \varphi$.*
2. *$\models \varphi$ iff φ is valid in all Kripke structures.*
3. *More generally, for any $T \subseteq \mathcal{L}_\infty(A)$, $T \models \varphi$ iff $T \models_{Kr} \varphi$.*

Later in this chapter we will give a sound and complete proof theory for the relation $T \models \varphi$, when these are in the finitary language $\mathcal{L}(A)$.

We now return to the result promised earlier by showing that nodes satisfying the same sentences in two different Kripke structures are related by a bisimulation of the structures.

Theorem 11.5 *Let a and b belong to $V_{afa}[A]$. If a and b satisfy the same sentences of $\mathcal{L}_\infty(A)$ then a and b are A-bisimilar and hence equal.*

We won't prove this result just now since we'll prove a more difficult result (Theorem 11.7) below. Once you see that proof you'll easily see how the proof works for Theorem 11.5.

This last result would not be true if we used only the finitary language $\mathcal{L}(A)$, as the following example shows.

Proposition 11.6 *Let ω be the set of natural numbers construed as finite ordinals (see page 13), and let $\omega^* = \omega \cup \{\omega^*\}$. Let $A \subseteq \mathcal{U}$ be arbitrary. Then ω and ω^* satisfy the same finitary modal sentences $\varphi \in \mathcal{L}(A)$, yet they are different sets.*

Proof We show that for all finitary φ, the following are equivalent:

(1) $\omega \models \varphi$.

(2) For all but finitely many n, $n \models \varphi$.

(3) For infinitely many n, $n \models \varphi$.

(4) $\omega^* \models \varphi$.

In the case of the atomic φ, this equivalence is easy: A pure set satisfies an atomic φ iff φ is T. So (1)–(4) are also equivalent for atomic sentences.

Assume now that (1)–(4) are equivalent for φ. We show the same thing for $\neg\varphi$. We'll prove four implications for $\neg\varphi$.

(1) $\Rightarrow$ (2): Assume that $\omega \models \neg\varphi$, so that $\omega \not\models \varphi$. Since (3) implies (1) for φ, we have the negation of (3). That is, there are only finitely many n such that $n \models \varphi$. So for all but finitely many n, $n \models \neg\varphi$. This is (2).

(2) $\Rightarrow$ (3): Assume that for all but finitely many n, $n \models \neg\varphi$. Then (3) fails for φ. Since (2) implies (3) for φ, we see that it is not the case that for all but finitely many n, $n \models \varphi$. So there must be infinitely many n which satisfy $\neg\varphi$. This is (3).

(3) $\Rightarrow$ (4): Assume that there are infinitely many n such that $n \models \neg\varphi$. Then (2) is false for φ. Since (4) implies (2) for φ, we have the negation of (4) for φ. This means that $\omega^* \not\models \varphi$. So $\omega^* \models \neg\varphi$.

(4) $\Rightarrow$ (1): Assume $\omega^* \models \neg\varphi$. Then $\omega^* \not\models \varphi$. So $\omega \not\models \varphi$. Therefore $\omega \models \neg\varphi$.

Now assume the equivalence of (1)–(4) for φ and ψ. We show (1) $\Rightarrow$ (2) for $\varphi \wedge \psi$; the other steps are as above. suppose $\omega \models \varphi \wedge \psi$. So we have N and M so that for $n \geq N$, $n \models \varphi$, and for $n \geq M$, $n \models \psi$. Then for $n \geq max(N, M)$, $n \models \varphi \wedge \psi$.

Finally, we show the equivalence for $\Diamond\varphi$, assuming it for φ. If $\omega \models \Diamond\varphi$, then there is some n so that $n \models \varphi$. But then all $m \geq n$ satisfy $\Diamond\varphi$, so (2) holds for $\Diamond\varphi$. As before (2) always implies (3). Assuming (3) for $\Diamond\varphi$, we have at least one n so that $n \models \varphi$. So $\omega^* \models \Diamond\varphi$ as well. This proves (3). Now assuming (4) for $\Diamond\varphi$, we prove (1) in two cases: either some $n \in \omega$ satisfies φ, or ω^* itself does. If some n satisfies φ, then $\omega \models \Diamond\varphi$. If $\omega^* \models \varphi$, then by induction hypothesis, some n satisfies φ. So $\omega \models \Diamond\varphi$.   ⊣

We know by Theorem 11.5 that the sets ω and ω^* are not equivalent in $\mathcal{L}_\infty(A)$. In fact, we can exhibit a sentence on which they differ. Let τ_n be defined by recursion so that $\tau_0 = $ T and $\tau_{n+1} = \Diamond\tau_n$. Thus $n \models \tau_n \wedge \neg\tau_{n+1}$. Now consider the sentence

$$\Diamond \bigwedge_{n \in N} \tau_n$$

This sentence is true in ω^* but is not true in ω.

From our point of view, the central point about this example is that ω and ω^* are not *finite*. The full infinitary language $\mathcal{L}_\infty(A)$ is expressive enough to separate them, but the finitary language $\mathcal{L}(A)$ is not. With this in mind, we make a general definition:

Definition Let $A \subseteq \mathcal{U}$. A set a is *hereditarily finite over* A if every set $b \in TC(\{a\})$ is finite and $support(a) \subseteq A$. The class of all sets which are hereditarily finite over A is denoted $HF^1[A]$.[1]

If we restrict attention to $HF^1[A]$, then different sets satisfy different sets of sentences.

Theorem 11.7 *Let a and b belong to $HF^1[A]$. If a and b satisfy the same sentences of $\mathcal{L}(A)$ then a and b are A-bisimilar and hence equal.*

Proof Write $\equiv$ for the relation on $HF^1[A]$ given by: $a \equiv b$ iff a and b satisfy the same sentences of $\mathcal{L}(A)$. We claim that $\equiv$ is a bisimulation on sets. Suppose that $a \equiv b$. Then clearly a and b contain the same urelements, since all of those would have to be in A, and hence in $\mathcal{L}(A)$.

Suppose that a' is a set in a. We need some set $b' \in b$ so that $a' \equiv b'$. Suppose that no such b' exists. Then for each $b' \in b$ there is some $\varphi = \varphi(b')$ so that a' and b' differ on φ. Since our language has $\neg$, we may assume that $a' \models \varphi(b')$ and $b' \models \neg\varphi'$. Now b is a finite set, so we have a sentence $\psi = \bigwedge_{b' \in b} \varphi(b')$ of $\mathcal{L}(A)$. We see that $a' \models \psi$, so $a \models \Diamond\psi$. But $b \not\models \Diamond\psi$, for otherwise there would be a $b' \in b$ which satisfies ψ, and in particular which satisfies $\varphi(b')$. This argument proves half of the bisimulation condition; the other half is similar. ⊣

The proof of Theorem 11.5 is quite similar, though simpler, since no cardinality considerations apply.

Theorem 11.7, combined with earlier results, give us the following corollary. Let us say that a Kripke structure G is *finitely branching* if for each $w \in G$, the set of children of w is finite. This condition clearly entails that the set $d[G]$ of decorations for G is a subset of $HF^1[A]$.

Corollary 11.8 *Let g and g' be nodes in Kripke structures G and G' over A, respectively. Then g and g' are related by a bisimulation of G and G' if and only if they satisfy the same sentences of $\mathcal{L}_\infty(A)$. If G and G' are finitely branching, then we can add to the equivalences the condition that they satisfy the same sentences of $\mathcal{L}(A)$.*

11.2 Characterizing sets by sentences

Theorem 11.5 shows that each set $a \in V_{afa}[A]$ is characterized by the proper class of all the infinitary sentences that it satisfies. In fact, we can do much better than this. To see what we mean, consider the following definition.

Definition Let $a \in V_{afa}[A]$, and let θ be a sentence in $\mathcal{L}(A)$ or $\mathcal{L}_\infty(A)$. We

---

[1]The class $HF^1[A]$ and related classes are studied in Chapter 18. Nothing from that later work is used here, though. You may be interested to know that if A is a set, then so is $HF^1[A]$. And $HF^1[A]$ is the largest class C with the property that $C = \mathcal{P}_{fin}(C \cup A)$.

say that θ *characterizes* a in $V_{afa}[A]$ provided that $b \models \theta$ iff $b = a$, for all $b \in V_{afa}[A]$.

The main work of this section is to prove the following result:

Theorem 11.9 *Every set* $a \in V_{afa}[A]$ *is characterizable in* $V_{afa}[A]$ *by some sentence* θ^a *of* $\mathcal{L}_\infty(A)$.

Since this result takes a little work, let's start with an example. The general proof will be an elaboration of this example.

Example 11.2 Let $A \subseteq \mathcal{U}$ and let us find a sentence θ^Ω characterizing the set Ω. Since Ω has no urelements, let $\varphi_0 = \bigwedge_{p \in A} \neg p$. For each natural number n define, inductively,

$$\varphi_{n+1} = \varphi_0 \wedge \Diamond \varphi_n \wedge \Box \varphi_n.$$

Finally, let $\theta = \bigwedge_{n=0,1,\ldots} \varphi_n$. An easy induction on n shows that $\Omega \models \varphi_n$, and hence $\Omega \models \theta$.

We claim that θ characterizes Ω in $V_{afa}[A]$. To see this let $a \in V_{afa}[A]$ be such that $a \models \theta$. Since $a \models \varphi_0$, a contains no urelements. Since $a \models \varphi_1$, a is non-empty and each of its elements satisfies φ_0. In general, we see that every element of the transitive closure of a is nonempty but contains no urelements. This implies that $a = \Omega$.

We might note that there are sets $b \notin V_{afa}[A]$ which also satisfy θ. For example, suppose that $q \notin A$ is an urelement, and let $b = \{q, b\}$. Then $b \models \theta$. This is why we must be a bit careful in stating characterization results. There is no single θ which characterizes Ω among all sets, but for each $A \subseteq \mathcal{U}$, there is a θ which works for $V_{afa}[A]$.

We now turn to the key definitions used in proving Theorem 11.9.

Definition For any set Φ of sentences, let $\triangle\Phi$ be the following sentence:

$$\bigwedge_{\varphi \in \Phi} \Diamond \varphi \wedge \Box \bigvee_{\varphi \in \Phi} \varphi$$

Notice that $a \models \triangle\Phi$ iff every member of Φ is true in some member of a and every member of a satisfies some member of Φ. Thus if Φ were a set of characterizing sentences, then $\triangle\Phi$ would characterize the set of sets characterized by some member of Φ. This observation is used in the following exercise.

Exercise 11.4 Let $A \subseteq \mathcal{U}$. Use the Induction Principle for $WF[A]$ to prove that for every $a \in WF[A]$, there is a single sentence $\psi_a \in \mathcal{L}_\infty(A)$ so that for all $b \in V_{afa}[A]$, $b \models \psi_a$ iff $b = a$. Moreover, if A is finite and $a \in HF^0[A]$, then ψ_a may be taken to belong to $\mathcal{L}(A)$.[2]

The following simple observation will be of help.

Lemma 11.10 *Let Φ and Ψ be sets of sentences. Assume that every $\psi \in \Psi$ is logically entailed by some $\varphi \in \Phi$ and that every $\varphi \in \Phi$ logically entails some $\psi \in \Psi$. Then $\triangle\Phi$ logically entails $\triangle\Psi$.*

Proof Assume $a \models \triangle\Phi$. We need to show that $a \models \triangle\Psi$. Let us first show that a is a model of the first conjunct of $\triangle\Psi$, namely

$$\bigwedge_{\psi \in \Psi} \Diamond\psi.$$

Pick any $\psi \in \Psi$. By the assumption, there is a $\varphi \in \Phi$ that entails ψ. Since $a \models \triangle\Phi$, $a \models \Diamond\varphi$. Hence there is a $b \in a$ such that $b \models \varphi$. Hence $b \models \psi$, and so $a \models \Diamond\psi$, as desired. Showing that a models the second conjunct is similar, using the other assumption of the proposition.

$\dashv$

We now come to the main definition.

Definition Let $A \subseteq \mathcal{U}$. For each $a \in V_{afa}[A]$ we define a transfinite sequence φ_α^a of sentences of $\mathcal{L}_\infty(A)$ by recursion on ordinals as follows:

$$\begin{aligned}
\varphi_0^a &= \bigwedge_{p \in A \cap a} p \wedge \bigwedge_{p \in A - a} \neg p \\
\varphi_{\alpha+1}^a &= \varphi_0^a \wedge \triangle\{\varphi_\alpha^{a_0} \mid a_0 \in a\} \\
\varphi_\lambda^a &= \bigwedge_{\alpha < \lambda} \varphi_\alpha^a \qquad \text{for } \lambda \text{ a limit ordinal}
\end{aligned}$$

(As usual, empty conjunctions here count as T, empty disjunction as F.)

We summarize some elementary observations about these sentences in the next lemma.

Lemma 11.11 *Let a and b belong to $V_{afa}[A]$, and let α and β be ordinals.*

1. $b \models \varphi_0^a$ iff $b \cap \mathcal{U} = a \cap \mathcal{U}$.
2. $a \models \varphi_\alpha^a$.
3. If $\alpha \geq \beta$ then $\models \varphi_\alpha^a \rightarrow \varphi_\beta^a$.
4. If φ_α^a and φ_α^b are jointly satisfiable then $\varphi_\alpha^a = \varphi_\alpha^b$.
5. If $\varphi_\alpha^a = \varphi_\alpha^b$ then $\varphi_\beta^a = \varphi_\beta^b$ for all $\beta < \alpha$.

Proof Part 1 follows easily from the fact that $a \cap \mathcal{U} \subseteq A$ and $b \cap \mathcal{U} \subseteq A$. Part 2 is an easy induction on α. So is part 3, given Lemma 11.10.

Part 4 is also proven by induction on α. It is obvious in the case of 0 and the induction step for limit ordinals is obvious, so let us suppose it holds for

---

[2]$HF^0[A]$ is the set of all wellfounded sets whose transitive closure is finite and with support contained in A. It will be discussed later in the book.

α and prove it for $\alpha + 1$. Suppose $\varphi^a_{\alpha+1}$ and $\varphi^b_{\alpha+1}$ are jointly satisfied by some set c. We want to prove that $\varphi^a_{\alpha+1} = \varphi^b_{\alpha+1}$. Given the form of these sentences, it clearly suffices to show that for each $a_0 \in a$ there is a $b_0 \in b$ such that $\varphi^{a_0}_\alpha = \varphi^{b_0}_\alpha$, and vice versa. By the inductive hypothesis, it suffices to show that for each $a_0 \in a$ there is a $b_0 \in b$ such that $\varphi^{a_0}_\alpha$ and $\varphi^{b_0}_\alpha$ are jointly satisfiable (and vice versa). Let $a_0 \in a$. Then since $c \models \varphi^a_{\alpha+1}$, $c \models \Diamond \varphi^{a_0}_\alpha$ (by the second conjunct) so that is a $c_0 \in c$ such that $c_0 \models \varphi^{a_0}_\alpha$. But since $c \models \varphi^b_{\alpha+1}$, $c \models \Box \bigvee_{b_0 \in b} \varphi^{b_0}_\alpha$ (by the third conjunct). Hence there is a b_0 in b such that $c_0 \models \varphi^{b_0}_\alpha$. But then $\varphi^{a_0}_\alpha$ and $\varphi^{b_0}_\alpha$ are jointly satisfiable, as desired. Part 5 follows immediately from the earlier parts. ⊣

This result shows us that we can think of the sentences φ^a_α as *canonical invariants* of a. Our aim is to show that eventually one of these invariants characterizes a. Theorem 11.9 follows immediately from the following result, whose proof will absorb us for the rest of this section.

Theorem 11.12 *For every set* $a \in V_{afa}[A]$ *there is an* α *such that* φ^a_α *characterizes* a.

Our original proof of this result was less elegant than the proof given below, which is due to Alexandru Baltag and is used with his kind permission.

First, we need the notion of a *regular cardinal number*. (You may wish to first review page 20 for the definition of a cardinal.) A cardinal κ is regular iff for every set I of cardinality less than κ, if a_i has cardinality less than κ for each $i \in I$, then $\bigcup_{i \in I} a_i$ has cardinality less than κ. For example, ω is a regular cardinal. And if κ is an infinite cardinal, then its successor cardinal κ^+ is always regular. (The successor cardinal of κ is the least cardinal greater than κ.) This last result is proved using the Axiom of Choice. It is easy to see that if κ is regular then whenever B is a set of size κ and $f : \kappa \to B$, then there is a subset $S \subseteq \kappa$ such that $\bigcup S = \kappa$ and f is constant on S. In other words, every map from κ to a set of smaller size is constant on an unbounded subset of κ.

Let us define $H_\kappa[A]$ to be the largest collection C of sets such that every c in C is a subset of $C \cup A$ and has size less than κ. For $\kappa = \omega$, $H_\kappa[A] = HF^1[A]$. Notice that $H_\kappa[A]$ is transitive.

Lemma 11.13 (Baltag's Lemma) *Let* κ *be an infinite regular cardinal. Define a relation* R *on* $V_{afa}[A]$ *by:*

$$a \, R \, b \text{ iff there is a set } c \in H_\kappa[A] \text{ such that } \varphi^a_\kappa = \varphi^c_\kappa = \varphi^b_\kappa.$$

Then R *is a bisimulation relation. Hence if* $a \, R \, b$ *then* $a = b$ *and* $a \in H_\kappa[A]$.

Before proving this lemma, let us show that it yields the theorem. Given a, pick κ so that κ is regular and $a \in H_\kappa[A]$. We claim that φ^a_κ characterizes a. For suppose $b \models \varphi^a_\kappa$. By Lemma 11.11.4, $\varphi^a_\kappa = \varphi^b_\kappa$. But since $a \in H_\kappa[A]$, we have $a \, R \, b$, where R is as in Baltag's Lemma. Hence $a = b$.

We now turn to the proof of Baltag's Lemma.

Proof Suppose that $a \, R \, b$. Pick a $c \in H_\kappa[A]$ such that $\varphi_\kappa^a = \varphi_\kappa^b = \varphi_\kappa^c$. It is clear that $a \cap A = b \cap A$ since φ_κ^c gives complete information about the urelements in $c \cap A$ (see Lemma 11.11.1). The rest of the proof has to do with the set part of the bisimulation condition.

Claim 1 For every set $a' \in a$ and $\alpha < \kappa$ there is some set $c' \in c$ such that $\varphi_\alpha^{a'} = \varphi_\alpha^{c'}$. For this c', there is some set $b' \in b$ such that $\varphi_\alpha^{c'} = \varphi_\alpha^{b'}$.

To prove the first assertion, note that $a \models \Diamond \varphi_\alpha^{a'}$, so also $c \models \Diamond \varphi_\alpha^{a'}$. Thus for some $c' \in c$, $c' \models \varphi_\alpha^{a'}$. So $\varphi_\alpha^{a'}$ and $\varphi_\alpha^{c'}$ are jointly satisfiable (by c'). Thus they are identical, by Lemma 11.11.4. The second assertion is proved in the same manner.

Claim 2 For any set $a' \in a$ there is a set $c' \in c$ such that $\varphi_\kappa^{a'} = \varphi_\kappa^{c'}$.

By Claim 1, for each α there is some $c' \in c$ such that $\varphi_\alpha^{a'} = \varphi_\alpha^{c'}$. At this point we use the regularity of κ: There are κ ordinals α, and each one may be associated to some $c' \in c$. Since c has fewer than κ elements, there must be some fixed $c' \in c$ so that for an unbounded set $S \subseteq \kappa$, whenever $\alpha \in S$, $\varphi_\alpha^{a'} = \varphi_\alpha^{c'}$. Since S is unbounded, this equation holds for *all* $\alpha < \kappa$. Since κ is a limit ordinal, $\varphi_\kappa^{a'} = \varphi_\kappa^{c'}$. This is our claim.

Claim 3 There is an ordinal $\alpha < \kappa$ such that for all sets $c', c'' \in c$, if $\varphi_\alpha^{c'} = \varphi_\alpha^{c''}$ then $\varphi_\kappa^{c'} = \varphi_\kappa^{c''}$.

To prove this claim, first note that for a fixed pair $\langle c', c'' \rangle$ such that $\varphi_\kappa^{c'} \neq \varphi_\kappa^{c''}$ there must be some α so that $\varphi_\alpha^{c'} \neq \varphi_\alpha^{c''}$. Let

$$ S \;\; = \;\; \{ \langle c', c'' \rangle \mid \varphi_\kappa^{c'} \neq \varphi_\kappa^{c''} \}. $$

S is a subset of $c \times c$, and it is a basic fact of set theory that the cardinality of $c \times c$ is again $< \kappa$. So the cardinality of S is $< \kappa$. By regularity, there is a fixed $\alpha < \kappa$ so that for all $\langle c', c'' \rangle \in S$, $\varphi_\alpha^{c'} \neq \varphi_\alpha^{c''}$.

Claim 4 Let $a' \in a$ be a set. Fix c' as in Claim 2. There is some set $b' \in b$ such that $\varphi_\kappa^{c'} = \varphi_\kappa^{b'}$.

To prove this last claim choose α as in Claim 3. By Claim 1, there is some $b' \in b$ such that $\varphi_\alpha^{c'} = \varphi_\alpha^{b'}$. We'll show that $\varphi_\kappa^{c'} = \varphi_\kappa^{b'}$. To show this, it suffices to show that $\varphi_\beta^{c'} = \varphi_\beta^{b'}$ whenever $\alpha < \beta < \kappa$. Fix such a β. By Claim 1 (actually the analogous result with b replacing a), there is some $c'' \in c$ such that $\varphi_\beta^{b'} = \varphi_\beta^{c''}$. Hence

$$ \varphi_\alpha^{c'} \;\; = \;\; \varphi_\alpha^{b'} \;\; = \;\; \varphi_\alpha^{c''}. $$

Since $\varphi_\alpha^{c'} = \varphi_\alpha^{c''}$ we have $\varphi_\kappa^{c'} = \varphi_\kappa^{c''}$ by Claim 3.

Of course, Claims 2 and 4 suffice for the proof of the result, since every $c' \in c$ belongs to $H_\kappa[A]$. $\dashv$

Definition For any set $a \in V_{afa}[A]$, we define the *degree* of a, written $\deg(a)$, be the least ordinal α such that φ_α^a characterizes a. We also let

$$\theta^a = \varphi_{\deg(a)}^a.$$

It follows from our results that every set $a \in V_{afa}[A]$ has a degree, and that $\deg(a)$ is at most the least infinite cardinal κ such that $a \in H_\kappa[A]$.

Corollary 11.14 *For all sets* $a, b \in V_{afa}[A]$, *the following are equivalent:*

1. $a = b$
2. $\theta^a = \theta^b$
3. $b \models \theta^a$

Proof It suffices to prove (3) $\Rightarrow$ (1), but this follows immediately from the fact that θ^a characterizes a. ⊣

Corollary 11.15 *For all sets* $a, b \in V_{afa}[A]$, *the following are equivalent:*

1. $a \in b$
2. $\models \theta^b \rightarrow \Diamond\theta^a$
3. $b \models \Diamond\theta^a$

Corollary 11.16 *For all* $a \in HF^1[A]$, θ^a *is a conjunction of sentences* φ_n^a, *each of which belongs to* $\mathcal{L}(A)$.

In general, suppose that κ is an infinite regular cardinal with the property that every element of $TC(a)$ has size $< \kappa$. Then the sentence θ^a is a conjunction of κ sentences, each of which is built up using infinitary conjunctions and disjunctions of size $< \kappa$.

Exercise 11.5 Consider $\mathcal{L}_\infty(A)$ as a proper class which is preordered by the relation $\varphi \leq \psi$ iff $\models \varphi \rightarrow \psi$. Write $\equiv$ for the equivalence relation determined by $\leq$, and $<$ for the strict preorder determined by it. Call a sentence φ *maximal* if $\varphi < \mathsf{F}$, and if whenever $\varphi < \psi$, then $\psi \equiv \mathsf{F}$. Prove that the maximal sentences are exactly those of the form θ^a for some set a.

11.3   Baltag's Theorems

In this section we want to prove two interesting theorems of Alexandru Baltag. They are actually two of many interesting results found by Alexandru Baltag in the course of the MOC Workshop; they will appear in Baltag (to appear). The basic question being addressed is: when is a set $a \in V_{afa}[A]$ characterizable by a single sentence of, or a theory of, the finitary language $\mathcal{L}(A)$?

Theorem 11.17 (Baltag) *Assume that A is finite. A set $a \in V_{afa}[A]$ is characterizable by a sentence of the finitary language $\mathcal{L}(A)$ iff $a \in HF^0[A]$, that if, iff a is wellfounded and $TC(\{a\})$ is finite.*

Define the *modal depth* of a sentence $\varphi \in \mathcal{L}_\infty(A)$ as follows:

$$\begin{aligned}
\mathrm{d}_\diamond(\varphi) &= 0 \text{ if } \varphi \text{ is atomic,} \\
\mathrm{d}_\diamond(\neg\varphi) &= \mathrm{d}_\diamond(\varphi), \\
\mathrm{d}_\diamond(\bigwedge\Phi) &= \sup\{\mathrm{d}_\diamond(\varphi) \mid \varphi \in \Phi\}, \quad \text{and} \\
\mathrm{d}_\diamond(\Diamond\varphi) &= \mathrm{d}_\diamond(\varphi) + 1.
\end{aligned}$$

A routine induction on α shows that for each a, $\mathrm{d}_\diamond(\varphi_\alpha^a) = \alpha$.

Proof The right to left direction follows easily from our results above (see Exercise 11.4). Going the other way, define for each set $a \in V_{afa}[A]$ sets a^n by recursion on n: $a^0 = A \cap a$; $a^{n+1} = \{b^n \mid b \in a\} \cup (A \cap a)$. Then an easy induction on n shows that if φ has modal depth n, $a \models \varphi$ iff $a^n \models \varphi$. It follows that if φ characterizes a, then $a = a^n$ for $n = \mathrm{d}_\diamond(\varphi)$. By induction on n we also see that there are only finitely many sets a^n as a ranges over $V_{afa}[A]$; this is where the assumption that A is finite is used. A final induction on n shows that a^n is wellfounded. It follows that $a^n \in HF^0[A]$ for all $a \in V_{afa}[A]$ and all n. $\dashv$

Before stating Baltag's second theorem, we state the following useful observation relating modal depth and sentences of the form φ_α^a.

Proposition 11.18 *For all $a, b \in V_{afa}[A]$, a and b satisfy the same sentences with modal depth at most α iff $\varphi_\alpha^a = \varphi_\alpha^b$.*

Proof The direction from left to right follows from the fact that $a \models \varphi_\alpha^a$ and the depth of this sentence is α. To prove the converse, we prove by double induction on α and χ that if $\mathrm{d}_\diamond(\chi) \leq \alpha$ and $b \models \varphi_\alpha^a$ then χ is true in a iff it is true in b. The only non-trivial step of this induction is when χ is of the form $\Diamond\psi$. Suppose that $b \models \chi$. Then $\mathrm{d}_\diamond(\psi) + 1 = \mathrm{d}_\diamond(\chi) \leq \alpha$, so $\mathrm{d}_\diamond(\psi) < \alpha$. Thus $b \models \varphi_{\mathrm{d}_\diamond(\psi)+1}^a$. Hence

$$b \models \Box \bigvee_{a_0 \in a} \varphi_{\mathrm{d}_\diamond(\psi)}^{a_0}.$$

Then there is a $b_0 \in b$ such that $b_0 \models \psi$. But then there is an $a_0 \in a$ such that $b_0 \models \varphi_{\mathrm{d}_\diamond(\psi)}^{a_0}$. By the induction hypothesis on sentences, $a_0 \models \psi$. So $a \models \Diamond\psi$; i.e., $a \models \chi$. The converse is similar. $\dashv$

Corollary 11.19 *Assume A is finite.*

1. *a and b satisfy the same sentences of $\mathcal{L}(A)$ iff $\varphi_\omega^a = \varphi_\omega^b$.*
2. *a is characterizable by a theory of $\mathcal{L}(A)$ iff $\deg(a) \leq \omega$.*

Proof The first part is immediate from Proposition 11.18. For the second assertion, suppose that a is characterizable by a set S of finitary sentences. Then if $b \models \varphi_\omega^a$, $b \models S$ by Proposition 11.18. So $b = a$. This proves the left to right half. The other direction is obvious, since φ_ω^a is a conjunction of finitary sentences. $\dashv$

We now state Baltag's second theorem.

Theorem 11.20 (Baltag) *Assume that A is finite. Let $a \in V_{afa}[A]$. a is characterizable by a theory of the finitary language $\mathcal{L}(A)$ iff $a \in HF^1[A]$.*

The right to left half will follow easily from results which we have already seen. Before proving the converse, though, it is useful to prove the following observation.

Proposition 11.21 *For all $b \in a \in V_{afa}[A]$, $\deg(b) \leq \deg(a)$. Moreover, if $\deg(a)$ is a successor, then $\deg(b) < \deg(a)$.*

Proof This is proved by induction on $\alpha = \deg(a)$. The case for $\alpha = 0$ is vacuous, since the sets of degree 0 are subsets of A. So suppose $\alpha = \beta + 1$. We will prove that φ_β^b characterizes b. Let c be a set such that $c \models \varphi_\beta^b$. Consider $a' = (a - \{b\}) \cup \{c\}$. But examining the structure of the sentence $\varphi_{\alpha+1}^a$ we see that $a' \models \varphi_{\alpha+1}^a$. Therefore $a' = a$. Since $b \in a$ we must therefore have $b = c$. (This trick of Baltag's is one we will use again below.) This completes the induction step for successor ordinals. The case where α is a limit is similar but simpler. ⊣

Using this result, we can prove the following lemma, a special case of the theorem we are after.

Lemma 11.22 *Assume that A is finite. If $a \in V_{afa}[A]$ is characterizable by a theory of $\mathcal{L}(A)$, then so is every $b \in a$, and hence every $b \in TC(a)$.*

Proof Immediate by Corollary 11.19 and Proposition 11.21. ⊣

We can now give the proof of Theorem 11.20.

Proof Let us first prove the easy half. Assume that $b \in HF^1[A]$. Then b is characterized by φ_ω^b, by Baltag's Lemma. But this is a conjunction of finitary sentences, hence a theory of $\mathcal{L}(A)$.

Assume that b is characterizable by a theory in $\mathcal{L}(A)$. We need to show that $b \in HF^1[A]$. By the above lemma, it suffices to show that b is finite. The proof will use a similar trick to the one used in the proof of Proposition 11.21. Namely, we will assume that b is not finite and construct a set c such that $b_1 = b - \{c\}$ and $b_2 = b \cup \{c\}$ both satisfy φ_ω^b, and hence the full $\mathcal{L}(A)$ theory of b. But $b_1 \neq b_2$ so b cannot be characterized by the full $\mathcal{L}(A)$ theory of b. To construct c, we use a well-known result that we will prove later in the chapter, namely, that the language $\mathcal{L}(A)$ is compact: given any set T of sentences, if every finite subset of T is satisfiable, then so is T.

We now define sets c_n by recursion on $n \in \omega$. Assume that we have c_i for all $i < n$. Let $c_n \in b$ be a set with the following property:

(1)   there are infinitely many $c \in b$ such that for all $i \leq n$, $\varphi_i^c = \varphi_i^{c_i}$.

Here is why c_n is well-defined: Given c_i for all $i < n$ as in (1), there is an infinite set S such that for all $c \in S$ and all $i < n$, $\varphi_i^c = \varphi_i^{c_i}$. Since A is finite, there are only finitely many sentences of the form φ_n^a as a ranges over the whole class $V_{afa}[A]$. Hence there must be an infinite subset $S_0 \subseteq S$ such that for $c, d \in S_0$, $\varphi_n^c = \varphi_n^d$. Then c_n may be taken to be any element of S_0. Let

$$ T = \{\varphi_0^{c_0}, \varphi_1^{c_1}, \ldots, \varphi_m^{c_m}, \ldots\}. $$

Since every finite subset of T is consistent, so is the whole set T, by compactness. Let c be a set that satisfies T. We claim that $\varphi_n^c = \varphi_n^{c_n}$ for every n. The reason is that $c \models \varphi_n^{c_n}$, so φ_n^c and $\varphi_n^{c_n}$ are jointly satisfiable. So the result follows from Lemma 11.11.5.

Let $b_1 = b - \{c\}$ and $b_2 = b \cup \{c\}$. It is easy to see that for all n, $b_1 \models \varphi_n^b$ and $b_2 \models \varphi_n^b$. The key point concerns b_1: Since infinitely many elements of b "look like" c, b_1 satisfies the same finitary sentences as b. Formally, one uses (1) to check that for all n, $b_1 \models \varphi_n^b$. The argument for b_2 is similar. We therefore see that $b_1 \models \varphi_\omega^b$ and $b_2 \models \varphi_\omega^b$. So $b_1 = b = b_2$. But this is impossible, since $c \in b_2 - b_1$. ⊣

11.4 Proof theory and completeness

We have semantical notions of validity and logical consequence, denoted $\models \varphi$ and $T \models \varphi$. More generally, given a collection W that is transitive on sets, we have notions $\models^W \varphi$ and $T \models^W \varphi$, with the full notion being the case where $W = V_{afa}[A]$. We have seen that for any Kripke structure, the set of sentences valid in that Kripke structure is the same as the set valid in the set obtained by decorating the Kripke structure.

We would like to develop a proof theory to go with these notions of consequence. That is, for an arbitrary collection W that is transitive on sets, it would be nice to have a way to have a definition of a relation of *provable*, written $\vdash_W$, with two properties. First, to say that $\vdash_W \varphi$ should guarantee that φ has been proved to be true in all sets in W; i.e., $\models^W \varphi$. And second, it should be the case that whenever $\models^W \varphi$, then $\vdash_W \varphi$.

The logic associated with W will, of course, be sensitive to the set-theoretic properties of W. There is, at this time, no known result which lets us carry out this task in anything like full generality. In this section we consider the special case where W is the universe of all sets. In the next section we look at some examples that can be obtained by transferring known results from modal logic. This leads to a number of interesting sub-universes of the universe $V_{afa}[A]$ of sets over A.

The System K of modal logic The most basic system of modal logic is the system K for the finitary language $\mathcal{L}(A)$. This is the system that is complete when we look at arbitrary sets in $V_{afa}[A]$. The system K has *axioms* and *rules of inference*. The axioms of K are of two forms: (1) all substitution instances of tautologies of classical propositional logic, and (2) all substitution instances of the sentences

$$\Box(\varphi \rightarrow \psi) \rightarrow ((\Box\varphi) \rightarrow (\Box\psi)).$$

(For an example of what we mean by a substitution instance, $\Diamond q \vee \neg(\Diamond q)$ is a substitution instance of $\varphi \vee \neg\varphi$. Since the latter is a tautology, the former is an axiom of K.)

There are two rules of inference of K. First, from φ, deduce $\Box\varphi$. In symbols:

$$\frac{\varphi}{\Box\varphi}$$

This rule is called *necessitation*.

The second rule is: from φ and $\varphi \rightarrow \psi$, deduce ψ. In symbols:

$$\frac{\varphi \quad \varphi \rightarrow \psi}{\psi}$$

This is called *modus ponens* and requires two sentences to apply.

Definition A sentence $\varphi \in \mathcal{L}(A)$ is *provable in K* if it is in the smallest set of sentences containing the axioms of K and closed under modus ponens and necessitation. We write $\vdash \varphi$ if φ is provable in K. More generally, φ is *provable from* a set T if there are $\psi_1, \ldots, \psi_n$ in T so that $\vdash \psi_1 \wedge \ldots \wedge \psi_n \rightarrow \varphi$. We write $T \vdash \varphi$ in this case.

Example 11.3 We give a proof in K of $\vdash \Box(\Box((\Diamond p) \vee \neg(\Diamond p)))$.

$((\Diamond p) \vee \neg(\Diamond p))$
$\Box((\Diamond p) \vee \neg(\Diamond p))$
$\Box(\Box((\Diamond p \vee \neg(\Diamond p)))$

(1) is a substitution instance of a tautology. (2) follows from (1) by necessitation, as does (3) from (2).

Exercise 11.6 Prove the following about K.
  1. For all φ and ψ, $\vdash \Box\varphi \wedge \Box\psi \rightarrow \Box(\varphi \wedge \psi)$.
  2. For all φ and ψ, $\vdash \Box\varphi \wedge \Diamond\psi \rightarrow \Diamond(\varphi \wedge \psi)$.

The main aim of this section is to show the following result.

Theorem 11.23 (Soundness and Completeness of K) *Let $T \subseteq \mathcal{L}(A)$ and $\varphi \in \mathcal{L}(A)$. Then the following are equivalent:*

(1) $T \vdash \varphi$
(2) $T \models_{Kr} \varphi$.
(3) $T \models \varphi$.

We have already proven that (2) and (3) are equivalent. The proof that (1) entails either of these is an easy induction on the length of proof.[3] It is called the *soundness* result, since it shows that the logic behaves in a reasonable way. So the main content of the above is that (2) entails (1). This is called a *completeness* result since it shows that the logic is capable of proving everything that it should. The remainder of this section will be devoted to building up the machinery needed to give this proof.

The basic idea is fairly simple. We are going to build a certain labeled graph Th. The nodes of Th consist of maximal, consistent subsets of K, that is, sets U of sentences which do not allow us to prove everything and are maximal in this regard: if you add any other sentence, the result is inconsistent. The edge relation of Th is defined by

$$U \to V \text{ iff for all } \varphi \in \mathcal{L}(A), \varphi \in V \text{ implies } \Diamond \varphi \in U.$$

This is equivalent to saying that if $\Box \varphi \in U$ then $\varphi \in V$.

Let us write $th(a)$ (or $th_\infty(a)$) for the set of all sentences $\varphi \in \mathcal{L}(A)$ (or $\varphi \in \mathcal{L}_\infty(A)$, respectively) such that $a \models \varphi$. This collection is called the *A-theory* (or infinitary *A-theory*) of a. The way to remember the meaning of the edge relation in Th is that if we have sets a and b, and if $a \in b$, then we want to have $th(b) \to th(a)$. However, this is a semantic guide to the definition, and at this point we do not know whether every maximal consistent T is of the form $th(a)$ for some a.

The labeling relation of Th is defined by having U labeled by the set of atomic sentences in U. This defines our labeled graph G. Since the set Th is an A-labeled graph, we can interpret $\mathcal{L}(A)$ on it. We will show that if U is any node of G (i.e., any theory), then $U \models_G \varphi$ iff $\varphi \in U$. This result, known as the "truth lemma," will be outlined below in Exercise 11.7. Using it, we can complete the proof of our result as follows.

Proof of Theorem 11.23 To prove that (2) implies (1), suppose that $T \not\vdash \varphi$. Then $T \cup \{\neg\varphi\}$ is consistent. By Zorn's Lemma, we can get a maximal consistent set U which includes $T \cup \{\neg\varphi\}$. (An example of the use of Zorn's Lemma may be found in the proof of Lemma 11.24 below.) Then we look at Th. By the Truth Lemma, $U \models T \cup \{\neg\varphi\}$. Hence $T \not\models_{Th} \varphi$. ⊣

So what remains is to show the Truth Lemma. In order to keep the presentation short, we assume familiarity with standard facts about maximal consistent subsets of propositional systems. These are usually studied in courses which cover the completeness results for propositional logic. For example, if U is maximal consistent, then for all φ, either φ or $\neg\varphi$ belongs to U. And if $\varphi \vee \psi$

---

[3]We did not actually define a notion of proof, so this induction is really on the notion of provable. One could, of course, introduce a notion of proof to go along with our notion of provable.

belongs to U, then either φ or ψ belongs to U. Finally U is closed under deduction.

Lemma 11.24 *Let U be a maximal consistent set, and suppose $\Diamond\varphi \in U$. Then there is a maximal consistent V so that $\varphi \in V$ and $U \to V$ in the system Th of maximal consistent sets.*

Proof Call a set $S \subseteq \mathcal{L}$ *good* if $\varphi \in S$, and if whenever S_0 is a finite subset of S, then $\Diamond \bigwedge S_0$ belongs to U. The idea is if $U \to V$, then V would be good, and that any good set is a partial description of some such V.

We want to use Zorn's Lemma on the collection of good sets. We can do this, since the union of every chain of good sets is easily seen to be good. Let V be any maximal good set. We first claim that V is consistent. If not, let $V_0 \subseteq V$ be a finite set such that $\vdash \bigwedge V_0 \to \mathsf{F}$. Then using necessitation, $\vdash \Box(\bigwedge V_0 \to \mathsf{F})$. By Exercise 11.6,

$$\vdash \Box(\bigwedge V_0 \to \mathsf{F}) \wedge \Diamond \bigwedge V_0 \to \Diamond\mathsf{F}.$$

Since the original U is maximal consistent, it contains this sentence and is closed under deduction. Therefore, U also contains $\Diamond\mathsf{F}$. But $\Box\mathsf{T}$ is provable, since we use necessitation on the tautology T. This means that U is inconsistent, and this is a contradiction.

Now that we know V is consistent, we show that it is maximal consistent. If not, there would be some ψ so that neither ψ nor $\neg\psi$ belonged to V. Then by maximality, neither $V \cup \{\psi\}$ nor $V \cup \{\neg\psi\}$ would be good. So we would have a finite $V_0 \subseteq V$ so that $\vdash \bigwedge V_0 \to \psi$ and $\vdash \bigwedge V_0 \to \neg\psi$. Then $\vdash \neg \bigwedge V_0$. So $\vdash \neg\Diamond \bigwedge V_0$ by necessitation. But since V is good, $\Diamond \bigwedge V_0 \in U$. Again we have the contradiction that U is inconsistent. ⊣

The Completeness Theorem for $\boldsymbol{K}$ has a number of important consequences. Among them is the following result.

Corollary 11.25 (Compactness Theorem) *Let $T \subseteq \mathcal{L}(A)$ be such that every finite subset of T is satisfied by some set. Then there is a set a such that $a \models T$.*

Proof If not, $T \models \mathsf{F}$. Hence $T \vdash \mathsf{F}$. Since proofs are finite, there is a finite $T_0 \subseteq T$ such that $T_0 \vdash \mathsf{F}$. But this implies that T_0 is not satisfiable. This contradicts our hypothesis. ⊣

Exercise 11.7 (Truth Lemma) Let $U \in Th$. Prove that $th(U) = U$. That is, that for all φ, $\varphi \in U$ iff $U \models \varphi$, where the satisfaction relation takes place in Th.

Notice that it follows from our results that every maximal consistent theory U is of the form $th(a)$, for some set a. First, Exercise 11.7 shows that U is of the form $th(x)$, where x is a node in some A-labeled graph G. (Namely, G is

Th, and x is U itself.) But then we can let $a = d(U)$ where d is the decoration of the labeled graph G. Our results show that $U = th(a)$

Exercise 11.8 A set S is *closed under* $\Box$ if whenever $\varphi \in S$, $\Box\varphi \in S$ also. Prove that $th(\Omega)$ is the only maximal consistent set which contains no urelements and is closed under $\Box$.

11.5 Characterizing classes by modal theories

The modal logic K considered in the previous section axiomatized the set of sentences of $\mathcal{L}(A)$ true in all sets. From the original perspective of modal logic, this is a very weak logic. In this section we want to examine various well-known logics extending K and see where they lead us when we look at them as carving out sub-universes of the universe of all sets.

Example 11.4 From basic intuitions about necessity and possibility, you might expect the following to be valid.

$$(T) \quad \Box\varphi \to \varphi$$
$$(D) \quad \Box\varphi \to \Diamond\varphi$$

However, neither of these is valid in the semantics we have given. The first is not valid since $\Box p$ is true in $a = \{\{p\}, \{p, q\}\}$, for example, but p is not true there. Similarly for the second; $\Box p \to \Diamond p$ is not valid since the $\Box p$ is true in $a = \{q, r\}$, for example, but $\Diamond p$ is not true there.

In traditional approaches to modal logic, when one wants a Kripke structure in which such schemes are valid, one typically achieves this by imposing some condition on the graph $\langle G, \to \rangle$ underlying the Kripke structure $\langle G, \to, l \rangle$. $\Box\varphi \to \varphi$ would be insured by having each node a child of itself, that is, by insisting that the edge relation is reflexive. $\Box\varphi \to \Diamond\varphi$ would be insured by the weaker condition of having every node have some child, that is, there be no terminal nodes in the graph.

In general, a *Kripke frame* is an unlabeled graph $\mathcal{G} = \langle G, \to \rangle$. We think of a frame $\mathcal{G}$ as the class of models obtained by adding a labeling. Similarly, we say that a sentence φ (or a theory T) is *valid* on $\mathcal{G}$ if for every labeling l, φ (or T) is valid on $\langle G, \to, l \rangle$.

It is fairly easy to see that each sentence of the form $\Box\varphi \to \varphi$ is valid on all reflexive frames. Furthermore, we have a converse: if all possible labelings of a frame $\mathcal{G}$ satisfy all sentences of the form $\Box\varphi \to \varphi$, then $\mathcal{G}$ must be reflexive. To see this, we check the contrapositive. Suppose that $g \not\to g$ in G. Let l label all children of g with $\{p\}$. Then $g \models \Box p \land \neg p$.

In other words, conditions on frames (such as reflexivity) correspond to schemes. There are numerous results of this type, and they are an important aspect are of modal logic. Now given that we have been more concerned with labeled graphs than frames, it would have been nice if the correspondence had been in terms of Kripke structures rather than frames. Half of the correspondence does go through: if $\langle G, \to, l \rangle$ is reflexive, then every instance of $\Box \varphi \to \varphi$ holds. But the converse fails: it is possible to have a non-reflexive labeled graph satisfy all of the axioms. For example, the graph could be based on the natural numbers, with

$$ 0 \quad \to \quad 1 \quad \to \quad \cdots \quad \to \quad n \quad \to \quad \cdots $$

and the empty labeling. Then this graph satisfies all sentences in our scheme. To see why, note that the graph is bisimilar to Ω. Then it is easy to check that Ω satisfies the scheme.

From our point of view, this means that there are *too many* labeled graphs to get a correspondence. So we ask the question: if one only looks at *canonical* Kripke structures, can we get a tighter correspondence?

However, we are getting ahead of ourselves a bit. Before going on, let's go back to the situation in standard modal logic. Since we cannot characterize structures by the sentences which hold in them, the question in traditional approaches is to *characterize sentences by models*.

Definition Let T be some set of sentences of $\mathcal{L}_\infty(A)$ and let $\mathcal{C}$ be a collection of Kripke structures. $\mathcal{C}$ is a *semantics for* T if every sentence in T is valid in every $G \in \mathcal{C}$. $\mathcal{C}$ is a *complete semantics for* T if, in addition, any sentence not provable from T in $\boldsymbol{K}$ is invalid in some $G \in \mathcal{C}$. In other words, T is a complete semantics for $\mathcal{C}$ if the following condition holds: $T \vdash \varphi$ iff φ is valid in $\mathcal{C}$.

In Figure 6, we give some examples of known complete Kripke semantics for various theories. In each case we list the defining condition on Kripke structures and then give the theory for which this class of structures provides a complete Kripke semantics. (All of the theories are understood as being additions to the basic modal logic $\boldsymbol{K}$.) The third column contains the customary name for the theory involved.

We might make a few remarks about the names used in the figure. In **WkDense**, the word "weak" comes from the fact that g_3 might be equal to g_1 or g_2. Also, the name (L) is for Löb (see Exercise 5.1).

We are not going to prove the correspondence results indicated in Figure 6. However, you should be able to find proofs in any text on modal logic.

The completeness results characterize the theories by the Kripke structures. From the perspective of set theory, we want to go the other way. Given a theory T, can we characterize the sets which appear in canonical structures satisfying T? That is, can we use motivations from modal logic to isolate interesting classes of sets? This will be of interest to us later because each of these is associated with a natural monotone operator, the basis for our work in Chapter 15.

Class	Defining Property			Axioms
Ne	no terminals: $(\forall g)(\exists h)\, g \to h$		(D)	$\Box\varphi \to \Diamond\varphi$
Refl	reflexive: $(\forall g)\, g \to g$		(T)	$\Box\varphi \to \varphi$
Trans	transitive: $(\forall g_1)(\forall g_2)(\forall g_3)$ if $g_1 \to g_2 \to g_3$ then $g_1 \to g_3$		(4)	$\Box\varphi \to \Box\Box\varphi$
Symm	symmetric: $(\forall g_1)(\forall g_2)$ if $g_1 \to g_2$ then $g_2 \to g_1$		(B)	$\varphi \to \Box\Diamond\varphi$
WkDense	weakly dense: $(\forall g_1)(\forall g_2)(\forall g_3)$ if $g_1 \to g_2$ then there is a g_3 such that $g_1 \to g_3 \to g_2$			$\Box\Box\varphi \to \Box\varphi$
Triv	all nodes related: $(\forall g)(\forall h)\, g \to h$		**S5**	(T) + $\Diamond\varphi \to \Box\Diamond\varphi$
WF	wellfounded: there are no infinite sequences $a_0 \to a_1 \to a_2 \to \cdots$ and also transitive		(L)	$\Box(\Box\varphi \to \varphi) \to \Box\varphi$

FIGURE 6 The correspondence of classes of Kripke frames and modal logics.

Definition

1. A set a is *reflexive* if $a \in a$.
2. The collection of *hereditarily reflexive sets* is the largest collection *HRefl* satisfying the following condition: If $a \in HRefl$ then a is a reflexive set and every set $b \in a$ is in *HRefl*.[4]
3. The collection of *hereditarily nonempty* sets is the largest collection *HNe* satisfying the condition: If $a \in HNe$ then a contains some set b and every set $b \in a$ is in *HNe*.

As examples, we note that Ω is hereditarily reflexive. So is $a = \{x, a, b\}$ where $b = \{y, b, \Omega\}$ and $x, y \in \mathcal{U}$.

Lemma 11.26 *HRefl* $\subset$ *HNe*.

Proof The inclusion holds since *HRefl* satisfies the condition defining *HNe*. The inclusion is proper, since the sets $a = \{p, b\}$ and $b = \{q, a\}$ are hereditarily non-empty but not reflexive.   ⊣

---

[4]Definitions involving the "largest collection" are explored in detail in Chapter 15 below. For now, we can state an equivalent form of the definition. The hereditarily reflexive sets are the sets a such that for all sets $b \in TC(\{a\})$, b is reflexive.

Proposition 11.27 *Let G be a Kripke structure and let W be the canonical Kripke structure bisimilar to it.*

1. *If the edge relation of G is reflexive then $W \subseteq HRefl$.*
2. *The edge relation of G has no terminals iff $W \subseteq HNe$.*

Proof We first prove (1). Assume that the edge relation is reflexive. Then W satisfies the following condition: If $a \in W$ then a is reflexive and every set $b \in a$ is in W. But *HRefl* is the largest collection satisfying this condition. The proof of the direction from left to right of (2) is similar, so let us prove the other direction. Actually, we will prove the contrapositive. So assume that G has a terminal node g. Then $d(g)$ has only urelements in it, so $d(g)$ does not contain any sets. Hence $W \nsubseteq HNe$. ⊣

Example 11.5 The converse of part (1) of Proposition 11.27 does not hold. Consider the set $a = \Omega$. This is clearly hereditarily reflexive. But we can find a graph for it which is not reflexive. The fact that $a = \{\{a\}\}$ suggests a non-reflexive graph. Namely, we have two nodes g_1, g_2 with edges $g_1 \to g_2 \to g_1$. Indeed, we can go further and find an unlabeled graph for Ω which is irreflexive, that is, has $g \nrightarrow g$ for all nodes g. Let the nodes of the graph consist of the natural numbers, with $n \to n + 1$.

We will show how to obtain completeness results for both collections of sets, *HRefl* and *HNe*. More generally, we will develop a machinery which allows us to transfer completeness results for known modal logics over to completeness results for sub-universes of the universe of all sets.

Definition Let $\mathcal{C}$ be a class of Kripke structures. The *hereditarily $\mathcal{C}$ sets* are those that appear as sets in some canonical Kripke structure in $\mathcal{C}$. That is,

$$H(\mathcal{C}) \quad = \quad \bigcup\{W \in \mathcal{C} \mid W \subseteq V_{afa}[A] \text{ is transitive on sets}\}.$$

Thus, for example, $HRefl = H(\mathbf{Refl})$ and $HNe = H(\mathbf{Ne})$.

Proposition 11.28 *If $\mathcal{C}$ is a semantics for T then $\models^{H(\mathcal{C})} \varphi$ for all φ such that $T \vdash \varphi$.*

This result is an immediate consequence of the definitions. What we want, of course, is the converse. To obtain this, we need the following definition.

Definition A collection $\mathcal{C}$ of Kripke structures is *closed under decoration* if for every $G \in \mathcal{C}$, the canonical Kripke structure bisimilar to G is also in $\mathcal{C}$.[5]

---

[5]We do not know the answer to the following natural question: What first-order classes of graphs are closed under decoration? Exercise 11.13 gives a class which is not so closed. There is a related

Now it turns out that all of the familiar (familiar to us, at least) classes of Kripke structures that provide complete semantics for modal logics are closed under decoration. In particular:

Proposition 11.29 *All of the classes* **Ne**, ..., **WF** *listed in Figure 6 are closed under decoration.*

Proof These all have simple proofs. We will show one by way of illustration, namely, **Trans**. Let G be a transitive Kripke structure and let d be its decoration mapping G onto the set W. We want to show that the edge relation on W is transitive, that is, if $a, b, c \in W$ and $a \in b \in c$ then $a \in c$. We have to start in the right order and apply the bisimulation condition. First, pick g_c such that $d(g_c) = c$. Then using the bisimulation condition, pick a child g_b of g_c such that $d(g_b) = b$. Now apply the bisimulation condition again to obtain a child g_a of g_b such that $d(g_a) = a$. Since G is transitive, we see that g_a is a child of g_c. But then $d(g_a) \in d(g_c)$, i.e., $a \in c$, as desired.   ⊣

Theorem 11.30 *Assume that $\mathcal{C}$ is a complete semantics for a theory T, and that $\mathcal{C}$ is closed under decoration. Then for all sentences $\varphi \in \mathcal{L}(A)$,*

$$\models^{H(\mathcal{C})} \varphi \text{ if and only if } T \vdash \varphi.$$

Proof Suppose that $T \nvdash \varphi$. By completeness of $\mathcal{C}$, there is a Kripke structure $G \in \mathcal{C}$ and a node $g \in G$ such that $g \nvDash \varphi$, even though every sentence in T holds in G. But then let W be the canonical Kripke structure bisimilar to G. Since $\mathcal{C}$ is closed under decoration, $W \in \mathcal{C}$. Hence $d(g)$, the set which decorates the node g, is in $H(\mathcal{C})$. And $d(g)$ makes all sentences in T true, but it does not make φ true.   ⊣

This theorem allows us to transfer many known completeness theorems from Kripke structures to canonical Kripke structures. And now we state a sharper form of the characterization question which we have raised several times in this section:

Question Suppose that T is a set of sentences closed under provability in $\mathbf{K}$, and that $\mathcal{C}$ is a complete semantics for T. Is $H(\mathcal{C})$ the collection of *all* sets making T true?

Notice that a positive answer would give a stronger result than Theorem 11.30. In any case, Theorem 11.30 characterizes sentences by classes, and we want to go the other way. The purpose of this section is to get characterization results of this kind for the theories axiomatized in Figure 6. As it happens, the answer depends partly on whether we are using the full infinitary language

---

result, due to van Benthem (1985). He studied the classes of frames *closed under p-morphisms*. We won't define p-morphisms here, but we mention that decorations are special kinds of p-morphisms. He obtained a syntactic characterization of the kind we are after.

$\mathcal{L}_\infty(A)$ or just the finitary version $\mathcal{L}(A)$. We'll illustrate this by considering the example of (T), the scheme of all sentences of the form $\Box\varphi \to \varphi$.

Let $\boldsymbol{T}_\infty$ be the closure under the rule of necessitation of all instances of (T) where φ is allowed to come from $\mathcal{L}_\infty(A)$.

Theorem 11.31 *For any set $a \in V_{afa}[A]$ the following are equivalent:*

1. *$a \in HRefl$*
2. *$a \models \boldsymbol{T}_\infty$*

Before proving this result, let's note the following observation which will help us with this and similar results.

Proposition 11.32

1. *Let W be transitive on sets. Then the set of all infinitary sentences φ such that $\overset{W}{\models} \varphi$ is closed under the rule of necessitation.*
2. *Let T be some set or class of infinitary sentences closed under the rule of necessitation, and let $W = \{a \in V_{afa}[A] \mid a \models T\}$. Then W is transitive on sets.*

Both proofs are routine. We now return to the proof of Theorem 11.31.

Proof of Theorem 11.31 Let us first prove that (1) implies (2). By Proposition 11.32.1 it suffices to show that every $a \in HRefl$ is a model of $\Box\varphi \to \varphi$ for every $\varphi \in \mathcal{L}_\infty(A)$. Assume $\models^{a} \Box\varphi$. Then every $b \in a$ is a model of φ. But a is a member of itself, so a is a model of φ. Hence a is a model of $\Box\varphi \to \varphi$, as desired.

The implication from (2) to (1) is an easy consequence of Theorem 11.14. Let $W = \{a \mid a \models \boldsymbol{T}_\infty\}$. We will show that W satisfies the following condition: If $a \in W$, then a is reflexive and every set $b \in a$ is in W. Our result will follow since *HRefl* is the largest collection satisfying this condition.

So assume that $a \in W$. Let's first prove that $a \in a$. Let $\varphi = \neg\theta^a$. Since $a \in W$, $a \models \Box\varphi \to \varphi$. But then $a \models \theta^a \to \neg\Box\neg\theta^a$; i.e., $a \models \theta^a \to \Diamond\theta^a$. Since $a \models \theta^a$, we have $a \models \Diamond\theta^a$. Hence there is a $b \in a$ such that $b \models \theta^a$. Now $b \in V_{afa}[A]$, since $support(b) \subseteq support(a)$. By Theorem 11.14, $b = a$.

Finally, we need to verify that if $a \in W$ then so is every set $b \in a$. This follows from Proposition 11.32.2 and the fact that $\boldsymbol{T}_\infty$ is closed under necessitation. ⊣

Having this result, it is natural to ask about the finitary version of T. This is just the theory whose axioms are the $\mathcal{L}(A)$-instances of the scheme (T) above. It follows from the known fact that **Refl** is a complete semantics for T and Theorem 11.30 that this theory is complete for the class *HRefl* of hereditarily reflexive sets; that is, anything not provable in T can be falsified in some set

$a \in HRefl.$[6] On the other hand, there are sets which are models of T that are not hereditarily reflexive.

Example 11.6 Let $p_0, p_1, \ldots$ be an infinite sequence of urelements. By the Solution Lemma, there are sets a_C as C ranges over the subsets of N so that the following equation holds:

$$a_C \quad = \quad \{p_n \mid n \in C\} \cup \{a_D \mid D \neq C\}.$$

Let $Q = \{a_C \mid C \subseteq N\}$. We first check that Q is not reflexive. Suppose toward a contradiction that $Q \in Q$. Then $Q = a_C$ for some C. Then as $a_C \in Q = a_C$, there is some $D \neq C$ such that $a_C = a_D$. But C and D differ on some urelement, so $a_C \neq a_D$. We also claim that for each finitary φ, $\not\models^Q \varphi$. The reasoning is as follows: we first prove by induction on φ that if F is the finite set of all p_n occurring in φ and $C \cap F = D \cap F$, then $a_C \models \varphi$ iff $a_D \models \varphi$. The proof of this is not hard, and we omit it. Then note that if $a_D \models \varphi$, we can consider $D = C \cup \{p_k\}$ for some k large enough so that p_k does not occur in φ and does not belong to C. Then $D \neq C$ so $a_D \in a_C$; also $a_D \models \varphi$. So $a_C \models \Diamond\varphi$. This argument shows that every instance of T is valid in Q but Q is not reflexive, let alone hereditarily reflexive.

Recall our experience with bisimulation and the finitary language in Theorems 11.7 and 11.5. These results suggest that the hereditarily finite sets are the appropriate site for a characterization result.

Theorem 11.33 *For any set* $a \in HF^1[A]$ *the following are equivalent:*

1. $a \in HRefl$
2. $a \models T$

Proof We only prove (2) $\Rightarrow$ (1). As in Theorem 11.31, we just show that if $a \models T$, then $a \in a$. However, this time we must do a bit more work. Recall that θ^a is an infinite conjunction $\bigwedge \varphi_n^a$, where each φ_n^a is finitary. Our hypothesis tells us that for each n, $a \models \varphi_n^a \to \Diamond\varphi_n^a$. So for each n, there is some child b of a so that $b \models \varphi_n^a$. Since a has only finitely many children, we have that for some fixed b, $b \models \varphi_n^a$ for all n. Thus $b \models \theta^a$. Since b, too, belongs to $V_{afa}[A]$, $b = a$. Thus $a \in a$. ⊣

This result is our main characterization result for the finitary T. We can re-work it a bit to have a result about Kripke structures themselves. We state it in such a way as to be only about structures and sentences, though our proof goes by way of sets and *AFA*.

---

[6]We can derive this result easily from our earlier work, too. Recall the labeled graph *Th* used in our completeness proof. Look at the subgraph consisting of those maximal consistent sets containing all instances of (T). It is easy to verify that this graph is reflexive. Hence the canonical Kripke structure bisimilar to it will consist of hereditarily reflexive sets.

Corollary 11.34 *Let G be a finitely branching Kripke structure, all of whose node labels belong to a set $A \subseteq \mathcal{U}$. Then the following are equivalent:*

1. *All instances of (T) in $\mathcal{L}(A)$ are valid in G.*
2. *G is bisimilar to a reflexive Kripke structure.*

Proof We only need to consider the canonical Kripke model to which G is bisimilar. Call this set a. Then a belongs to $HF^1[A]$ and a and G satisfy the same sentences so the result follows from 11.33. ⊣

Although we have stated the results on the past few pages just for (T), they hold for all of the other classes listed in Figure 6. The proofs of the equivalence in all of the cases except the last are similar to our work for (T) and **Refl**. For (L), a slightly more involved argument is needed, and you are invited to work this out in Exercises 11.11 and 11.12.

We refer you to Barwise and Moss (1996) for a discussion of the relation of modal correspondences with the set theoretic results that we have obtained in this chapter. We should mention, though, that the logic D is a bit different from the others on our list; one need not pass to the infinitary language to get the characterization result. This is the content of our first exercise.

Exercise 11.9 This exercise shows that the modal logic T is different from the modal logic D with regard to its infinitary theory. Consider the modal logic D whose axioms consist of those of K plus the axiom $\Diamond T$. Since K contains the rule of necessitation, this logic also allows us to prove $\Box \Diamond T$, $\Box \Box \Diamond T$, $\Box \Box \Box \Diamond T$, ... Show that

$$HNe \quad = \quad \{a \in V_{afa}[A] \mid a \models \Diamond T, \Box \Diamond T, \Box \Box \Diamond T, \Box \Box \Box \Diamond T, \ldots\}$$

Exercise 11.10 Let T be any class of finite or infinitary sentences, and let $T^\Box$ be the closure of T under necessitation. Let

$$W \quad = \quad \{a \in V_{afa}[A] \mid a \models T^\Box\}.$$

Prove that

$$W \quad = \quad \bigcup \{a \mid a \text{ is transitive on sets and } \models^{\underline{a}} T\}.$$

The point of the last exercise is that it allows many of our characterization results to be recast, avoiding the closure under necessitation but involving a unary union operation instead. For example, in parallel to Theorem 11.31, we see that

$$HRefl \quad = \quad \bigcup \{a \mid a \text{ is transitive on sets and } \models^{\underline{a}} T'_\infty\},$$

where T'_∞ is the class of all instances of (T) in $\mathcal{L}_\infty(A)$.

Exercise 11.11 Let $A \subseteq \mathcal{U}$, and let $WF[A]$ be the class of all sets which are wellfounded and which have their support in A. Let $\boldsymbol{L}_\infty(A)$ be the closure under necessitation of all instances of (L), where φ comes from $\mathcal{L}_\infty(A)$. Prove that

$$WF[A] \;=\; \{a \in V_{afa}[A] \mid a \models \boldsymbol{L}_\infty(A)\}.$$

Exercise 11.12 *As in the previous exercise, let $A \subseteq \mathcal{U}$. Prove that for all $a \in HF^1[A]$,*

$$a \in WF[A] \quad \textit{iff} \quad a \models \boldsymbol{L}(A).$$

Exercise 11.13 Show that the class of irreflexive Kripke structures is not closed under decoration. [Hint: a solution to this exercise is contained in one of the examples in this section.]

Exercise 11.14 Show that $H(\mathbf{Triv})$ is the union of all sets W of sets satisfying the following conditions: (1) W is transitive on sets, and (2) if $a, b \in W$ $a \in b$. Give an example of such a set W containing at least two members.

Historical Remarks

Modal logic is a staple of philosophical logic. However basic the connection of infinitary modal logic to sets is, it was only first explored by Aczel (at the time he was writing his 1988 book). He knew the pure set version of Theorem 11.7. The connections with non-wellfounded sets were rediscovered independently a few years later by Abramsky (1988) and Moss. All of them worked only in the pure-set case, without urelements. As far as we know, the fact that *AFA* allows ordinary modal logic to be developed via what we call canonical Kripke structures is new here. The countable case of Theorem 11.14 appears in van Benthem and Bergstra (1995). Both the countable case and the general one are similar to what is known as Scott's Theorem in infinitary logic. The general result, and some of Baltag's results, seem to be very close in spirit to results in Fagin (1994). Fagin works in a different setting, but it seems likely that one could obtain our results via his or vice versa. Fagin also has a relatized notion of degree which seems quite interesting and suggests the following. Given a transitive set or class W and a set $a \in W$ define $deg_W(a)$ to be the least α such that a is the only member of W satisfying φ_α^a. It would be interesting to investigate the behavior of this function in this setting.

Modal logic also plays a role in theoretical computer science. To understand the connection on a general level, look back at the work on nondeterministic labeled transition systems and streams from Chapter 3. Consider a set *Act* of atomic actions, and make modal sentences by using new modalities $\Box_a$ and $\Diamond_a$ for all $a \in Act$. If $\mathcal{T} = \langle S, \delta \rangle$ is an lts over *Act*, we can interpret this language on $\mathcal{T}$ using the following clause in the semantics:

$$s \models_{\mathcal{T}} \Diamond_a \varphi \qquad \text{iff} \qquad \text{for some } b \text{ such that } \langle a, t \rangle \in \delta(s),$$
$$t \models_{\mathcal{T}} \Diamond_a \varphi$$

In Chapter 19 we will study this logic in a more general setting, where we will see a general way to develop such logics by building on the work done in the next part of the book. The current discussion is intended to give a hint that all kinds of phenomena involving transitions and changes can be studied using appropriate versions of modal logic.

12

Games

My favorite paradox of all is known as *hypergame*. It is due to the mathematician William Zwicker.

Raymond M. Smullyan (1983)

In Section 5.3, we considered an informal notion of game in connection with the Hypergame Paradox. This chapter presents a more formal account, and it shows what hypersets can contribute to a resolution of this paradox. In order to do that, we first take up the general issue of modeling games in Section 12.1. The games we are concerned with are two-person games of perfect information, like chess and checkers, in that both players have complete information of each other's previous moves. On the other hand, we do not include many features of real two-person games, for example, features like an initial deal of a deck of cards or the ability to bet. We also formalize what it means for one of the players to have a winning strategy in a game. This leads to some classical results which we present at the end of the section.

We turn in Section 12.2 to applications of games. These include the connection of games to quantifiers, and also a game-theoretic formulation of bisimulation. We return to the Hypergame Paradox in the final section of this chapter.

12.1 Modeling games

To provide a set-theoretic model of the notion of a two person game, we will think of a play of a game as producing a sequence of moves in a set M of possible moves. A given play might end, so we need to allow for sequences that come to an end. We also want to model the fact that the moves available to a player at a given stage in a game depend on what has happened earlier in the game.

Let M be some set of *moves*. Let $M^{<\infty}$ be the set of finite sequences from M. To be precise, we take the set ω of finite natural numbers (see page 13) and then define

$$M^{<\infty} \;=\; \bigcup \{p \mid p : n \to M \text{ for some } n \in \omega\}.$$

Similarly, we let $M^{\leq\infty}$ be the set of finite and infinite sequences from M, defined analogously.[1]

One special finite play is ϵ, the empty play. Formally, ϵ is the empty function and hence is $\emptyset$, but we use a different symbol to remind ourselves of its special role.

There are two parts to our model of a game: the rule for how to play the game and a method for determining who has won a given play of the game. We begin by modeling the notion of a rule. For this, we need a few definitions concerning plays.

We define the *length* of a play, $\mathrm{lh}(p)$, to be the number of terms in the sequence, if p is a finite sequence, or ω if p is an infinite sequence. So, for example, $\mathrm{lh}(\langle a, b, a\rangle) = 3$, $\mathrm{lh}(\epsilon) = 0$, and $\mathrm{lh}(\langle a, a, a, \ldots\rangle) = \omega$. If n is a natural number and $n \leq \mathrm{lh}(p)$ then we define the *n-th move of p*, $\mathrm{mv}(p, n)$, to be the n-th term in p. Hence $\mathrm{mv}(p, n) \in M$. We define the *truncation* $\mathrm{tr}(p, n)$ of a play p at n, for $n \leq \mathrm{lh}(p)$, to be the play obtained by chopping off everything after the n-th move. For example, $\mathrm{tr}(p, 0) = \epsilon$, and $\mathrm{tr}(p, 1) = \langle 1^{st}p\rangle$.

Finally, we say that a play p *extends* a play q if $q = \mathrm{tr}(p, \mathrm{lh}\, q)$. Note that p extends all of its truncations.

Given a finite play p and a move m we write $p^\frown m$ for the play that starts with p and ends with m. Note that for all m, $\epsilon^\frown m = \langle m\rangle$.

Definition A *rule* on M is a function R with domain a nonempty subset of $M^{<\infty}$ taking subsets of M as values. The elements of $dom(R)$ are called the *legal finite plays*. The elements of $R(p)$ are called the *moves allowed after p*. If p is a legal play and there are no legal moves after p (that is, $R(p) = \emptyset$) then p is said to be a *terminal play*. R must satisfy the following condition: for all finite plays p and moves m, p is a legal play and m is allowed after p if and only if $p^\frown m$ is a legal play; that is, $p \in dom(R)$ and $m \in R(p)$ iff $p^\frown m \in dom(R)$.

A (finite or infinite) play $p \in M^{\leq\infty}$ is *legal* according to rule R if for all

---

[1]The details concerning the coding of sequences are not critical to the results of this chapter. Indeed, we would have preferred to use streams in connection with $M^{\leq\infty}$ but decided to make this chapter accessible at this point of the book.

$n \in \omega$, $\mathrm{mv}(p, n+1) \in R(\mathrm{tr}(p,n))$. For example, if $p = \langle m_1, m_2, m_3, \ldots \rangle$, then p is legal iff $m_1 \in R(\epsilon)$, $m_2 \in R(\langle m_1 \rangle)$, $m_3 \in R(\langle m_1, m_2 \rangle)$, etc.

So far, we have only dealt with plays from a set M, and we have been quite abstract. For games, we always think of a play as being made alternately by two players. These players will be called I and II. (In other contexts, they might be given more descriptive names.)

To remind ourselves of who is playing what in response to what, we sometimes depict a legal infinite play $\langle m_1, m_2, m_3, \ldots \rangle$ as follows:

I	II
$m_1 \in R(\epsilon)$	
	$m_2 \in R(\langle m_1 \rangle)$
$m_3 \in R(\langle m_1, m_2 \rangle)$	
	$m_4 \in R(\langle m_1, m_2, m_3 \rangle)$
$\vdots$	$\vdots$

or more briefly just as

I	II
m_1	
	m_2
m_3	
	m_4
$\vdots$	$\vdots$

Now that we have moves, plays and players, we can speak about games. Suppose we are given a rule R on a set M of moves. If p is a terminal play in the game, then it means that the player whose turn it is to move has no legal move. In this case we declare that player the loser. In other words, if $R(p) = \emptyset$, and p has odd (even) length then p is said to be a *winning play* for I (for II respectively). But what about the plays that go one forever? How do we decide who wins such a play? The answer is that this must be built into the specification of the game.

Definition A *game* is a triple $\mathcal{G} = \langle M, R, W \rangle$ such that R is a rule on M and W is a subset of the infinite legal plays of R.

Given such a game, we define the set W_I of winning plays for I to be the set of finite winning plays for I together with all plays in W. We define the set W_{II} of winning plays for II to be the set of finite winning plays for II together with all infinite legal plays not in W.

Example 12.1 [Chess] We first give a model of the game of chess.[2] We identify the white player with I and the black player with II. We take M to

---

[2]If you do not know how to play chess, you will not really be at a loss in what follows. The

be the set of all possible board positions. Let R be such that for all finite plays $p = \langle m_1, \ldots, m_k \rangle$, $R(p)$ is the set of board positions that could be reached by taking m_k and moving one piece in accordance with the rules of chess.

We first stipulate that in chess there are no infinite winning plays for I. This would mean that the only way for I to win is to checkmate II in finitely many moves. If the game were to go on forever, then II would win that play. There would be no possibility of a draw.

There is a rule of chess that says that if a board position is repeated three times in the course of a game, then a draw is declared. Now our formalism cannot accommodate draws (and draws will not be needed in the mathematical examples). But we could modify R so that if a position occurs three times in a sequence p, then whichever player was responsible for the third repetition automatically loses.

Example 12.2 [The Membership Game] Suppose we are given some set a. We can use a to define what we will call the *membership game* $\mathcal{G}_a$ on a as follows. Player I starts by picking a member m_1 of a. Player II responds by picking a member m_2 of m_1. Player I then picks some $m_3 \in m_2$, and so on. Player I wins if II cannot move at some stage; otherwise II wins. Suppose for example, $a = \{0, 1, a\}$. Player I can win by playing 0. If I plays 1 or a, though, then player II can win by playing 0.

Exercise 12.1 Let a be a set, and consider the membership game $\mathcal{G}_a$. Figure out what M, R, and W_I are in this game.

The following are important kinds of games.

Definition A game $\mathcal{G}$ is *wellfounded* if every legal play is finite. $\mathcal{G}$ is *open* if for every infinite winning play p for I there is a finite initial segment q of p such that all legal infinite extensions of q are wins for I; $\mathcal{G}$ is *closed* if for every infinite winning play p for II there is a finite initial segment q of p such that all legal infinite extensions of q are wins for II. Notice that wellfounded games are both open and closed.

The membership game $\mathcal{G}_a$ is a wellfounded game iff a is a wellfounded set. Note too that $\mathcal{G}_a$ is an open game whether or not a is wellfounded. Concerning chess, the rule that says that three repeats is a draw implies that $\mathcal{G}$ is wellfounded. It is here that we use the fact M is finite. So in any infinite play, some position must be repeated infinitely often, hence at least three times. Without some rule of this type, a board game like chess would not be wellfounded.

Proposition 12.1 *There is no set of all wellfounded games. Hence there is no set of all games.*

---

actual rules of chess are far too long for us to spell out completely, and our point is just that any rule-based game *could* be modeled according to our definitions.

Proof If $a \neq b$ then $\mathcal{G}_a \neq \mathcal{G}_b$. Hence, since there is a proper class of wellfounded sets, there is a proper class of wellfounded games. ⊣

Example 12.3 The reason for the "open (closed) game" terminology comes from topological considerations. Suppose, for example, that

$$M = \{0, 1, 2, 3, 4, 5, 6, 7, 8, 9\}$$

and that the infinite plays $p = \langle m_1, m_2, m_3, \ldots \rangle$ are thought of as generating a decimal number

$$r_p \quad = \quad .m_1 m_2 m_3 \ldots \quad = \quad \sum_{n=1}^{\infty} \frac{m_n}{10^n}.$$

Each such r_p is a number in $[0, 1]$. Let $W \subseteq [0, 1]$. We define a game $\mathcal{G}_W$ by taking as wins for I all of the infinite sequences p such that $r_p \in W$. The connection to topology is that W is an open (or closed) set of real numbers iff $\mathcal{G}_W$ is an open (closed) game.

Suppose, for example, that W is the open interval $(.25, .75)$. Then $\mathcal{G}_W$ is an open game. Any real number in this interval is there because some finite initial segment of it forces the end result to be between $.25$ and $.75$. For example, the number $.2500000222 \cdots$ is forced in by the finite play corresponding to $.25000002$.

We now want to model the notions of a strategy and a winning strategy for a game. Intuitively, a strategy σ for one of the players should be a method for that player to determine their next move, at any stage of play. So σ will be a function. For example, a strategy for I should have as its domain a set of finite legal plays p of even length, and for such p, if $\sigma(p)$ is defined, then $\sigma(p) \in R(p)$. Furthermore, to be a winning strategy, it should be the case that no matter how II plays, if I follows σ, then the resulting play will be a win for I.

Definition Let $\mathcal{G} = \langle M, R, W_I \rangle$ be a game.

1. A *strategy* σ for I (or II) in $\mathcal{G}$ is a function from the finite legal plays of even (odd) length into M.
2. If σ is a strategy for I, then a play p is played *in accord with* σ if I's move at each stage is given by σ; that is, if for each odd integer $2n + 1 \leq \ln p$, $\sigma(\text{tr}(p, 2n)) = \text{mv}(p, 2n + 1)$. If σ is a strategy for II, then p is in accord with σ if II's move at each stage is given by σ, that is, if for each even integer $2n + 2 \leq \ln p$, $\sigma(\text{tr}(p, 2n + 1)) = \text{mv}(p, 2n + 2)$
3. A strategy σ for I (or II) is a *winning strategy* if every play which is played in accord with σ is a winning play for I (for II, resp.).

4. A game $\mathcal{G}$ is *determined* if one of the two players has a winning strategy.[3]

Note that it is not possible for both players to have winning strategies in a game, since if we pit the two strategies against one another, both players would win, contradicting the fact that W_I and W_{II} are disjoint. The following is a classical and important result in the theory of two-person games.

Proposition 12.2 (Gale-Stewart Theorem) *Every open game is determined.*

Proof Let $\mathcal{G} = \langle M, R, W_I \rangle$ be an open game. Suppose that I has no winning strategy. We will find a winning strategy for II. Actually, the strategy is the common-sense one: play defensively. That is, always play so that I does not get into a position from which I has a winning strategy. We need to see (1) that this is always possible, and (2) that if II follows this strategy, the resulting play is a win for II. (It is only in (2) that the assumption that the game is open is relevant.)

To prove (1), suppose that after a given finite play p of $\mathcal{G}$, it is II's turn to move, II has a legal move, but no matter what move $m \in R(p)$ player II makes, there is a winning strategy σ_m for I in the resulting game. Then I already has a winning strategy at position p, namely: play σ_m in response to m. Since I has no winning strategy at the start of the game, this shows that II can always move so as to insure that I still has no winning strategy.

To prove (2), assume p_1 be some completed play of the game played where II follows the strategy. If I wins this play, then there is a finite stage p at which the fact that I is going to win is decided, by the definition of open game. But that means that at that stage I has a winning strategy: play anything that is legal. ⊣

Corollary 12.3 *Every wellfounded game is determined.*

Proof Every wellfounded game is open (and closed). ⊣

Corollary 12.4 *For every set a, the game $\mathcal{G}_a$ is determined.*

Proof Every such game is open. ⊣

Exercise 12.2 This result gives us a natural way to divide the universe of sets into two parts: the class *Odd* of sets a such that I has a winning strategy in $\mathcal{G}_a$ and the class *Even* of those for which II has a winning strategy. For example, $\emptyset$ and Ω are both even, so any set containing either of them is odd.

Actually, the class *Even* divides nicely into two subclasses $Even_1$ and $Even_2$. $Even_1$ is the class of sets a such that II has a winning strategy σ in $\mathcal{G}_a$ with the

---

[3]It might be better to call this a game with a guaranteed outcome, since the word "determined" makes it sound like the players have no choice in how to play. But the word "determined" is quite standard in the theory of these sorts of games.

additional property that in all plays, II has won after a finite number of moves. $Even_2$ is the class of sets a such that II has a winning strategy, but for which there is some play by I which results in an infinite play p such that none of the finite truncations of p are wins for II. For example $\emptyset \in Even_1$, and every urelement also belongs to $Even_1$. On the other hand, $\Omega \in Even_2$. Note that $Even_1 \cap Even_2 = \emptyset$.

Let a be a pure set (i.e., $support(a) = \emptyset$). Prove the following:

1. $a \in Odd$ iff some $b \in a$ belongs to $Even_1$.
2. $a \in Even_1$ iff every $b \in a$ belongs to Odd.
3. $a \in Even_2$ iff every $b \in a$ belongs to $Odd \cup Even_2$, and some $b \in a$ belongs to $Even_2$.

(Similar results hold for arbitrary sets, but the formulations are a bit more complicated due to the urelements.)

Exercise 12.3 Use the previous exercise to show that the classes Odd and $Even_1$ are proper classes. Assuming *AFA*, prove that $Even_2$ is also a proper class.

Corollary 12.5 *Every closed game is determined.*

Proof Let $\mathcal{G}$ be a closed game. Define $\neg\mathcal{G}$ to be the same as $\mathcal{G}$ except that the positions that count as wins for I are the complement of those in $\mathcal{G}$. Then $\neg\mathcal{G}$ is open, so it is determined. But then if one of the players has a winning strategy in $\neg\mathcal{G}$, the other player has one in $\mathcal{G}$. ⊣

There are determined games which are neither open nor closed. The standard proofs of this go by considering the real-number games of Example 12.3. It is not hard to find such a countable set W which is neither open nor closed; for example, the rational numbers in $[0, 1]$ would work. The next exercise shows that $\mathcal{G}_W$ would be determined, but neither open nor closed.

Exercise 12.4 Let W be a countable subset of $[0, 1]$. Prove that II has a winning strategy in $\mathcal{G}_W$.

In addition, assuming the Axiom of Choice, one can construct games which are not determined. It is even possible to do this using simple sets of possible plays, such as $\{0, 1\}$. However, the details of the construction would take us too far afield. We encourage the interested reader to take up the subject in another book, for example Moschovakis (1980).

12.2 Applications of games

From quantifiers to games

There is a close relationship between games and quantifiers. This connection was first exploited by the logicians Ehrenfeucht and Fraïsse in order to prove results about the expressive power of languages. As a result, certain games are known as Ehrenfeucht-Fraïsse games. Although the results of this section are not needed for any work later in this book, we feel that they are so fundamental that everyone should know about them.

To give an example, consider some sentence

$$(1) \qquad (\forall x)(\exists y)(\forall z)(\exists w) \ R(x, y, z, w)$$

about numbers. We can think about this assertion in terms of a game played by two players. In this context, the players are named $\forall$ and $\exists$, rather than I and II, and the game works as follows: first $\forall$ picks some number x. After seeing this number, $\exists$ responds with a number y. After this, $\forall$ picks z. Finally, $\exists$ gets to pick one last number w. The sequence $\langle x, y, z, w \rangle$ of choices constitutes one play of the game. The play is won by $\exists$ if and only if $R(x, y, z, w)$.

Of course, there are many possible plays of the game. A winning strategy for $\exists$ would be a function which tells $\exists$ a way to play that results in a win each time. Such a function would be defined on single numbers (corresponding to the initial choice make by $\forall$) and on pairs of numbers (corresponding to $\forall$'s first two moves). When would $\exists$ have a winning strategy? Just in case the sentence is true.

The upshot is that a complicated sentence in first-order logic is re-interpreted as an assertion about winning strategies in a game. The point of doing this is that the original sentence (1) is too complicated to be understood directly, but by reformulating it in terms of games, we make it easier to work with.

From games to quantifiers

Going the other way around, the existence of a winning strategy for one of the players can be expressed using quantifiers. Given any game $\mathcal{G}$, we can informally express the existence of a winning strategy for Player I by means of the infinite "expression"

I	II
$\exists m_1 \in R(\epsilon)$	
	$\forall m_2 \in R(\langle m_1 \rangle)$
$\exists m_3 \in R(\langle m_1, m_2 \rangle)$	
	$\forall m_4 \in R(\langle m_1, m_2, m_3 \rangle)$
$\vdots$	$\vdots$

$$\langle m_1, m_2, m_3, \ldots \rangle \in W_I$$

Dually, the existence of a winning strategy for II can be expressed by:

I	II
$\forall m_1 \in R(\epsilon)$	
	$\exists m_2 \in R(\langle m_1 \rangle)$
$\forall m_3 \in R(\langle m_1, m_2 \rangle)$	
	$\exists m_4 \in R(\langle m_1, m_2, m_2 \rangle)$

$$\vdots \qquad\qquad\qquad \vdots$$

$$\langle m_1, m_2, m_3 \rangle \in W_{II}$$

Once again, we have an infinite expression. By the way, the usual de Morgan operations between negations and quantifiers do not apply to infinite strings of alternating quantifiers. (For if they did, then for every game $\mathcal{G}$, the two infinite assertions above would be negations of one another. So if I had no winning strategy, then II would have one. This contradicts the existence of non-determined games.) They would apply if the game $\mathcal{G}$ were determined.

The bisimulation game

The condition of bisimulation is fairly complicated. In order to check that a given pair of objects (systems of equations, sets, graphs, streams, what-have-you) is related by a bisimulation, we may well need to see whether other pairs are related, and so on. This is the kind of definition that admits a nice reformulation in terms of games. In this section we will carry out this reformulation in the case of pointed systems of equations. The reader will be able to see that the same idea applies much more generally.

So we'll devise a game played between pointed systems $\mathsf{e} = \langle \mathcal{E}, x \rangle$ and $\mathsf{f} = \langle \mathcal{F}, y \rangle$.[4] The idea is that I tries to show that the systems are not bisimilar, and II tries to respond with parts of a reason why they are.

We'll first describe the game informally, and then turn to formal definitions. To fix notation, we'll write $\mathcal{E}$ as $\langle X, A, e \rangle$, $\mathcal{F}$ as $\langle Y, B, e' \rangle$. To make life simpler, we'll assume that $X \cap Y = \emptyset$. (Exercise 12.5 will ask you to think about removing that assumption.) If $\langle X, A, e \rangle$ is a flat system, we define functions $indets : X \to \mathcal{P}(X)$ and $atoms : X \to \mathcal{P}(A)$ given by $indets(e, x) = e_x \cap X$ and $atoms(e, x) = e_x \cap A$.

A play of the game always starts $\langle x, y \rangle$ in mind. If $atoms(e, x) \neq atoms(e', y)$, then I has won and there is no point in playing further. Next, I either picks some $z \in indets(e, x)$ or some $z \in indets(e', y)$. Now II must respond with an indeterminate in the other system: if I played $z \in indets(e', y)$, then II must respond with some $w \in indets(e, x)$. In all cases, II must respond so that the final two points have the same atoms.

---

[4]You might want to look back at our notations and definitions concerning pointed systems from Section 9.1. In addition to the notion of a pointed system, $\mathsf{e} = \langle \mathcal{E}, x \rangle$, we'll need the notion of a bisimulation between pointed systems (see page 105).

Because of I's ability to play from either side, we'll use letters like u and v to range over $X \cup Y$ in what follows.

The players continue like this, with I picking from either side, and II responding from the other side. If at any point I is faced with $u \in X$ and $v \in Y$ such that $atoms(e, c) \neq atoms(e', v)$, then player I is declared the winner of the play, and so II has lost. Concerning infinite plays, we stipulate that II wins on all of them. This implies that the bisimulation game is open, and hence determined.

We now turn to a formalization of this game. In order to simplify the notation a bit, let's agree to use u and v for letters from *either* X *or* Y, and when we do this, we'll write $atoms(u)$ to mean $atoms(e, u)$ when $u \in X$ and $atoms(e, u)$ when $u \in Y$. To formalize the bisimulation game, we take the set M of possible moves to be $X \cup Y$. We need to specify R and W_I.

First, we need to handle the requirement that if $atoms(e, x) \neq atoms(e, y)$, then I wins automatically. To do this, we set $R(\epsilon) = \{x, y\}$, and

$$R(\langle u \rangle) \quad = \quad \begin{cases} \{x\} & \text{if } u = y \text{ and } atoms(e, x) = atoms(e', y) \\ \{y\} & \text{if } u = x \text{ and } atoms(e, x) = atoms(e', y) \\ \emptyset & \text{if } atoms(e, x) \neq atoms(e', y) \end{cases}$$

This way, if the starting values x and y do not have the same atoms, then II will be unable to play. If $atoms(e, x) = atoms(e, y)$, then the first two moves will either be $\langle x, y \rangle$ or $\langle y, x \rangle$.

For each sequence of positions of even length $n \geq 2$,

$$(2) \qquad\qquad p \quad = \quad \langle u_1, v_1, u_2, v_2, \ldots, u_n, v_n \rangle,$$

we take

$$R(p) \quad = \quad indets(u_n) \cup indets(v_n).$$

Again, this says that I may choose an indeterminate associated with either of the last two moves. (Note that if both $indets(e, u_n)$ and $indets(e, v_n)$ are empty, then II has won.)

Finally, here is the part of R that governs the moves by II:

$$R(p^\frown u_{n+1}) = \{z \in indets(v_n) \mid atoms(z) = atoms(u_{n+1})\}.$$

So II must respond to each play by I with an indeterminate from the opposite side, an II must always arrange that u_n and v_n have the same atoms. This completes the specification of R.

For W_I, the set of infinite winning plays of I, we take $\emptyset$. Thus, II wins if the game goes on forever. (Of course, II can additionally win if I arrives at a position p with $R(p) = \emptyset$.)

Example 12.4 Let p and q be urelements. Let x, y, $z \in \mathcal{U}$. We'll depart from our usual usage and let X, Y, Z, W also denote urelements. Suppose we have systems $\mathbf{e} = \langle \mathcal{E}, x \rangle$ and $\mathbf{f} = \langle \mathcal{F}, X \rangle$ satisfying the following equations:

$$
\begin{array}{rcl}
x & = & \{x,y\} \\
y & = & \{p,z\} \\
z & = & \{y\}
\end{array}
\qquad\qquad
\begin{array}{rcl}
X & = & \{X,Y,Z\} \\
Y & = & \{p,W\} \\
Z & = & \{p,W\} \\
W & = & \{Y\}
\end{array}
$$

We claim that II has a winning strategy in $\mathcal{G}(\mathsf{e},\mathsf{f})$. To show this we'll give a table that gives a strategy for II. In the first column we list positions, including the starting position (x,X). In the next columns headed by (1) and (2) we list pairs of moves. Whenever I makes a move in one of these columns, II makes the corresponding move in the other column. (Sometimes II will have more than one possible move.) Notice that every position that can arise by following this strategy, that is, every pair of sets from (1) and (2), appears in the Position column.

Position	(1)	(2)
	x	X
(x,X)	y	Y
	y	Z
(y,Y)	z	W
(y,Z)	z	W
(z,X)	y	Y
	y	Z
(z,W)	y	Y

For example, suppose that in the position (x,X) I plays X. Then to play according to the table, II must play by x. If I plays Y or Z, then I must play x.

It is easily checked that this strategy always gives a legal move for I. By following this strategy a play of the game will either go on forever, or if it ends after finitely many rounds, then II wins that play. In this way, II has a winning strategy.

Now the table above is essentially a bisimulation between the original pointed systems. This is the whole point: the existence of the bisimulation relation is equivalent to the existence of a winning strategy in the associated game. We now make this precise.

For any pointed systems $\mathsf{e} = \langle \mathcal{E},x\rangle$ and $\mathsf{f} = \langle \mathcal{F},y\rangle$, we have defined a game $\mathcal{G} = \mathcal{G}(\mathsf{e},\mathsf{f})$. Define the relation $R_{\mathcal{G}}$ on pointed systems by

$$R_{\mathcal{G}}(\mathsf{e},\mathsf{f}) \quad \text{iff} \quad II \text{ has a winning strategy in } \mathcal{G}(\mathsf{e},\mathsf{f})$$

Once again, each bisimulation game is open, hence determined. So $R_{\mathcal{G}}(\mathsf{e},\mathsf{f})$ iff I does not have a winning strategy in $\mathcal{G}(\mathsf{e},\mathsf{f})$.

Theorem 12.6 $R_{\mathcal{G}}$ *is exactly the relation of bisimulation on pointed systems. That is,* $\langle \mathcal{E}, x \rangle \equiv \langle \mathcal{F}, y \rangle$ *iff* $\langle \mathcal{E}, x \rangle R_{\mathcal{G}} \langle \mathcal{F}, y \rangle$.

Proof We continue to assume that $X \cap Y = \emptyset$. First, let S be a bisimulation between $\langle \mathcal{E}, x \rangle$ and $\langle \mathcal{F}, y \rangle$. We show how to get a winning strategy σ for II in the game.

II's strategy σ is to always play according to S. That is, to respond given the position $p ^\frown u_{n+1}$, II should play as follows: First, see whether u_{n+1} belongs to X or to Y. Let's suppose that $u_{n+1} \in X$. Then since $\langle \mathcal{E}, u_n \rangle \equiv \langle \mathcal{F}, v_n \rangle$, there is some $v_{n+1} \in Y$ such that $\langle \mathcal{E}, u_{n+1} \rangle \equiv \langle \mathcal{F}, v_{n+1} \rangle$. II should play any such v_{n_1}. This defines the strategy σ.

To prove that σ is a winning strategy, we show that if II plays by σ, then when faced with a play of even length as in (2), say with $u_n \in X$ and $v_n \in Y$, it will be the case that $\langle \mathcal{E}, u_n \rangle \equiv \langle \mathcal{F}, v_n \rangle$. The proof of this is by induction on n. So II can never lose playing by σ. Since I has no infinite winning plays, σ is a winning strategy.

Going the other way, we must prove that if II has a winning strategy, say σ, then the pointed systems $\langle \mathcal{E}, x \rangle$ and $\langle \mathcal{F}, y \rangle$ are bisimilar. Let $S \subseteq X \times Y$ be as follows:

$$S(u, v) \quad \text{iff} \quad \begin{array}{l} \text{either } u = x \text{ and } v = y, \\ \text{or there is a finite sequence } p \text{ obtained from} \\ \text{a play of the game where } II \text{ plays by } \sigma \\ \text{whose last two terms are } \langle u, v \rangle \text{ or } \langle v, u \rangle.. \end{array}$$

Note that in this definition, it might be that u was played before v, or vice-versa. Then we check that S is a bisimulation. Suppose that $S(u, v)$. Since σ is a winning strategy, $atoms(u) = atoms(v)$. Also, $R(x, y)$ since σ is a winning strategy. Further, general properties of winning strategies imply that S is a bisimulation.   ⊣

Exercise 12.5 Figure out what would have to change in the bisimulation game when $X \cap Y \neq \emptyset$. Be sure that Theorem 12.6 still holds.

Exercise 12.6 Suppose that we change the rules of the bisimulation game so as to force all plays to be finite. We might consider a new game $\mathcal{G}'$ defined like $\mathcal{G}$ except that at the beginning, player I starts with a natural number n. A play of the game continues for at most n rounds, and after that, if I has not won, then II is declared the winner. Of course, I may declare different n in different plays of the game. Give an example of two pointed systems e and f which are not bisimilar but for which the II has a winning strategy in wins $\mathcal{G}'(\mathsf{e}, \mathsf{f})$.

12.3   The Hypergame Paradox resolved

We now want to apply our machinery to examine the construction used in the Hypergame Paradox first discussed in Section 5.3.

One quick way for someone to evade the Hypergame Paradox would simply be to adopt the modeling of games that we have, use the Foundation Axiom, and thereby conclude that no game can be a play in itself. On this way of looking at things, the Hypergame simply does not exist. As you must know by now, we do not agree with this solution. We feel that this position goes wrong on a few counts: it does seem perfectly reasonable to have a game be a possible move in itself. On the other hand, there must be some sort of lesson in the Hypergame Paradox. So the ban on circularity diverts our attention from the real source of the paradox.

In a sense, Proposition 12.1 is the technical point made by Zwicker (1987) in his resolution of the Hypergame Paradox. If a game has to have a definite set of possible first moves, then there are too many wellfounded games to constitute the set of first moves of the hypergame. As he puts it, "Hypergame does not exist, so don't even think about playing it." While this solution is technically correct, it does not feel very satisfying, since it does not make any use of the reasoning of the Hypergame Paradox.

It seems we should be able to take any set S of games and use it to define a new game, the first move of which was to choose a game from S. After that, play proceeds in the game chosen. Actually, the idea is to let S be the set of all wellfounded games, but we have already seen that there is no such set. But let's see what happens if we try to formalize the construction when we do start with a *set S* of games.

We want to pass from S to a game S' whose first move might be any of the wellfounded games in S; and if S' is wellfounded, then S' should be a possible first move of itself. The paradox shows that there is *no game S'* with all of these features. We'll return to this point at the end of this section.

In the meantime, we will examine two constructions that do work, and which have interesting properties. These both pass from a set S of games to a bigger set. We will call the first the "supergame." In that construction, the first move of the new game must be one of the games in S; it need not be a wellfounded game. In our second construction, the first move might be the hypergame itself, but we do not ask whether hypergame itself is wellfounded in the definition of the game. We name this construction "hypergame," since it is the one that makes use of hypersets in a crucial way.

The Supergame Construction

Let S be a set of games. The *supergame over S*, written S^+, is defined informally as follows. Player I starts by playing one of the games $\mathcal{G} \in S$. Then the players switch sides: that is, II plays first in the game $\mathcal{G}$, and the remainder of the play is a play of $\mathcal{G}$. For example, if $S = \{\text{checkers, chess}\}$ then I can choose to play either checkers or chess as the first move of the supergame over S. If I chooses chess, then the remainder of a play of the hypergame over S is a play of chess, with player II taking the white pieces and player I the black.

It should be clear that for every set S of games, there is a supergame S^+. That is, we can define S^+ precisely, using S. (Shortly we'll turn to the hypergame construction, and we'll see a more involved construction in more detail.)

Proposition 12.7 *Let S be a set of games.*

1. *If S is a set of wellfounded games then S^+ is wellfounded.*
2. *If S is a set of wellfounded games S^+ is not in S.*

Proof (1) just the reasoning used in the first half of the reasoning of the Hypergame Paradox. (2) This is the reasoning used in the second half of the paradox. That is, if $S^+ \in S$, then both players could take turns playing S^+ forever. ⊣

As an immediate consequence of this we obtain a different proof of the fact there is no set of all wellfounded games. For if S were such a set, then S^+ would be a wellfounded game not in S.

The Hypergame Construction

Given a set S of games, there is another game S^* closely related to the supergame S^+. We dub it the *hypergame over S* and define it informally as follows. The allowable first moves by I in S^* are the games in S or the game S^* itself. If I picks some $\mathcal{G} \in S$ (other than S^* itself, if it happens to belong to S) then the players switch sides, and the remainder of the play is a play of $\mathcal{G}$, with player II beginning (and so I takes on the role of the second player in this game). Player I wins if the play from the second move on counts as a win for II in the game played in the first move. We also need to say that happens if I plays S^* itself. Then player II become like player I was in the first place. If II plays some $\mathcal{G} \in S$, and then the rest is a play p of $\mathcal{G}$. In this case, I wins the overall play in S^* iff I wins p considered as a play of $\mathcal{G}$. But of course II could also play S^*. If either player ever plays a game in S, then roles switch as above. But of course they could both keep playing S^* forever. In this case, we declare II the winner.

For example, if $S = \{\text{checkers}, \text{chess}\}$, then player I can again start by playing checkers or chess. But I can also start with S^*. In that case, II would get to begin a match of the hypergame over S^*. In particular, II could begin checkers, chess, or S^* itself. And the players could go round and round like this, forever.

We first prove that our mathematical framework is rich enough to model this intuitive description of the hypergame over S, provided S is a set of wellfounded games.

Proposition 12.8 *For any set S of wellfounded games, the hypergame S^* is a well-defined game.*

Proof Unlike the supergame, this result makes crucial use of hypersets. The proof uses the General Solution Lemma. Before we begin the proof, we recall the modeling of sequences suggested at the beginning of the chapter. One could prove the result with other modeling of sequences, but the proof would be more tedious. For notation convenience, we write a typical game $\mathcal{G} \in S$ as

$$\mathcal{G} \;=\; \langle M_{\mathcal{G}}, R_{\mathcal{G}}, W_I^{\mathcal{G}} \rangle.$$

Let x, y, z, and w be urelements outside of *support*(S). We are going to consider a system of equations

$$(3) \qquad \begin{aligned} x &= \langle y, z, w \rangle \\ y &= M \\ z &= R \\ w &= W \end{aligned}$$

M, R, and W are sets which we specify below. After this, we'll take a solution s to this system. The game S^* we are after will be s_x.

Let $M = S \cup \{x\} \cup \bigcup_{\mathcal{G} \in S} M_{\mathcal{G}}$. Our construction will arrange that all moves of S^* will come from $M[s]$.

R is the function defined on a subset of $M^{<\infty}$ as follows: For each $p \in M^{<\infty}$, if $p = \epsilon$, or $p = \langle x, x, \ldots, x \rangle$, then $R(p) = S \cup \{x\}$. If there is a game $\mathcal{G} \in S$ and a finite legal play q of $\mathcal{G}$ such that $p = \langle \mathcal{G}, q \rangle$ or $p = \langle x, \ldots, x, \mathcal{G}, q \rangle$, then $R(p) = R_{\mathcal{G}}(q)$. Otherwise $p \notin dom(R)$.

Finally, W consists of those plays $p \in M^{\leq \infty}$ such that for some $\mathcal{G} \in S$, either

1. $p = \langle x, x, \ldots, x, \mathcal{G}, q \rangle$, with the length of the x sequence some even number (possibly 0) and q a win for II in $\mathcal{G}$
2. $p = \langle x, x, \ldots, x, \mathcal{G}, q \rangle$, with the length of the x sequence some odd number and q a win for I in $\mathcal{G}$

Let s be the solution to the system in (3). Then s_x is a game which we call S^*. We need to check that it satisfies the condition of the hypergame. For this, we'll check that the infinite play $\langle S^*, S^*, \ldots \rangle$ is not in $W[s]$. Note that S^* is not a wellfounded game. So $S^* \notin S$. This implies that the infinite play of S^* is not a win for I. ⊣

We do not know whether S^* exists for all sets S of games, wellfounded or not. We believe it likely, however.

Proposition 12.9 *Let S be any set of wellfounded games. Then S^* is not wellfounded, so $S^* \notin S$.*

Proof We saw above that the sequence $\langle S^*, S^*, \ldots, \rangle$ is a legal play of S^*. No initial piece of it is a win for either player. It follows that S^* is not wellfounded, so $S^* \notin S$. ⊣

Thus, a different way of looking at what is going on in the Hypergame Paradox is that the informal description of Hypergame is ambiguous between the construction S^+ and the construction S^*. Suppose S is a set of wellfounded games. If by the hypergame on S one means S^+, then the hypergame is indeed wellfounded. But it is not in S so is not a legal first move of itself. On the other hand, if by the hypergame on S one means S^*, then the hypergame is not wellfounded but is a legal first move of itself.

Yet another choice for the meaning of the informal description of Hypergame is that it refers to T^*, where T is the set of wellfounded games in S. But note that T^* is not a wellfounded game, so it does not completely work as a model of the Hypergame. We return to a related point below.

Exercise 12.7 Let S be a set of determined games.

1. Prove that S^+ is determined.
2. Prove that S^* is determined.

There is yet one more insightful way of looking at the Hypergame Paradox. The informal version of the Hypergame Paradox seems to call on us to build a game S' so that the set $R_{S'}(\epsilon)$ of legal first moves is the set

$$\{\mathcal{G} \in S \cup \{S'\} \mid \mathcal{G} \text{ is wellfounded}\}.$$

Neither of the games S^+ or S^* satisfy this condition. Indeed, no game can, as the paradox shows.

Here is a simpler, purely set-theoretic, version of this fact.

Proposition 12.10 *Let s be any set. There is no set a that satisfies the "equation"*

$$a \quad = \quad \{b \in s \cup \{a\} \mid b \text{ is wellfounded}\}.$$

Proof Suppose such a set a exists. By the right-hand side, a is a set of wellfounded sets, and so is wellfounded. But then

$$a \quad \in \quad \{b \in s \cup \{a\} \mid b \text{ is wellfounded}\}.$$

But then $a \in a$ and so a is not wellfounded, which is a contradiction. ⊣

The moral of this story is that one must be careful. In writing down what might look like equations we may not have something that fits our model of an equation. In such a case, there is no guarantee that it will have a solution. The Solution Lemma lets us solve certain kinds of equations, like

$$a \quad = \quad \{b \mid b \in s \cup \{a\}\}.$$

The Separation Axioms also let us solve certain other kinds of equations, like

$$c \;=\; \{b \in s \cup \{a\} \mid \varphi(b)\}$$

for any condition φ at all. But we cannot do both things simultaneously. Attempting to define a set a by, say,

$$a \;=\; \{b \in s \cup \{a\} \mid \varphi(b)\}$$

can lead one into a truly vicious circle.

Historical Remarks

As we remarked earlier, the Hypergame Paradox was first formulated in Zwicker (1987). Zwicker noted the similarity of the paradox to the Burali-Forti Paradox (that there is no set of all ordinals). The only other paper which we know of which discusses it is d'Agostino and Bernardi (to appear). That paper is concerned with recursion-theoretic aspects of the paradox. The key point here is new: with *AFA* there are games with the property that one of the players can play that very game.

The classes *Even* and *Odd* in Exercises 12.2 and 12.3 were inspired by classes I and II discussed in section 1.2 of Forster (1995).

The connection of games and quantifiers is very important in fields like descriptive set theory; see Moschovakis (1980) for example. The connection between games and bisimulation is part of the folklore of the subject.

13

The semantical paradoxes

The theory of hypersets has many applications in computer science, as we will show in the course of this book. We believe it also has potential applications in a number of other fields. In this chapter we want to initiate what seems to us a potentially rich subject, one we call "anti-founded model theory." Anti-founded model theory is the study of models whose domains can contain other models, in a possibly non-wellfounded manner. We shall be particularly interested in "reflexive" models, that is, models that are elements of their own domain. Such models allow us to develop a new approach to the analysis of the semantical paradoxes of self-reference. Our aim here is not to "solve" the paradoxes but to set up a framework where we can see quite clearly just what the assumptions are that give rise to the paradox. One can then examine these assumptions and decide which, if any, are unwarranted.

The material in this chapter is not used in the rest of the book, so can be skipped without loss of continuity.

13.1 Partial model theory

In this section we review some standard material about partial models and Kleene's three-valued logic.

We assume there are disjoint sets Rel of relation symbols, $Const$ constant symbols, and Var variables. We assume that $Rel \cup Const \cup Var \subseteq \mathcal{U}$, though nothing crucial hangs on this decision. Each $R \in Rel$ is assumed to have some arity, a natural number $n > 0$. We assume that Var is infinite. We will be making additional assumptions about Rel as we go along.

We assume the reader is familiar with the notion of a sentence built up from these symbols by the first-order operations $\neg, \wedge, \vee, \rightarrow, \forall x$, and $\exists x$. For economy of exposition, we will take $\neg, \wedge$ and $\exists$ as primitive, the others being defined in the usual manner.

Definition A *partial model* M is a tuple $\langle D_M, L_M, Ext_M, Anti_M, d_M, c_M \rangle$ satisfying the following conditions:

1. D_M is a non-empty set, called the *domain* of M.
2. $L_M \subseteq Rel \cup Const$ is a set of relation symbols and constant symbols, called the *language* of M.
3. Ext_M and $Anti_M$ are functions with domain $L_M \cap Rel$ such that for each n-ary symbol $\mathsf{R} \in L_M$, $Ext_M(\mathsf{R})$ and $Anti_M(\mathsf{R})$ are disjoint n-ary relations on D_M, called the *extension* and *anti-extension* of R in M, respectively.
4. d_M is a function with domain $L_M \cap Const$ taking values in D_M; if $d_M(\mathsf{c}) = b$ then b is said to be the *denotation of* c in M.
5. c_M is a function with domain a subset of Var taking values in D_M, called the *context* of M. If $c_M(\mathsf{v}) = b$ then b is said to be the *denotation* of the variable v in M.

We also define $den_M(\mathsf{t})$ to be $d_M(\mathsf{t})$ if t is a constant and to be $c_M(\mathsf{t})$ if t is a variable in the domain of c_M; $den_M(\mathsf{t})$ is called the *denotation* of t.

A *total model* is a model M such that for each n-ary relation symbol R of L_M and every n-tuple $m_1, \ldots, m_n$ from the domain of M, either $\langle m_1, \ldots, m_n \rangle \in Ext_M(\mathsf{R})$ or $\langle m_1, \ldots, m_n \rangle \in Anti_M(\mathsf{R})$, and such that $den_M(\mathsf{c})$ is defined for every constant symbol c of L_M.

We say that a sentence φ of L_M is *defined in* M if every constant and variable of φ has a denotation in M. We write $Def(M)$ for the set of sentences defined in M.

We say that an object $b \in D_M$ is *named* in M if it is in the range of den_M. For any such b we will often use $\ulcorner b \urcorner$ to indicate an arbitrary constant or variable that denotes b in M. We must use this notation with some care, since for a given b, just what $\ulcorner b \urcorner$ is depends on what model M we are considering. And even if M is fixed, $\ulcorner b \urcorner$ need not be unique.

Since we are dealing with partial models, there are two notions of enlarging a model, one where the domain is fixed, one where the domain is allowed to grow. We begin with the former.

Definition A model M_2 is an *extension* of a model M_1, written $M_1 \subseteq M_2$, if (i) the domain D_{M_1} of M_1 is the same as the domain D_{M_2} of M_2, (ii) the language L_{M_1} of M_1 is a subset of the language L_{M_2} of M_2, (iii) for each relation symbol $R \in L_{M_1}$, the extension and anti-extension of R in M_1 are subsets of the extension and anti-extension of R in M_2, respectively, (iv) the denotation function d_{M_1} of M_1 is a subfunction of the denotation function d_{M_2}, and (v) the context c_{M_1} is a subfunction of the context c_{M_2} of M_2.

We want to use total models to represent some situation, or the world, and partial models to represent partial information about that situation or world. In this regard, the following is trivial but important.

Proposition 13.1 *Every model M has an extension M_{tot} which is a total model.*

We now want to say what it means for a sentence of our language to be true on the basis of a partial model. For total models, this should agree with the standard definition, and we want to make sure that any sentence that is true in a partial model is true in any extension of it, since more information should not invalidate anything made true by a partial model. The basic idea is that we must keep track not just of the sentences that are made definitely true, but also those that are made definitely false. This is captured by the two relations defined below. To facilitate the definition, we write $M_1 =_x M_2$ to indicate that the models M_1 and M_2 are identical, with the possible exception as to the value of the contexts on the variable x, where either one might be undefined, or they might have different values.

Definition [Kleene evaluation] We define $M \models \varphi$ and $M \models^- \varphi$, for models M and sentence φ defined in M as follows:

Atomic Sentences If φ is atomic, say $R(x, y, z)$, then

$$M \models \varphi \quad \text{iff} \quad \langle den_M(x), den_M(y), den_M(z) \rangle \in Ext_M(R)$$
$$M \models^- \varphi \quad \text{iff} \quad \langle den_M(x), den_M(y), den_M(z) \rangle \in Anti_M(R)$$

Compound Sentences

$$M \models \neg\varphi \qquad \text{iff} \quad M \models^- \varphi$$
$$M \models^- \neg\varphi \qquad \text{iff} \quad M \models \varphi$$
$$M \models (\psi_1 \wedge \psi_2) \quad \text{iff} \quad M \models \psi_1 \text{ and } M \models \psi_2$$
$$M \models^- (\psi_1 \wedge \psi_2) \quad \text{iff} \quad M \models^- \psi_1 \text{ or } M \models^- \psi_2$$
$$M \models \exists x\psi \qquad \text{iff} \quad \text{for some } M' \text{ such that } M' =_x M,$$
$$M' \models \psi$$
$$M \models^- \exists x\psi \qquad \text{iff} \quad M' \models^- \psi \text{ whenever } M' =_x M$$
$$\text{and } \psi \in Def(M')$$

We note that an atomic sentence $R(x, y, z)$ might be undefined in M; this would be the case if either $R \notin L_M$ or if one of the variables were not in the domain of c_M. If φ is atomic, then $\varphi \in Def(M)$ iff either $M \models \varphi$ or $M \models^- \varphi$. This is not so for compound sentences. It is possible for $M \models \exists x\varphi$ even when $x \notin dom(c_M)$.

The defined symbols work exactly as you would expect. (See Lemma 11.1 for a parallel result concerning the modal logic.)

Proposition 13.2 *If M is a total model and $\varphi \in Def(M)$ then $M \models \neg\varphi$ iff $M \not\models \varphi$.*

Proof A straightforward proof by induction on sentences establishes $M \not\models \varphi$ iff $M \models^- \varphi$. But by definition $M \models \neg\varphi$ iff $M \models^- \varphi$. ⊣

A straightforward induction on sentences also establishes the following:

Proposition 13.3 *Let $M_1 \subseteq M_2$. If $M_1 \models \varphi$ then $M_2 \models \varphi$. Similarly, if $M_1 \models^- \varphi$ then $M_2 \models^- \varphi$.*

Corollary 13.4 *There is no model M and sentence φ such that $M \models \varphi$ and $M \models^- \varphi$.*

Proof Suppose $M \models \neg\varphi$ and $M \models \varphi$. Let N be a total extension of M. Then $N \models \neg\varphi$ and $N \models \varphi$, by the proposition. But for total models N, we have $N \models \neg\varphi$ iff $N \not\models \varphi$. ⊣

For partial models M, there is an important gap between $M \not\models \varphi$ and $M \models \neg\varphi$. For example, if the constant symbols c does not denote in M, then the sentence R(c) is certainly not true in M, but neither is it false in M in the sense of $M \models^- R(c)$, this latter being equivalent to $M \models \neg R(c)$.

Definition A sentence φ is *not true* in the model M if $M \not\models \varphi$. In contrast, φ is *false* in M if $M \models \neg\varphi$.

If φ is false in M then it is not true in M, by the above. The converse holds in the case of total models, but not in general.

In addition to the notion of an extension of a model, we also need the notion of an expansion of a model. A model M_2 is an *expansion* of a model M_1, written $M_1 \sqsubseteq M_2$, if (i) the domain D_{M_1} of M_1 is a subset of the domain D_{M_2} of M_2, and all the other conditions for $M_1 \subseteq M_2$. (That is, (ii) the language L_{M_1} of M_1 is a subset of the language L_{M_2}, (iii) for each relation symbol R $\in L$, the extension and anti-extension of R in M_1 are subsets of the extension and anti-extension of R in M_2, respectively, and (iv) each term which denotes in M_1 also denotes in M_2, and denotes the same thing.) The following is trivial but important.

Proposition 13.5 *Let M_2 be an expansion of M_1 and φ be an existential sentence defined in M_1. If $M_1 \models \varphi$ then $M_2 \models \varphi$.*

Proof An existential sentence is a sentence in which every existential quantifier is in the scope of an even number of negations, and every universal quantifier is in the scope of an odd number of negations. By using de Morgan's laws, every such sentence is equivalent (true in the same models) as a sentence with no universal quantifiers and where each negation applies to an atomic sentence. For sentences in this form, the result is proved by an easy induction. $\dashv$

We will also use the notion of a principal expansion of a model. We say that M' is a *principal expansion of* M *by* a if $M \sqsubseteq M'$ and $D_{M'} = D_M \cup \{a\}$. We write $M(a)$ to indicate a model which is a principal expansion of M by a. We have not insisted that $a \notin D_M$, so that our definition permits the possibility that $M = M(a)$. We will say that $M(a)$ is a *proper* principal expansion of M if $a \notin D_M$.

Note that in the cases where $a \notin D_M$, there are in general many different principal expansions of M by a, depending on how the extensions and anti-extensions of the predicates get extended on new tuples containing the element a. Thus we must not think of the notation "$M(a)$" as suggesting function application.

13.2 Accessible models

The elements of the domain of a model can be anything at all. In particular, they could be urelements, sentences of L, and models of L, among other things. This allows us to specialize our language to include semantical predicates, predicates like the truth predicate or the reference predicate, and explore their behavior in problematic cases. Motivated by the notion of an accessible world in Chapter 11 and its connection to the notion of a reflexive set, we make the following definition.

Definition A model N is *accessible* in the model M if $N \in D_M$. The model M is *reflexive* if M is accessible in itself.

As the first application of the General Solution Lemma in this chapter, let's prove that every model can be expanded to a total, reflexive model.

Proposition 13.6 *For every model M, there is a principal expansion $M^+ = M(M^+)$ such that M^+ is reflexive and total.*

Proof To simplify notation, let's suppose that the language L_M of M contains only one binary symbol R, and no constants. Given the model M, let M^+ be the model defined as follows:

$$
\begin{aligned}
M^+ &= \langle D, L_M, Ext_M, Anti_{M^+}, d_M, c_M \rangle \\
D &= D_M \cup \{M^+\} \\
Anti_{M^+} &= \{\langle \mathsf{R}, R^- \rangle\} \\
R^- &= D \times D - Ext_M(\mathsf{R})
\end{aligned}
$$

Such a model always exists, by the General Solution Lemma. In general, we will be content to define models in this way, but since this is the first such proof, let's go into a bit more detail. There are four parts of this model that are new, so let's introduce four indeterminates, x, y, z and w, and write the following system of equations:

$$\begin{aligned} x &= \langle y, L_M, Ext_M, z, d_M, c_M \rangle \\ y &= D_M \cup \{x\} \\ z &= \{\langle R, w \rangle\} \\ w &= (D_M \cup \{x\})^2 - Ext_M(R) \end{aligned}$$

Notice that the right hand side of each equation really is a set. By the General Solution Lemma, there is a unique solution s to this system. If we let $M^+ = s_x$, $D = s_y, Anti_{M^+} = s_z$ and $R^- = s_w$, we get a structure satisfying the equations given above. ⊣

Notice that M and M^+ do not in general have the same domain, since we have thrown a model into the domain of M^+ which may not have been in M. This means that quantification means different things in M and M^+. This is not as big a problem as it might seem, though. If U is some unary predicate, then a U-bounded sentence is one in which all quantifiers range over the extension of U, that is, are of one of the forms:

$$\forall x\, (U(x) \rightarrow \ldots)$$

$$\exists x\, (U(x) \wedge \ldots)$$

Let's assume that the extension of U is the domain of M in M, and hence also in M^+. It is easy to see that if φ is U-bounded, for any U, then $M^+ \models \varphi$ iff $M \models \varphi$. So, for example, if we had a sentence about natural numbers, or people, for example, in M, we could treat it as a bounded sentence, and it would have the same meaning in M^+.

The construction of M^+ from M can violate the intended meaning of some predicates, though. This might happen if we had a predicate Mod whose extension is intended to be all models. In the above construction, we have thrown M^+ into the anti-extension of Mod. We would have the undesirable result that $M^+ \models^- Mod(M^+)$. Also, as we have constructed M^+, there will be no way to refer to M^+ in M^+. We will be quantifying over it, but it will not be the denotation of any term.

Actually, there is in general a great deal of flexibility in the construction of the model M^+ in the previous result, flexibility that we can use on a case-by-case basis to solve this sort of problem. To indicate this flexibility, we generalize the result as follows:

Proposition 13.7 (Principal Expansion Lemma) *Let M be any model and let $M(u)$ be a proper principal expansion of M, where u is an urelement not in the support of M. There is a reflexive model M^+ such that $M^+ = M(u)[s]$ where s is the substitution $u \mapsto M^+$. In particular, $M^+ = M(M^+)$.*

Proof Consider the following equation:

$$(1) \qquad\qquad u \;=\; M(u)$$

Let $M^+ = s_u$ be the value assigned to u by the unique solution s to (1). Then $M^+ = M(u)[s]$. It is easy to check that M^+ is a principal expansion of M by M^+. $\qquad\qquad\dashv$

To see how we might use the flexibility built into the lemma, let's think again about the problem mentioned above.

Example 13.1 If our language had a predicate Mod for being a model and we wanted to make sure that M^+ satisfied this predicate in itself, we could insure this by forming a proper principal expansion $M(u)$ putting u into the extension of Mod, and then apply The Principal Expansion Lemma. Another problem with our previous construction was that we constructed a model that was not denoted by anything in our language. This is easily rectified. If we want to make sure that we can refer to M^+ in M^+ by the previously unused variable (or constant) x, we can just add the pair $\langle \mathsf{x}, u \rangle$ to the context (or denotation function d, respectively) in $M(u)$.

It could happen that the model M^+ is the same as the model M, for example, if M is reflexive and u is treated as a copy of M in $M(u)$. In this case, M^+ is not isomorphic to $M(u)$ as one might expect. Thus it is instructive to see under what conditions we can guarantee that these models are isomorphic.

Corollary 13.8 *In the Principal Expansion Lemma, if M does not contain any model which is a principal expansion of itself, then $M(u)$ is isomorphic to M^+ under the isomorphism which is the identity on D_M and sends u to M^+.*

Proof Using the notation from the proof of Proposition 13.7, let f be the restriction of $[s]$ to $D_M \cup \{u\}$. By the choice of u, the function f is the identity on D_M, sends u to M^+. Since M^+ is a principal expansion of M and M was assumed to contain no such models as elements, M^+ is not in D_M so f is one-one. It is easy to check that f is an isomorphism of $M(u)$ and M^+. $\qquad\dashv$

13.3   Truth and paradox

In Chapter 5 we discussed a number of paradoxes, all of which somehow involved circularity. These paradoxes were a significant psychological hindrance to the development of the theory of non-wellfounded sets. But now that the theory has been developed for other purposes, it seems like an interesting idea to turn the tables on the paradoxes by using hypersets to develop a framework in which to explore the paradoxes. We are going to concentrate on the semantical paradoxes having to do with truth and reference. The framework presented here is inspired by Barwise and Etchemendy (1987), their book *The Liar*, but develops things in a rather different way.

In general, paradoxes are a very fruitful area of investigation. Gödel's Theorem, of course, is a mathematically precise result of the first magnitude directly inspired by the Liar Paradox. Similarly for Tarski's Theorem on the undefinability of truth. A fruitful way of studying the semantical paradoxes is to find a reasonable mathematical setting in which we can analyze the notions of truth and reference and use the reasoning of the paradoxes to prove theorems that shed light on the implicit assumptions behind these notions, assumptions which give rise to the paradoxes.

In this section, we explore paradoxes involving the truth predicate. We concentrate on the Liar Paradox and some of its close relations, as discussed in Chapter 5. Our aim is to try to develop an understanding of the pre-theoretic intuitions that give rise to it. We do this by presenting a consistent mathematical framework in which we can express the Liar sentence and speak about truth, and see what the argument of the Liar allows us to prove within this framework.

Adding a truth predicate to the language

In this section we assume that one of the predicates of L is a truth predicate True(x, y) which is intended to express the condition that the sentence denoted by x is true in the model denoted by y. With a fixed M in mind, we write $True(a, b)$ to mean that $\langle a, b \rangle$ belongs to the extension of True in M, and we write $False(a, b)$ to mean that $\langle a, b \rangle$ belongs to the anti-extension of True in M.

We are thus thinking of models whose domains contain, among other things, sentences and models. To stress the background role of the model in which the truth predicate is evaluated, we sometimes write this as $\text{True}_y(x)$ rather than True(x, y). Since we want this predicate to mean "true" we need to impose some conditions to make this so. The question is, what are the right conditions?

If M is a total model, then it is clear that the condition we want is the following: for all $\varphi, N \in D_M$,

(T0) $True(\varphi, N)$ if and only if N is a model, $\varphi \in Def(N)$ and $N \models \varphi$.

Condition (T0) is a version of Tarski's famous T-scheme, once we make the structures in which the sentences are being evaluated explicit.

Many people working on the paradoxes have argued that the paradoxes force us to partial models. The argument is that the assumption that the Liar is true leads to contradiction, as does the assumption that it is false, so it must be neither. One of our aims is to explore this argument and see whether its conclusion is inescapable. It is for this reason that we have included partial models in our account.

For partial models, we need to consider what the right analogue of (T0) would be. Let (T1) be the left-to-right half of (T0):

(T1)   If $True(\varphi, N)$, then N is a model, $\varphi \in Def(N)$ and $N \models \varphi$.

Condition (T1) seems unproblematic as a condition on the extension of True. But what about the anti-extension of True? Two possibilities suggest themselves, (T2) and (T3).

(T2) If N is a model, $\varphi \in Def(N)$, and $False(\varphi, N)$, then $N \models^- \varphi$.

(T3) If N is a model, $\varphi \in Def(N)$, and $False(\varphi, N)$, then $N \not\models \varphi$.

If we restrict attention to total models M and N, then (T2) and (T3) are each equivalent to the converse of (T1). In other words, restricted to total models, (T0) is equivalent to the conjunction of (T1) and either (T2) or (T3). But when we look at partial models, the two conditions are quite different, and capture different intuitions. One makes sense if we want the anti-extension of True to capture the notion of "false in N" while the other makes sense if we want it to capture the notion of "not true in N" in the sense defined in the previous section.

Which of these we accept will makes a big difference. To see why, let us look at what they say about a model M satisfying (T1) and containing in its domain a partial model N and a sentence φ that is defined in N but neither true nor false in N. Condition (T2) says that for $\langle \varphi, N \rangle$ to be in the anti-extension of True, N must make φ definitely false, not leave it undetermined. Thus (T2) precludes the pair $\langle \varphi, N \rangle$ from being in the anti-extension of True in M, just as (T1) precludes it from being in the extension of True in M. Consequently, (T2) forces an essential incompleteness on models that have partial models in their domain. The pair $\langle \varphi, N \rangle$ cannot be in either the extension or the anti-extension of True.

By contrast, (T3) does not do this. Under the (T3) conception, all that it means for $\langle \varphi, N \rangle$ to be in the anti-extension of True in M is that N does not make φ true. It could be either false or undetermined. Thus there is nothing in (T3) that seems to force M to be partial just because it has partial models in its domain.

The book *The Liar* (Barwise and Etchemendy 1987) presents and relates two different accounts of the paradoxes, what we call there the Russellian and the Austinian accounts. (It would perhaps have been fairer to call the former the Kripkean account.) The difference between these two accounts corresponds very roughly to which of (T2) and (T3) one adopts. The Russellian account corresponds to (T2) while the Austinian corresponds to (T3).

Definition Let M be a model.

1. M is *truth-correct* if it satisfies conditions (T1) and (T3) for all sentences $\varphi \in D_M$ and all models $N \in D_M$.

2. M is *truth-complete* if it is truth-correct and satisfies the converses of (T1) and (T3). In other words, whenever $N \in M$ and $\varphi \in Def(N)$, then $N \models \varphi$ implies $True(N, \varphi)$ and $N \not\models \varphi$ implies $False(N, \varphi)$.

Thus M is truth-complete iff it satisfies condition (T0) above. If M has only total models in its domain, then (T2) and (T3) are equivalent, so in this case, the two notions agree. However, for models that contain non-total models in their domain, the two notions diverge.

It is worth comparing these notions with what is captured by Tarski's T-scheme as it is usually expressed:

The sentence $\ulcorner S \urcorner$ is true if and only if S.

Here $\ulcorner S \urcorner$ refers to a sentence S of English. Let's formulate this in our language so that we can see that it is completely silent about the difference between (T2) and (T3).

To express this in our language, we need be able to talk about truth in a model M in the very model M we are evaluating our sentences in. The T-scheme implicitly assumes that the world is reflexive, and that we have a way to talk about the world in the world. In other words, this scheme implicitly assumes that we have some term $\ulcorner M \urcorner$ to refer to the background model M itself. Given such a term, let us agree to write $\mathsf{True}(\ulcorner \varphi \urcorner)$ as an abbreviation of $\mathsf{True}_{\ulcorner M \urcorner}(\ulcorner \varphi \urcorner)$.

Proposition 13.9 *Let M be a reflexive model with a term $\ulcorner M \urcorner$ denoting itself. If M satisfies (T1) and its converse then:*

$$\mathsf{True}(\ulcorner \varphi \urcorner) \leftrightarrow \varphi$$

for each sentence $\varphi \in Def(M)$ that has a name $\ulcorner \varphi \urcorner$ in M.

Proof Immediate. ⊣

The proposition shows us that in order to capture the content of the T-scheme in a model M of the world, M needs to be reflexive and to satisfy (T1) and its converse; the latter is the case if M is truth-complete (or even just truth-correct and satisfying the converses of (T1) and (T2)). The proposition also shows that the T-scheme makes no useful claim about what it takes to make a sentence *false* in M, as opposed to just not true. Later we will see several examples of models satisfying the hypothesis of Proposition 13.9. Some will satisfy (T2), some (T3).

Exercise 13.1 As long as we are going to introduce a truth predicate, it really makes sense to add predicates for more primitive syntactic and semantic notions. For example, it would be natural to have a predicate **Sent** whose extension (in total models) is the set of sentences in the domain, a predicate **Mod** whose extension is the set of models in the domain, a relation symbol **Def** whose extension is the set of pairs $\langle \varphi, N \rangle$ such that $\varphi \in Def(N)$, and a denotation predicate **Denotes** whose extension is just the set of pairs $\langle x, y \rangle$ such that $den_M(x) = y$. With one exception, none of this seems particularly problematic, so we have not gone into it here.

1. What conditions analogous to (T1) and (T3) would one want to impose on the interpretation of these predicates? Show that with these conditions, the following would hold in all total, truth-correct, reflexive models:

$$\forall x \: [(\mathsf{Sent}(x) \wedge \mathsf{Def}(x, \ulcorner M \urcorner) \wedge \mathsf{Denotes}(\ulcorner \varphi \urcorner, x)) \rightarrow [\mathsf{True}(x) \leftrightarrow \varphi]]$$

2. What else would we need to add to the language in order to be able to state things like the following:

$$\forall x, y, z \quad [(\mathsf{Sent}(x) \wedge \mathsf{Sent}(y) \wedge \mathsf{Mod}(z)) \rightarrow$$
$$[\mathsf{True}(x \wedge y, z) \leftrightarrow \mathsf{True}(x, z) \wedge \mathsf{True}(y, z))]]$$

3. In the case of reflexive models we would really like to be able to state the truly universal statement:

$$\forall x \: [(\mathsf{Sent}(x) \wedge \mathsf{Def}(x, \ulcorner M \urcorner)) \rightarrow [\mathsf{True}(\ulcorner x \urcorner) \leftrightarrow x]]$$

Notice that this is not readily expressible in English. How could one extend the language so that this is a well-formed expression that is true in all total, truth-correct, reflexive models?

13.4 The Liar

Without further ado, let's see what the Liar Paradox tells us about the notions we now have at our disposal. A *Liar sentence* is any sentence of the form

$$\neg \mathsf{True}_h(\mathsf{this})$$

where this and h (for "here") are either constants or variables of our language.

Theorem 13.10 (The Liar) *Let λ be a Liar sentence* $\neg\mathsf{True}_h(\mathsf{this})$. *If M is a truth-correct model then at least one of the following must fail:*

1. this *denotes λ in M.*
2. h *denotes M in M.*
3. $M \models \lambda \vee \neg\lambda$.

In particular, if (1) and (2) both hold, then M is not total.

Proof To prove this, we just repeat the reasoning that leads to the Liar Paradox when treated informally. Assume (1), (2), and (3), and let's get a contradiction. Since $M \models \lambda \vee \neg\lambda$, we know that either $M \models \lambda$ or $M \models \neg\lambda$. First, suppose $M \models \lambda$, that is, $M \models \neg\mathsf{True}_h(\mathsf{this})$. Since this denotes λ and h denotes M, the definition of $\models$ shows us that $\langle \lambda, M \rangle$ is in the anti-extension of True. But then by (T3), $M \not\models \lambda$ after all. Thus $M \not\models \lambda$. Now suppose that $M \models \neg\lambda$, that is, that $M \models \mathsf{True}_h(\mathsf{this})$. But then $\langle \lambda, M \rangle$ is in the extension of True, so by (T1), $M \models \lambda$, which contradicts $M \models \neg\lambda$.[1]   ⊣

---

[1] It is worth noting that (T1) was used only to show that $M \not\models \neg\lambda$ and (T3) was used only to show $M \not\models \lambda$.

Most "solutions" to the Liar Paradox that have been proposed can be seen as abandoning one of the assumptions (1), (2) and (3). Tarski's proposal not to allow a language to contain its own truth predicate makes (1) impossible. That's not very interesting and has pretty much been abandoned after Kripke's work. Proposals which advocate "truth gaps" give up (3). They see the Liar as somehow requiring us to give up the idea that the world is total, that a claim either is true or it isn't.

Proposals which see the problem in some sort of indexicality or context sensitivity can be seen as giving up (2). The idea, roughly, is that the act of asserting the Liar about the whole world results in some sort of pragmatic shift in context, a shift that is overlooked in the reasoning that seems to lead to paradox. Put crudely, here (as referred to by h) before the claim and here after the claim are slightly different. Let's see how this shift might happen by looking at some examples.

Example 13.2 Let's first construct a truth-correct model M_0 that is as close to what is ruled out by the Liar as possible, but which gives up (2). We will construct M_0 so as to satisfy the following conditions:

1. this denotes $\lambda = \neg \mathsf{True_h}(\mathsf{this})$ in M_0.
2. M_0 is reflexive and has a name $\ulcorner M_0 \urcorner$ for itself.
3. M_0 is total.

(By the theorem, all that keeps this from being impossible is that it will be some term other than h that denotes M_0.)

Let M_0 be defined as in Figure 7:

M_0	$= \langle D, L, Ext,$		M_1	$= \langle D', L', Ext',$
	$\quad Anti, d, c\rangle$			$\quad Anti', d', c'\rangle$
D	$= \{\lambda, M_0\}$		D'	$= \{\lambda, M_0, M_1\}$
L	$= \{\mathsf{m_0}, \mathsf{True}\}$		L'	$= \{\mathsf{m_0}, \mathsf{True}\}$
Ext	$= \{\langle\mathsf{True}, \emptyset\rangle\}$		Ext'	$= \{\langle\mathsf{True}, \emptyset\rangle\}$
$Anti$	$= \{\langle\mathsf{True}, D \times D\rangle\}$		$Anti'$	$= \{\langle\mathsf{True}, D' \times D'\rangle\}$
d	$= \{\langle\mathsf{m_0}, M_0\rangle, \langle\mathsf{this}, \lambda\rangle\}$		d'	$= \{\langle\mathsf{m_0}, M_0\rangle, \langle\mathsf{this}, \lambda\rangle\}$
c	$= \emptyset$		c'	$= \{\langle\mathsf{h}, M_0\rangle\}$

FIGURE 7 Models M_0 and M_1

(We have put a second model M_1 above, too, for ease of comparison with M_0 in our next example.) This model M_0 clearly has the claimed properties. (The model M_0 is rather impoverished, having only two elements, λ and itself, but that's not a problem. We could make it bigger and richer, containing tables and chairs and other sentences, but no additional insight would be gained.)

Here we are with this nice, tidy little world M_0. Note that h has no denotation in M_0. So λ is neither true nor false in this world. Note that $\langle \lambda, M_0 \rangle$ is in the anti-extension of True, too, which is where it would have to be by (T1).

Having seen that λ is not true, some troublesome character might want to say exactly this. There are a couple of ways of doing this. One way is by means of the Liar sentence $\neg\text{True}_{m_0}(\text{this})$. No problem, since

$$M_0 \models \neg\text{True}_{m_0}(\text{this}).$$

Why is there no paradox? Because we have a slightly different Liar sentence, $\lambda' = \neg\text{True}_{m_0}(\text{this})$, not the old sentence λ from before, and the term this denotes λ in M_0.

This example is very reminiscent of the Strengthened Liar. It is not quite the Strengthened Liar, though, since the sentence changed from λ to λ'. To model the Strengthened Liar we would like to make the same claim with the original Liar sentence λ. We can do this as follows.

Example 13.3 Let's create a reflexive, truth-correct model M_1 with the following properties:

1. this denotes λ in M_1.
2. h denotes M_0 in M_1.
3. M_1 is total.

This will represent a case where our troublesome character uses the original Liar λ to say of the original world M_0 that λ is not true. The model is indicated in the right-hand side of Figure 7. Note that the model $M_0 \in D'$ is the model described in the left-hand side of the figure.

Notice how similar the two models are, though. Indeed, were it not for putting $\langle h, M_0 \rangle$ into the context, these models would be bisimilar and hence identical! That is, the only difference between them is a slight change in the context.[2]

Moral of the Liar Our discussion suggests that, contrary to appearances, the Liar Paradox does not force one into abandoning the intuitive idea that the world is total and that any claim either is true or it is not true. What it does is forces one to be extraordinarily sensitive to subtle shifts in context. This seems to us quite a plausible explanation for the intuitive reasoning behind the paradox: it just fails to be sufficiently sensitive to such subtleties.

---

[2]Notice that this pair of examples shows that there is a hole in the argument of those who would use the Strengthened Liar to embarrass proponents of truth gaps (see page 87). Recall that the punch line of that argument was that the Liar and the Strengthened Liar are making the very same claim, that the Liar is not true. These examples show how one can deny that they are making the very same claim.

The fact that the world is total, and that one *can* refer to the whole world, no way implies that one *must* refer to the whole world, that is, that one cannot make claims about parts of the world that are not total if one is so inclined.

Example 13.4 We construct a model M_3 which satisfies the following:

1. this denotes $\lambda = \neg \mathsf{True_h}(\mathsf{this})$ in M_3.
2. h denotes M_3 in M_3.
3. M_3 is truth-correct (but not total of course, as this is precluded by the Liar).

Let M_3 be defined as follows:

$$
\begin{aligned}
M_3 &= \langle D, L, Ext, Anti, d, c \rangle \\
D &= \{\lambda, M_3\} \\
L &= \{\mathsf{True}, \mathsf{this}\} \\
Ext &= \{\langle \mathsf{True}, \emptyset \rangle\} \\
Anti &= \{\langle \mathsf{True}, \emptyset \rangle\} \\
d &= \{\langle \mathsf{this}, \lambda \rangle\} \\
c &= \{\langle \mathsf{h}, M_3 \rangle\}
\end{aligned}
$$

Notice that M_3 is (vacuously) truth-correct, and reflexive. As predicted by the Liar, this model makes neither λ nor $\neg\lambda$ true.

Let's take a second look at the Strengthened Liar, this time in relation to M_3.

Example 13.5 There is a model M_4 which satisfies the following:

1. this denotes λ in M_4.
2. h denotes M_3 in M_4.
3. M_4 is truth-correct and total.

There is no problem since the Liar here is not about the model M_4, but about the original partial model M_3. On this way of modeling things, the Liar is not true, but the Strengthened Liar is true. That is, λ is not true in M_3 (and neither is its negation) but it is true (about M_3) in M_4. There is a shift in context between the Liar and the Extended Liar; they are both about the same model, M_3, but they are evaluated in different models, M_3 and its expansion M_4.

Exercise 13.2 Construct the model M_4 as in Example 13.5. [Hint: the main difference between the defining equations for M_3 and for M_4 comes in the anti-extension of **True**.]

Notice that the shifts in context forced on us by the Liar are the exception. Most sentences can be asserted about the whole world without these shifts taking place, even with circular reference. (See Exercise 13.6.)

The reader with an imagination will think of many interesting questions raised by this section. Since this book is about applications of non-wellfounded sets, not about the paradoxes, we resist the temptation to develop this framework further here, contenting ourselves with giving a few exercises. Some open problems and suggestions for further research are given on page 329 in the concluding chapter.

Exercise 13.3 Prove that every truth-correct model M has a principal expansion which is a reflexive, truth-correct model M'.

Exercise 13.4 Show that every truth-correct model M has an extension to a total, truth-correct model M_{tot}.

Exercise 13.5 Show that the following is false: every truth-correct model has a principal expansion to a total, truth-correct, reflexive model.

Exercise 13.6 Let τ be the *truth-teller*, the sentence $\mathsf{True_h(this)}$. Construct a pair of total, truth-correct models M_1, M_2 such that this denotes τ in each, h denotes M_i in M_i, but such that τ is true in M_1 and false in M_2.

Exercise 13.7 Let R be some unary predicate, let c be some constant symbol, and let γ be the *contingent Liar*, the sentence $\mathsf{R(c)} \wedge \neg\mathsf{True_h(this)}$.

1. Show that in any total, truth-correct model M in which γ is defined, if this denotes γ and h denotes M, then $M \models \neg R(c)$.
2. Construct a total, truth-correct model M such that this denotes γ and h denotes M in M.

13.5 Reference and paradox

The line of development taken on the Liar Paradox in the preceding section can be developed for other semantical paradoxes. To make this point, we discuss one more example, namely, the paradox of reference discussed in Chapter 5. In that discussion we considered two expressions, t_0 and t_1.

$$(t_0) \quad \begin{cases} 1 & \text{if this denotes 0} \\ 0 & \text{if this denotes 1} \end{cases}$$

$$(t_1) \quad \begin{cases} 1 & \text{if this denotes 0} \\ 0 & \text{if this denotes 1} \\ 0 & \text{if this does not denote anything} \end{cases}$$

The first of these could be used to refer to either 0 or 1, depending on the referent of "this," and it could also be used in ways which left it without a referent. The term t_1 was similar, but more problematic. Intuitively, it seems that if we use "this" to refer to t_1, then we have a contradiction on our hands. In this section we want to see how the above framework can be adapted to understand what lurks behind this sort of puzzle.

Definition by cases

We first extend the model-theoretic framework reviewed in Section 13.1 to allow for terms that express Definition by Cases (DC), also known as "if ... then ... else ..." in computer science. Thus, we add a new formation rule to our language:

> For any $n > 0$, any terms $t_0, \ldots, t_{n-1}$ and any sentences $\varphi_0, \ldots, \varphi_{n-1}$ there is a term $DC(t_0, \varphi_0; \ldots; t_{n-1}, \varphi_{n-1})$.

We will define the semantics of our language so that this term denotes the same thing that t_0 denotes if φ_0 is true, the same thing that t_1 denotes if φ_0 is false and φ_1 is true, etc. These terms are to be allowed to appear in sentences, so we need a simultaneous induction of the set of terms and sentences.

A term t is said to be a *term of M*, written $t \in Term(M)$, iff all the relation symbols, constants, and variables occurring in t denote in M. This does not entail that the term t denotes anything in M, though.

Definition Let M be a model. We extend the denotation function den_M to a subset of $Term(M)$ by recursion as follows. Let t be the term

$$DC(t_0, \varphi_0; \ldots; t_{n-1}, \varphi_{n-1})$$

and suppose that $t \in Term(M)$. If

$$M \models \varphi_0$$

and $den_M(t_0)$ is defined, then $den_M(t) = den_M(t_0)$. If

$$M \models \neg\varphi_0 \wedge \varphi_1$$

and $den_M(t_1)$ is defined, then $den_M(t) = den_M(t_1)$. ... If

$$M \models \neg\varphi_0 \wedge \ldots \wedge \neg\varphi_{n-2} \wedge \varphi_{n-1}$$

and $den_M(t_{n-1})$ is defined, then $den_M(t) = den_M(t_{n-1})$. Otherwise $den_M(t)$ is not defined.

To do this completely rigorously, we would need to define $M \models \varphi$, $M \models^- \varphi$, and $den_M(t) = a$ simultaneously. This is routine. The only point that might get overlooked is that if φ is atomic, then in order for either $M \models \varphi$ or $M \models^- \varphi$ to hold, $den_M(t)$ must be defined for all the terms t in φ.

Parallel to the persistence of truth under extension, we have the following simple result.

Proposition 13.11 *If $M_1 \subseteq M_2$ and $den_{M_1}(t) = a$ then $den_{M_2}(t) = a$.*

Proof And easy induction. ⊣

In this section, we are going to be using the identity predicate. Rather than assign it an extension and anti-extension, we simply always interpret it as actual identity, and say no more about it.

Adding a denotation predicate to the language

We want to introduce a denotation predicate Denotes(x,y,z), read "x denotes y in z." To stress the background role of the model in which the denotations are determined, we write this as $\text{Denotes}_z(x, y)$. Since we want this predicate to mean "denotes" we need to impose some conditions to make this so, in a manner parallel to our conditions (T1) and (T3) insuring that the truth predicate behaved properly.

For total models we want:

(D0) $\langle t, a, N \rangle$ is in the extension of Denotes if and only if N is a model, $t \in Term(N)$ and $den_N(t) = a$.

What should the general conditions be? Let (D1) be the left to right half of (D0):

(D1) If $\langle t, a, N \rangle$ is in the extension of Denotes then N is a model and $den_N(t) = a$.

(D1) seems clearly right as a condition on the extension of Denotes. But what about the anti-extension of Denotes? Again, there are two possibilities, (D2) and (D3).

(D2) If $\langle t, a, N \rangle$ is in the anti-extension of Denotes, and N is a model then $den_N(t) = b$ for some $b \neq a$.

(D3) If $\langle t, a, N \rangle$ is in the anti-extension of Denotes, and N is a model then $den_N(t)$ is either undefined or else is b for some $b \neq a$.

Things are a bit different than in the case for truth, since even for total models, (D2) is too strong. It is possible to have a term t in a total model M which does not denote in that model. Again, we need to assume the generally weaker (D3).

Definition A model M is *denotation-correct* if it satisfies conditions (D1) and (D3) for all models $N \in D_M$, all terms t and all $a \in D_N$.

Thus, for total models, M is denotation-correct iff it satisfies condition (D0) above.

There is an informal D-analogue of Tarski's T-schema. Instances of it are

The noun phrase "the morning star" denotes Venus if and only if the morning star is Venus.
The name "Cicero" denotes Tully if and only if Cicero is Tully.

More abstractly, the D-scheme would say

The term $\ulcorner t \urcorner$ denotes a if and only if $t = a$.

If we are working in a reflexive model M we can capture the D-scheme as follows. Let Denotes(x, y) be an abbreviation of $\text{Denotes}_{\ulcorner M \urcorner}(x, y)$.

Proposition 13.12 *If M is a reflexive, denotation-correct total model, then it satisfies the following:*

$$\forall x\,(\text{Denotes}(\ulcorner t\urcorner, x) \leftrightarrow t = x)$$

for all terms t that have a name in M.

Paradoxical terms

The informal paradoxes of denotation implicitly assume the world we live in is reflexive, denotation-correct, and total, and that we have some context insensitive way of referring to the world as a whole. But these assumptions conflict with the paradoxical terms in just the same way that the corresponding assumptions conflict with paradoxical sentences like the Liar. Recall the terms t_0 and t_1 from page 191. (The relationship between these two is similar to the relationship between the Liar and the extended Liar.)

We first want formal analogs of these terms. We assume that we have constants 0 and 1, and are only going to consider models in which they denote 0 and 1, respectively. We take

$$\text{DC}(1, \text{Denotes}_h(\text{this}, 0); 0, \text{Denotes}_h(\text{this}, 1))$$

Similarly, let t_1 be

$$\text{DC}(1, \text{Denotes}_h(\text{this}, 0); 0, \text{Denotes}_h(\text{this}, 1); 0, \neg\exists x\,\text{Denotes}_h(\text{this}, x))$$

We want to explore the behavior of these terms in denotation-correct models. As we expect from our earlier discussion, t_0 is unproblematic. If we are working in a denotation-correct model M where 0 denotes 0, 1 denotes 1, and this denotes t_0, then t_0 simply does not denote anything. Let's double-check that this is consistent by constructing a total, reflexive, denotation-correct model M containing t_0 where this denotes t_0. Here are the defining equations for our model:

$$
\begin{aligned}
M &= \langle D, L, Ext, Anti, den \rangle \\
D &= \{0, 1, 2, t_0, M\} \\
L &= \{\text{Denotes}\} \\
Ext &= \{\langle \text{Denotes}, De^+ \rangle\} \\
De^+ &= \emptyset \\
Anti &= \{\langle \text{Denotes}, De^- \rangle\} \\
De^- &= D \times D \times D \\
den &= \{\langle 0, 0 \rangle, \langle 1, 1 \rangle, \langle \text{this}, t_0 \rangle, \langle \text{h}, M \rangle\}
\end{aligned}
$$

It is clear that this model is total and reflexive. [3] It is also denotation-correct. To see this, first note (D1) is vacuously satisfied. To check (D3), we only need

---

[3]Note that the definition of a total model does not change when we add a denotation predicate. It must be the case that $De^+ \cup De^- = D \times D \times D$. But it need not be the case that every term denotes.

to worry about the term t_0, since it is the only term in the domain. (We could have added other terms to the domain.) The informal argument we gave in Chapter 3 shows that t_0 does not denote in M. So (D3) is satisfied.

The term t_1, though, is more interesting. Let us first see what happens if we add it to the model just constructed. Redefine M to be:

$$
\begin{aligned}
M &= \langle D, L, Ext, Anti, den \rangle \\
D &= \{0, 1, 2, t_0, t_1, M\} \\
L &= \{\mathsf{Denotes}\} \\
Ext &= \{\langle \mathsf{Denotes}, De^+ \rangle\} \\
De^+ &= \{\langle t_1, 0, M \rangle\} \\
Anti &= \{\langle \mathsf{Denotes}, De^- \rangle\} \\
De^- &= D \times D \times D - \{\langle t_1, 0, M \rangle\} \\
den &= \{\langle 0, 0 \rangle, \langle 1, 1 \rangle, \langle \mathsf{this}, t_0 \rangle, \langle \mathsf{h}, M \rangle\}
\end{aligned}
$$

Notice that this still denotes t_0, not t_1. We gave an informal argument in Chapter 3 that, under these circumstances, t_1 denotes 0. And this is how we have built our model. It is easy to check that this model is denotation-correct. It is obviously total and reflexive.

If we try to let *this* refer to t_1, though, we get problems. Indeed, exactly parallel to Theorem 13.10, we have the following:

Theorem 13.13 *Let M be any denotation-correct model in which* 0 *and* 1 *denote distinct objects. Then one of the following must fail:*

1. this *denotes* t_1*;*
2. h *denotes* M*;*
3. $M \models \exists \mathsf{x}\, \mathsf{Denotes}_\mathsf{h}(t_1, \mathsf{x}) \vee \neg \exists \mathsf{x}\, \mathsf{Denotes}_\mathsf{h}(t_1, \mathsf{x})$.

In particular, if (1) and (2) obtain, then M is not total.

The proof of this is just the working out of the informal paradox we discussed at the beginning of this section. So again, we see the same kind of trilemma that arose in the case of the Liar. We take the moral to be the same as that in the Liar, namely, that some uses of some terms force a shift in context that subtly changes what we are talking about.

Exercise 13.8 Construct a parallel discussion for the term t_1 as we did for λ in Examples 13.2 and 13.3.

Historical Remarks

The material in this section is new, but the overall idea of using hypersets in connection with the semantical paradoxes goes back to Barwise and Etchemendy (1987). Tarski (1956) is Tarski's seminal paper on the concept of truth. Kripke (1975) is perhaps the first source to suggest a cogent alternative to Tarski's theory. For a fairly recent (and different) treatment of the circularity of truth we refer to the reader to Gupta-Belnap (1993), where an excellent bibliography can also be found.

Bartlett and Suber (1987) is a collection of essays by different authors on aspects of self-reference in various fields. Many of the papers are descriptive rather than technical, and so they present areas where *AFA* might possibly be useful in modeling. In addition, the book has a very large bibliography. Sainsbury (1988) is another excellent source of paradoxes ripe for modeling using *AFA*.

CHAPTER 13

14

Streams

In this chapter we use non-wellfounded sets to develop a theory of streams over some alphabet A. We do this as an illustration of the techniques developed in Part II, and as an introduction to the theory of uniform operators to be developed in Chapter 16. While streams are interesting in their own right, the real aim of the chapter is to get us used to the important methods of coinduction and corecursion in a simple setting before looking at them in the general case.

14.1   The set A^∞ of streams as a fixed point

Let A be some set and let us consider the streams over A. For any particular stream $(a_1, (a_2, (a_3, \ldots)))$, the General Solution Lemma gives us an immediate way to model it as a set. We just solve the system of equations

$$
\begin{aligned}
x_1 &= \langle a_1, x_2 \rangle \\
x_2 &= \langle a_2, x_3 \rangle \\
&\vdots
\end{aligned}
$$

But what about the set of *all* streams over A?

Our discussion of streams in Section 3.1 suggests that the set of all streams should satisfy the equation

(1) $$ Z = A \times Z $$

But unfortunately this is not the kind of equation to which any version of the Solution Lemma applies. As it happens, if the equation were $Z = \{A \times Z\}$, we could solve it (see Section 16.6). But we have no general results which directly handle equation (1).

This equation clearly has at least one solution, namely, $Z = \emptyset$. But this solution is not of any interest in modeling the set of all streams since we think there really are streams. What we want is not just any solution to the equation, but the *largest* such solution, since it is the one that is guaranteed to contain all streams.

We will show how to solve this equation here, as an introduction to the solution of such equations. Our solution to $Z = A \times Z$ is quite explicit; we are able to write a fairly explicit representation of the elements of the solution set. However, it is not always necessary to do this. Later we shall study solutions to many other equations, like $Z = \mathcal{P}(A \times Z)$ and $Z = A \times Z \times Z$. We want to know that in each case there is a largest solution Z, and also some general properties about Z. And we would prefer to do this based on general principles rather than ad hoc representations.

Theorem 14.1 *For every set A there is a largest set Z such that $Z \subseteq A \times Z$. This set Z in fact satisfies the equation $Z = A \times Z$. Moreover, if $A \neq \emptyset$, then $Z \neq \emptyset$ as well.*

Proof Before reading the proof below, the reader should review Exercise 2.11. The difference between that result and this one is that the earlier result starts with A and builds Z isomorphic to $A \times Z$; here we assume *AFA* and get actual equality.

Let F be the set of functions f from N to A. For each such f, we have another such function f^+ given by

$$f^+(n) \quad = \quad f(n+1) \,.$$

Now for each $f \in F$, let x_f be an indeterminate. We want these indeterminates to be *new for A*. This means that none should belong to *support*(A). We discuss this point further in Section 14.2 below, but for now, the only thing worth mentioning is each $a \in A$ is unaffected by substitution operations $[s]$ which come from substitutions defined only on urelements outside of *support*(A). That is, $a[s] = a$ whenever s is a substitution that ignores the urelements in *support*(A).

Consider the system of equations $\mathcal{E}$ given by

$$x_f \quad = \quad \langle f(0), x_{f+} \rangle \,.$$

Note that when we write $f(0)$ in one of these equations, we really have one of the elements of A there. There is one equation for each element of F. Now $\mathcal{E}$ is a general system of equations, and so it has a solution s. For each f, we write s_f instead of $s(x_f)$. Then our equations above tell us that for all f, $s_f = \langle f(0), x_{f+} \rangle[s]$. Moreover, $[s]$ commutes with pairs, so we see that for all f, $s_f = \langle f(0)[s], s_{f+} \rangle$. Since we used indeterminates x_f that were new for A,

$a[s] = a$ for all $a \in A$. Hence $s_f = \langle f(0), s_{f+} \rangle$. This means that each s_f is a pair consisting of an element of A and an element of the form $s_{f'}$. The upshot is that we may take $Z = \{s_f \mid f \in F\}$ to solve $Z \subseteq A \times Z$.

We also want to see that Z is the largest set W such that $W \subseteq A \times W$. (You should see that there is real issue here, since *AFA* does not promise that there will be a *unique* solution to the "subset equation" $W \subseteq A \times W$.) Let $w \in W$. Let $f_w : N \to A$ be such that

$$f_w(n) = 1^{st}(2^{nd^n}(w)).$$

This f_w is the natural function on N associated with the stream W. Then s_{f_w}, as defined above, belongs to Z. We claim that for all $w \in W$, $s_{f_w} = w$. In order to avoid double subscripts, we'll write w^* for s_{f_w}. Notice first that $f_{2^{nd}(w)} = (f_w)^+$; the reason is that $2^{nd^{n+1}}(w) = 2^{nd^n}(2^{nd}(w))$ for all n. It follows that

$$\begin{aligned} w^* &= s(x_{f_w}) \\ &= \langle 1^{st}(w), s(x_{2^{nd}(w)}) \rangle \\ &= \langle 1^{st}(w), (2^{nd}(w))^* \rangle \end{aligned}$$

Now we get a bisimulation R relating each w to the corresponding w^*. The definition of R should be familiar to you from Section 7.3. We take

$$\begin{aligned} R = \quad &\{\langle w, w^* \rangle \mid w \in W\} \\ &\cup \{\langle \{a\}, \{a\} \rangle \mid a \in A\} \\ &\cup \{\langle a, a \rangle \mid a \in A\} \\ &\cup \{\langle \{a, w\}, \{a, w^*\} \rangle \mid a \in A, w \in W\} \end{aligned}$$

Since each w^* belongs to Z, this proves that $w \in Z$. Hence $W \subseteq Z$ as desired.

At this point we know that Z is the largest set such that $Z \subseteq A \times Z$. Now we check that $A \times Z \subseteq Z$. Let $a \in A$ and $b \in Z$. By the definition of Z, $b = s(x_f)$ for some $f : N \to A$. Let $g : N \to A$ be defined so that $g(0) = a$ and $g(n + 1) = f(n)$. Then $g^+ = f$, and so $s(x_g) = \langle a, s(x_f) \rangle = \langle a, b \rangle$ belongs to Z.

Finally, if $A \neq \emptyset$, then the set F of functions from N to A is non-empty. So $Z \neq \emptyset$.   ⊣

We use A^∞ as a notation for the set of streams over A, and we speak of it as a *fixed point*. Here is what this means. If C is any set or class and $\Gamma : C \to C$, then c is a *fixed point of* Γ if $\Gamma(c) = c$. If we let C be the class of all sets, and take $\Gamma(c) = A \times c$, then A^∞ is a fixed point of Γ. Indeed, A^∞ is the *greatest* fixed point, where the order we use is the inclusion order on sets.

In Chapter 15, we'll see that the view of streams as a fixed point generalizes to other examples. One of the key points is that properties of the operation Γ imply properties of its fixed points. For the operation $\Gamma(c) = A \times c$, the most important property is *monotonicity*. In general, this means that if $c \subseteq c'$, then

$\Gamma(c) \subseteq \Gamma(c')$. In the particular case of cartesian product with A, this just says that $A \times c \subseteq A \times c'$. On the other hand, consider the operator $\Gamma'(c) = c \to c$. This operator is not monotone, since if c is a proper subset of c', then a function from c to itself is not a (total) function on c'.

As we'll see later, monotone operators have greatest fixed points, and these have important properties. The non-monotone operator Γ' happens to have two fixed points, as we show in Exercise 14.3 below. At this time, there is no general theory of such fixed points. In contrast to this, there is a theory of fixed points of monotone operators which is quite useful.

We close this section with a series of exercises pertaining to Theorem 14.1.

Exercise 14.1 Is A^∞ the only nonempty set Z such that $Z = A \times Z$?

Exercise 14.2 Let A and B be sets. What is the largest set Z such that $Z = B \cup (A \times Z)$?

Exercise 14.3 Suppose that Z is a solution to $Z = Z \to Z$. Show that either $Z = \emptyset$ or $Z = \Omega$. In fact, show that if $Z \subseteq Z \to Z$, then either $Z = \emptyset$ or $Z = \Omega$. [Hint: Recall that if f is a function from Z to Z, then

$$f \quad = \quad \{\langle z, f(z) \rangle \mid z \in Z\}.$$

Now take a solution to $Z = Z \to Z$ and write down a system corresponding to this observation.]

Exercise 14.4 In contrast, suppose $a \in \mathcal{U}$. Find an infinite set Z such that $Z \subseteq \{a\} \cup (Z \to Z)$.

14.2 Streams, coinduction, and corecursion

In logic we are used to dealing with sets defined recursively, as the smallest set closed under certain operations. The standard method of proving things about recursively defined sets is the method of proof by induction. When dealing with circularity, it is also common to define sets by "corecursion," that is, as the largest sets such that every member of the set satisfies some condition. The corresponding method of proof for these sets is the method of "proof by coinduction."

The study of streams and operations on streams provides many simple yet natural examples where coinduction is needed, so we begin our discussion there and then consider the general case in Chapter 15.

Let A^∞ be the set of streams over A. As we have just seen, A^∞ is the largest solution to $A^\infty = A \times A^\infty$. Let $f : A \to A$ be arbitrary. Then f naturally induces a function $map_f : A^\infty \to A^\infty$ by "corecursion":

(2) $\qquad map_f(s) \quad = \quad \langle f(1^{st}(s)), map_f(2^{nd}(s)) \rangle.$

For example, if $A = N$, $f(n) = 2n$, and $s = \langle 3, \langle 6, \langle 9, \ldots \rangle \rangle \rangle$, then

$$map_f(s) \quad = \quad \langle 6, \langle 12, \langle 18, \ldots \rangle \rangle \rangle.$$

This is just one of a large number of natural functions on streams defined in a similar manner. Note that in contrast with with ordinary recursive definitions, there is no "base case" in (2) since streams do not have a last element. Consequently, equations like (2) are fundamentally different from definitions by recursion. Nevertheless, there are many similarities; if you understand recursion and induction, you can easily master corecursion and coinduction. We will prove that equations like (2) define functions on streams, and then show how some of our machinery on stream bisimulations can be used to prove various identities involving those functions.

How can we define the function map_f? For each stream s, consider $map_f(s)$ to be a new indeterminate. (It might help if you wrote $x_{map_f(s)}$ for a moment.) Then the collection of instances of (2), as a runs through A and s through A^∞, gives a system of equations. The unique solution to this system yields sets $map_f(s)$ for all $s \in A^\infty$. So this defines the function map_f.

At this point, we know that each value $map_f(s)$ is some set. But we need to make sure that each value is a stream. Here is where coinduction comes in.

Method of Proof [Coinduction Principle for Streams] To show that a set Z is a subset of A^∞, show that $Z \subseteq A \times Z$.

This method of proof is valid because A^∞ is the largest X such that every member of X is an ordered pair $\langle a, s \rangle$ such that $a \in A$ and $s \in X$; that is, the largest set X such that $X \subseteq A \times X$. Indeed, this is just a restatement of Theorem 14.1.

Returning to map_f, let

$$Z = \{map_f(s) \mid s \in A^\infty\}.$$

To show that $Z \subseteq A^\infty$ we need only show that $Z \subseteq A \times Z$. Take an element of Z, say $map_f(s)$, where s is some stream. Since $A^\infty = A \times A^\infty$, there are a and s' such that $s = \langle a, s' \rangle$. By (2), $map_f(s) = \langle f(a), map_f(s') \rangle$. This set is a pair composed of an element of A (namely $f(a)$) followed by another element of Z. This shows that $map_f(s) \in A \times Z$. Since we started with an arbitrary element of Z, we have shown that $Z \subseteq A \times Z$. Therefore Z is a set of streams.

Let's recall where we are with map_f. We saw a way to define it, using new urelements. Then we had the problem of showing that the definition gave a map *into the streams*, and for this we used coinduction.

It is convenient to have a concise method of defining such functions. For as useful as the general tools are, one quickly tires of referring to them. The way to streamline the whole process is to use *corecursion*.

Theorem 14.2 (Corecursion Principle for Streams) *Let C be an arbitrary set. Given any functions $G : C \to A$ and $H : C \to C$ there is a unique function*

$F : C \to A^\infty$ *satisfying the following, for all $c \in C$:*

$$(3) \qquad\qquad F(c) \quad = \quad \langle G(c), F(H(c)) \rangle$$

This formula is called the corecursion equation *for F.*

We can justify this definition by turning this corecursion equation into a system of equations in our formal sense, as we will see.

As an example of how this is used, consider again the case of $F = map_f$. We obtain this function by taking $C = A^\infty$, $G(c) = f(1^{st}(c))$, and $H(c) = 2^{nd}(c)$.

Proof Here is a method for justifying such a corecursive definition. It is important not only in its own right but because we will generalize it to give a general method for defining other kinds of functions by corecursion in a later chapter.

(a) Take a new set of urelements X in correspondence with C, writing x_c for the urelement corresponding to c.

(b) Form the system of equations

$$(4) \qquad\qquad x_c \quad = \quad \langle G(c), x_{H(c)} \rangle$$

(c) Take the solution s to this system, and write $F(c)$ instead of $s(x_c)$. This will define a unique function $F : C \to A^\infty$

(d) Apply $[s]$ to both sides of the system of equations (4). This will give a system of equations

$$F(c) \quad = \quad \langle G(c), F(H(c)) \rangle$$

Therefore F satisfies (3). $\qquad\qquad\qquad\qquad\qquad\qquad\qquad\qquad \dashv$

Applying this result, we know how to define a function map_f satisfying the system given by (2). How do we know map_f is unique? Well, suppose that we had two such functions, map_f and m. Then the set of pairs of corresponding images:

$$\{ \langle map_f(s), m(s) \rangle \mid s \in A^\infty \}$$

would be a stream bisimulation. And thus it would be a subrelation of the identity. This tells us that for all s, $map_f(s) = m(s)$.

Here is another way to get the uniqueness, an argument which also works in the general case. Notice the similarity of (3) and (4). Indeed, every function F satisfying (3) gives a solution to (4) when we set $s(x_c) = F(c)$. And every solution to the system gives a function F. Since the system has a unique solution, there must be a unique function as well.

For another example of a function defined by corecursion, consider the function $c : A \to A^\infty$ given by

$$c_a \quad = \quad \langle a, c_a \rangle.$$

This map c takes a and gives the constant stream $\langle a, \langle a, \langle a, \ldots \rangle \rangle \rangle$. We can define a map c on A by this equation, and then prove that its image $c[A]$ is a set of streams. The argument is entirely parallel to the one we gave for map_f.

Let us now prove some simple theorems about streams. As a first example, consider

(5) $$map_f(c_a) \;=\; c_{f(a)}$$

A moment's thought should convince you that (5) is quite plausible. But how can we prove it? We show that the set of pairs

$$R \;=\; \{\langle map_f(c_a), c_{f(a)} \rangle \mid a \in A\}$$

is a stream bisimulation, hence a subrelation of equality. So for all $a \in A$, $map_f(c_a) = c_{f(a)}$.

To begin, then, let $a \in A$ be fixed. Note that

$$1^{st}(map_f(c_a)) \;=\; f(1^{st}(c_a)) \;=\; f(a),$$

and also that $1^{st}(c_{f(a)}) = f(a)$. So the streams $map_f(c_a)$ and $c_{f(a)}$ have the same first component. As for their second components,

$$2^{nd}(map_f(c_a)) \;=\; map_f(2^{nd}(c_a)) \;=\; map_f(c_a),$$

and $2^{nd}(c_{f(a)}) = c_{f(a)}$. Thus the pair $\langle 2^{nd}(map_f(c_a)), 2^{nd}(c_{f(a)}) \rangle$ belongs to R. In this way, R is a stream bisimulation.

Toward a generalization Before leaving this discussion of definition by corecursion, let us make a remark that will be important later. As we have formulated corecursion for streams, the "givens" consist of two functions G and H, one mapping C into A, the other mapping C into C. This is, of course, equivalent to having a single function J mapping C into $A \times C$. (Given G, H we can define $J(c) = \langle G(c), H(c) \rangle$. Conversely, given J we can extract G and H by taking first and second coordinates.) This is significant since A^∞ is the largest C such that $C = A \times C$, and J maps into $A \times C$. When we turn to the more general case, it is this observation which suggests the proper generalization of Definition by Corecursion to other settings. We will see that the method of proof used to justify our result lifts in a straightforward manner to the general case.

Example 14.1 For another example of a function defined by corecursion, there is a unique function

$$zip : A^\infty \times A^\infty \to A^\infty$$

so that for all $a, b \in A$ and all $s, t \in A^\infty$,

$$zip(\langle a, s \rangle, \langle b, t \rangle) \;=\; \langle a, \langle b, zip(s, t) \rangle \rangle.$$

As its name suggests, this operation acts like a zipper. To define zip, pretend that each value $zip(s, t)$ is a new indeterminate and write a big system of equations.

The solution defines *zip* as a function on streams, but it does not show that the values themselves *are* streams.

Exercise 14.5 Prove that for all $s, t \in A^\infty$, $zip(s,t) \in A^\infty$. In other words, find an argument using coinduction which shows that A^∞ is closed under *zip*.

Exercise 14.6 How would we prove uniqueness of *zip*?

Example 14.2 We show how to prove one of the laws of *zip*; namely, that for all $a \in A$ and all $s, t \in A^\infty$ that

$$zip(\langle a, s \rangle, t) \quad = \quad \langle a, zip(t, s) \rangle.$$

To do this, we again use Lemma 7.4 on Stream Bisimulations. Let R be defined such that

$$zip(\langle a, s \rangle, t) \quad R \quad \langle a, zip(t, s) \rangle$$

for all a, s and t, and also such that the converse pairs are related. We claim that R is a stream bisimulation. Suppose that sRt. Then there are a, s', and t' such that $s = zip(\langle a, s' \rangle, t')$ and $t = \langle a, zip(t', s') \rangle$. (It is also possible that $t = zip(\langle a, s' \rangle, t')$ and $s = \langle a, zip(t', s') \rangle$, since we threw the converse pairs into R. This other case is handled similarly.) Note that

$$1^{st}(s) \quad = \quad 1^{st}(\langle a, s' \rangle) \quad = \quad a \quad = \quad 1^{st}(t).$$

Suppose that $t' = \langle c, t'' \rangle$. Then $2^{nd}(s) = \langle c, zip(s', t'') \rangle$, and $2^{nd}(t) = zip(\langle c, t'' \rangle, s')$. But since we included the converse pairs in R, we see that

$$\langle c, zip(s', t'') \rangle \quad R \quad zip(\langle c, t'' \rangle, s').$$

This completes the verification.

Exercise 14.7 Let $f : A \to A$. Use the corecursion principle to prove that there is a unique function

$$iter_f : A \to A^\infty$$

such that $iter_f(a) = \langle a, iter_f(f(a)) \rangle$ for all $a \in A$. (*iter_f* stands for "iterate f".) Thus, for example, if $f : N \to N$ is given by $f(n) = 2n$, then $iter_f(7) = \langle 7, \langle 14, \langle 28, \ldots \rangle \rangle \rangle$.

It turns out that

$$iter_f(a) \quad = \quad \langle a, map_f(iter_f(a)) \rangle.$$

Let us prove this using the techniques of bisimulation. To do this, let R be such that

$$iter_f(fa) \quad R \quad map_f(iter_f(a))$$

for all $a \in A$. We check that this R is a stream bisimulation. Two streams related by R have the same first components. And the second components of the pairs above are $iter_f(f(a))$ and $map_f(iter_f(fa)))$, and these streams are again related by R. This proves that for all a, $iter_f(fa) = map_f(iter_f(a))$. Thus

$$iter_f a = \langle a, iter_f(fa) \rangle = \langle a, map_f(iter_f(a)) \rangle.$$

Exercise 14.8 Prove the following laws:

1. $zip(c_a, c_a) = c_a$.
2. $iter_f(a) = zip(iter_g(a), map_f(iter_g(a)))$, where $g(a) = f(f(a))$.

One interesting fact, pointed out to us by Paul Miner and Jean-Yves Marion, is that one can obtain all instances of (3) from page 202 by composing instances of map_f and $iter_f$. Indeed, given $G : C \to A$ and $H : C \to C$, consider $map_G \circ iter_H$. Notice that

$$
\begin{aligned}
(map_G \circ iter_H)(c) &= map_G(\langle c, iter_H(H(c)) \rangle) \\
&= \langle G(c), map_G iter_H(H(c)) \rangle \\
&= \langle G(c), (map_G \circ iter_H)(H(c)) \rangle
\end{aligned}
$$

So $map_G \circ iter_H$ is the function F from (3).

14.3 Stream systems

At this point, we have seen several examples of functions which were shown to map into the class of streams. Before going on to a general approach, we must return to a point much earlier in this section.

When we defined map_f, we spoke of turning instances of (2) into a system of equations. Let's have another look at this. We'll say that a *stream system* is a set $X \subseteq \mathcal{U}$ together with a map $e : X \to A \times X$. This *is* a general system of equations, and we could try to prove that the solution set to every stream system is a set of streams.

Unfortunately, there is a problem with this, and this is why we mentioned "new indeterminates" when we began the discussion of map_f. For suppose that A itself were a set of urelements. We wouldn't want to use the elements from A as indeterminates in any stream system. For if we did, we could end up with a system like

$$
\begin{aligned}
a &= \langle a, b \rangle \\
b &= \langle b, b \rangle
\end{aligned}
$$

This solution is $a = b = \Omega$. However $\Omega \notin A^\infty$. Something has gone wrong here, since we want the solution to stream systems to always be streams. The problem here is that set of indeterminates $X = \{a, b\}$ clashes with the set A.

Fortunately, there is a way around this problem. For a fixed set A, let's say that $X \subseteq \mathcal{U}$ is *new for* A if $X \cap support(A) = \emptyset$. Since we are assuming the Strong Axiom of Plenitude on $\mathcal{U}$, this is not a big restriction: if $X \cap support(A)$ did happen to be non-empty, we could just use the axiom to move to a different set of urelements in forming our system.

Definition A *flat stream system over* A is a pair $\mathcal{E} = \langle X, e \rangle$, consisting of set $X \subseteq \mathcal{U}$ which is new for A, together with a map $e : X \to A \times X$.

Here is why this approach works.

Proposition 14.3 *Let A be any set.*

(a) If X is new for A and s is a substitution defined on X, then for all $a \in A$, $a[s] = a$. Hence $A[s] = A$.

(b) Furthermore, if b is any set, and s is a substitution defined on X, then $(A \times b)[s] = A \times b[s]$.

(c) If X is not new for A, then there is some substitution s and some $a \in A$ such that $a[s] \notin A$. This s has the property that for any nonempty set b, $(A \times b)[s] \neq A \times b[s]$.

Proof For (a), let X be disjoint from the support of A. Let s be a substitution with domain X. Let a be a set belonging to A. Since $support(a) \subseteq support(A)$, $[s]$ is constant on the support of a. So $a[s] = a$. Also, if $y \in A$ is an urelement, then $y \notin A$; hence $y[s] = y$. Since $A[s] = \{p[s] : p \in A\}$, it follows that $A[s] = A$.

For (b), recall that substitution commutes with the ordered pair operation. Thus

$$
\begin{aligned}
(A \times b)[s] &= \{\langle a, b\rangle[s] \mid a \in A\} \\
&= \{\langle a[s], b[s]\rangle \mid a \in A\} \\
&= \{\langle a, b[s]\rangle \mid a \in A\} \quad \text{by (a)} \\
&= A \times b[s]
\end{aligned}
$$

Turning to (c), suppose that $x \in X \cap support(A)$. Let $a \in A$ be such that $x \in support(a)$. Let $y \in \mathcal{U} - support(A)$; so no element of A has y in its support. Consider the substitution s which interchanges x and y. Then $support(a[s])$ contains y. Therefore $a[s] \notin A$. For the last assertion, note that $A[s] \neq A$. ⊣

Theorem 14.4 *Let $\mathcal{E} = \langle X, e \rangle$ be a flat stream system over A. Then the solution set of $\mathcal{E}$ is a set of streams over A.*

Proof Let s be the solution to $\mathcal{E}$, and let S be the solution set of $\mathcal{E}$. We show that $S \subseteq A \times S$; so it follows from Theorem 14.1 that $S \subseteq A^\infty$. For each $x \in X$, write e_x as $\langle a_x, y_x \rangle$. Note that each a_x belongs to A, and each y_x belongs to X. Now

$$
\begin{aligned}
S &= \{s_x \mid x \in X\} \\
&= \{e_x[s] \mid x \in X\} \\
&= \{\langle a_x[s], y_x[s]\rangle \mid x \in X\} \\
&= \{\langle a, y_x[s]\rangle \mid x \in X\} \quad \text{by Proposition 14.3(a)} \\
&\subseteq A \times S
\end{aligned}
$$

⊣

Now at this point, we would like a stronger result, one that will allow us to see immediately that many more systems have streams for their solutions.

Definition Let A and X be any sets. The set of *parametric A-streams (over X)* is the largest set H such that every member of H is an ordered pair $\langle a, b \rangle$ where $a \in A$ and $b \in X \cup H$. If $X = \emptyset$, then $H = A^\infty$, and (for emphasis) we call these the *nonparametric A-streams*.

Example 14.3 If $a \in A$ and $x \in X$, then the following are parametric A-streams: the stream $s = \langle a, s \rangle$, $\langle a, \langle a, x \rangle \rangle$, $\langle a, \langle b, \langle a, \langle b, \dots \rangle \rangle \rangle \rangle$. The first and last of these are nonparametric, but the second is parametric. (The calculation of the greatest fixed point here is made easier by Exercise 14.2.)

We will be studying this kind of definition further in various places later in the book. For now, let F be the set of "finite" parametric expressions; i.e., those like x, $\langle a, x \rangle$, and $\langle a, \langle b, \langle c, x \rangle \rangle \rangle$. Note that $F \cup A^\infty$ meets the condition on the set H above. So $F \cup A^\infty \subseteq H$. As it happens, $F \cup A^\infty = H$. (Why do you think this holds?) We use this fact in the next result.

Theorem 14.5 (Solution Lemma for Streams) *Let $\mathcal{E} = \langle X, e \rangle$ be a general system of equations in which each right-hand side e_x is a parametric A-stream over X. Then the solution set of $\mathcal{E}$ is a set of nonparametric A-streams.*

Proof First, using the characterization $F \cup A^\infty = H$ mentioned above, we may "flatten" $\mathcal{E}$ by adding more indeterminates such that each right hand side e_x is either a stream or an element of $A \times X$. Let s be the solution to $\mathcal{E}$. Since s is the identity on A, and since pairing commutes with substitution, we see that for all $x \in X$,

$$s_x \;=\; e_x[s] \;=\; \langle a, y[s] \rangle$$

where a is some element of A, and y is either an element of X or a parametric A-stream over X. Let Z be the solution set of $\mathcal{E}$. Then $Z \subseteq A \times (Z \cup A^\infty)$. Since $A^\infty = A \times A^\infty$, $(Z \cup A^\infty) \subseteq A \times (Z \cup A^\infty)$. So $Z \cup A^\infty \subseteq A^\infty$, and therefore $Z \subseteq A^\infty$. $\quad\dashv$

Theorem 14.5 gives us a general approach to the kinds of problems we have been solving. For example, it implies at once that map_f, c, zip, and $iter_f$ map into A^∞. To review this point, let's consider zip. For every pair (s, t) of streams over A, take an urelement $x_{s,t}$. Of course, we have to make sure that these do not belong to the support of A. Let $X = \{x_{s,t} \mid s, t \in A^\infty\}$. Let $\mathcal{E}$ be the general system given by

$$x_{s,t} \;=\; \langle \mathrm{i}^{st}(s), \langle 1^{st}(t), x_{2^{nd}(s), 2^{nd}(t)} \rangle \rangle.$$

Theorem 14.5 tells us that each value in the solution of this system is a stream. Now $zip(s, t)$ is the value of the solution on $x_{s,t}$. This proves that zip maps into the streams over A.

However, consider the function dm_f (for *doublemap$_f$*) given by

$$dm_f(s) \quad = \quad \langle f(1^{st}(s)), dm_f(dm_f(2^{nd}(s))) \rangle.$$

How can we show that dm_f is a well-defined map from A^∞ to A^∞? Before reading on, you might try the following exercise.

Exercise 14.9 Let $A = N$, and let $f(n) = n + 1$. Calculate

$$dm_f(\langle 0, \langle 1, \langle 2, \ldots \rangle \rangle \rangle).$$

As you can see from the solution, the way to see that dm_f is well-defined is to write a big system which defines $dm_f^n(s)$ for all $n \geq 1$ and all $s \in A^\infty$ simultaneously. So we take X to be a fresh set of urelements in correspondence with the formal terms of the form $dm_f^n(s)$, and then we have a system of equations of the appropriate type.

At this point, we have looked at streams and functions defined on them. As interesting as this example is, we are even more concerned with general tools to look at similar examples. In Chapters 15–17 of the book, we generalize all of our results on streams to a much wider context.

Historical Remarks

Stream functions like *map$_f$* are often considered in the literature on streams, but our work on streams and *AFA* seems to be new.

Part V

Further Theory

15

Greatest fixed points

Really, universally, relations stop nowhere, and the exquisite problem of
the artist is eternally but to draw, by a geometry of his own, the circle within
which they shall happily appear to do so.

Henry James, Prefaces

Mathematics and logic are full of sets that are defined recursively, as the
"smallest" set closed under certain operations. The notions of natural number, formula, and proof are three of the most obvious. The standard method
of proving things about recursively defined sets is the method of proof by
induction.

When dealing with circularity, it is also common to define sets by "corecursion," that is, as the largest sets such that every member of the set satisfies
some condition. We have seen examples of this with the set A^∞ of streams
over a set A, and the class of canonical states over a set Act of actions. These
were the greatest solutions of the "equations"

$$X \quad = \quad A \times X$$

and

$$X \quad = \quad \mathcal{P}(Act \times X)$$

respectively.

The usual method of proof for sets and classes defined as greatest fixed
points is the method of "proof by coinduction." In this chapter we will discuss
these methods in some generality. In the next chapter, we'll generalize the
notion of system of equation and relate solution sets to greatest fixed points.

As you might expect, we'll find the idea of solving systems of equations coming to the fore. We round out this part with a discussion of corecursion in Chapter 17.

The ideas behind all three chapters of the book were foreshadowed in Chapter 14. It might be a good idea to glance back at that work from time to time, since what we'll be doing will be a generalization of the stream work.

15.1 Fixed points of monotone operators

An *operator* is an operation

$$\Gamma : V_{afa}[\mathcal{U}] \to V_{afa}[\mathcal{U}].$$

That is, Γ is an operation taking sets to sets. A *monotone operator* is an operator Γ that satisfies the condition that for all sets a and b: if $a \subseteq b$ then $\Gamma(a) \subseteq \Gamma(b)$.

Here are some monotone operations we will use to illustrate the general discussion. All but the fourth has come up before in our discussion.[1]

$\Gamma_1(a) = \mathcal{P}(a)$

$\Gamma_2(a) = \mathcal{P}_{fin}(a)$

$\Gamma_3(a) = A \times a$ for some fixed set A.

$\Gamma_4(a) = \{f \mid f \text{ is a partial function from } a \text{ to } a\}$. We write this set as $a \rightharpoonup a$.

$\Gamma_5(a) = \mathcal{P}(Act \times a)$ for some fixed set Act.

Exercise 15.1 Verify that these operators are monotone.

The following simple observation allows us to build complex operators out of simple ones. For example it shows that $\Gamma(a) = \Gamma_1(\Gamma_3(\Gamma_2(a)))$ is a monotone operator.

Proposition 15.1 *The composition of monotone operators is monotone.*

Proof This means that if Γ and Δ are both monotone, we get a new monotone operator $\Gamma \circ \Delta$ defined by $\Gamma \circ \Delta(a) = \Gamma(\Delta(a))$. To check that this is monotone, suppose that $a \subseteq b$. Then $\Delta(a) \subseteq \Delta(b)$. Now we apply monotonicity to these two sets, and conclude that $\Gamma(\Delta(a)) \subseteq \Gamma(\Delta(b))$. ⊣

A monotone operator on sets can always be extended to a monotone operator on classes in a natural way. Simply define

(1) $$\Gamma(C) = \bigcup_{a \subseteq C} \Gamma(a).$$

for any proper class C. The a's in (1) range over sets. For example, if V_{afa} is the class of pure sets, then $\Gamma_3(V_{afa})$ would be the collection of all sets of the

---

[1]These example operators will be used throughout the remainder of the book. For this reason, we list them in an Appendix on page 335. In that Appendix, we also summarize the conditions on operators which we study. So you might like to have a look there from time to time.

form $\langle p, b \rangle$, where $p \in A$ and $b \in V_{afa}$. It is not hard to verify that when we extend a monotone Γ to an operator on classes, it is monotone as an operator on classes.

Exercise 15.2 Let C be $V_{afa}[\mathcal{U}]$, the class of all sets. Which of the following is true: $C \subseteq \mathcal{P}(C)$, or $\mathcal{P}(C) \subseteq C$?

We can think of a monotone operator Γ imposing one of two conditions on some set or class G: that everything present in G have a certain form, or as demanding that things of certain forms be present in G. These two conditions are captured in the notions of Γ-correct and Γ-closed sets. Intuitively, a set G is Γ-correct if everything in G is of some form allowed by Γ. By contrast, G is Γ-closed if everything demanded by Γ when applied to G is already present in G. We make these intuitions precise as follows:

Definition Let Γ be a monotone operator.

1. A set or class G is called Γ-*correct* if $G \subseteq \Gamma(G)$.
2. By contrast, G is called Γ-*closed* if $\Gamma(G) \subseteq G$.
3. G is a *fixed point* of Γ if it both closed and correct, that is, if $\Gamma(G) = G$.
4. G is the *greatest* (*least*) fixed point of Γ if G is a fixed point and if for every fixed point H of Γ, $H \subseteq G$ ($G \subseteq H$).

The first question we want to ask is whether a monotone operator Γ has fixed points, and if so, what they are like. We are especially interested in the least and greatest fixed points of monotone operators. Let's start by analyzing the fixed points of the operators Γ_1, Γ_3 and Γ_4. A generalization of Γ_2 is studied in depth in Chapter 18, and Γ_5 is treated in Exercise 15.3.

We will see later that for the operators that concern us, we can compute $\Gamma(a)$ by computing $\Gamma(X)$ on sets X of urelements in one-one correspondence with a and then substituting. Thus, let us call the elements of $\Gamma(X)$ the Γ-*forms* over X.

Example 15.1 Consider Γ_1, where $\Gamma_1(a) = \mathcal{P}(a)$. The Γ_1-forms are just the sets of urelements. By Cantor's Theorem, no set can be Γ_1-closed. Nevertheless, Γ_1 does have fixed points among the proper classes. The class *WF* of pure, wellfounded sets is the least fixed point. *WF* is Γ_1-correct since every set a is an element of some powerset, e.g. $\mathcal{P}(a)$. *WF* is Γ_1-closed since every set of pure, wellfounded sets is again a pure, wellfounded set. To see that *WF* is the least fixed point, let Y be such that $\Gamma_1(Y) \subseteq Y$. We use the principle of $\in$-induction for *WF* to show that $WF \subseteq Y$. Let $x \in WF$, and assume that every $y \in x$ belongs to Y. Then $x \subseteq Y$, so $x \in \mathcal{P}(Y) \subseteq Y$. So we have successfully shown what we need to use the $\in$-induction principle, and we conclude that $WF \subseteq Y$.

Γ_1 also has a greatest fixed point; namely, the class V_{afa} of all pure sets. Let's check that V_{afa} is a fixed point. First, let us check that V_{afa} is Γ_1-correct, that is, that $V_{afa} \subseteq \bigcup_{a \subseteq V_{afa}} \mathcal{P}(a)$. Let $b \in V_{afa}$. Then $b \in \mathcal{P}(b)$. Also, $b \subseteq V_{afa}$, since every element of a pure set is pure. So $b \in \bigcup_{a \subseteq V_{afa}} \mathcal{P}(a)$ as desired. Next, let us check that V_{afa} is Γ_1-closed, that is, $\mathcal{P}(V_{afa}) \subseteq V_{afa}$. Let $b \in \mathcal{P}(V_{afa})$. Then $b \in \mathcal{P}(a)$ for some $a \subseteq V_{afa}$. But then $b \subseteq a$. So b is a set of pure sets, and therefore is itself pure. That is, $b \in V_{afa}$.

The above shows that V_{afa} is a fixed point of Γ_1. Actually, V_{afa} is the greatest fixed point. In fact, V_{afa} is the greatest Γ_1-correct class. To see this, let $C \subseteq \mathcal{P}(C)$. But then C is transitive. Furthermore, since $\mathcal{P}(C)$ contains only sets, not urelements, C and every element of C is pure. So $C \subseteq V_{afa}$.

Note that whether the least and greatest fixed points of Γ_1 are equal depends on which axioms of set theory we take. For example, if we take *FA* and the assertion that there are no urelements, then the fixed points of Γ_1 coincide. If we adopt *AFA*, then they differ.

Example 15.2 Consider Γ_3 as above: $\Gamma_3(a) = A \times a$. The Γ_3-forms over X are simply pairs of the form $\langle a, x \rangle$ for $a \in A$ and $x \in X$. Since the empty set is a fixed point, it is clearly the least fixed point. Assuming *AFA*, the greatest fixed point is the set A^∞ of streams over A, and there are other fixed points as well. Assuming *FA*, the greatest fixed point is $\emptyset$ again.

Example 15.3 Recall that $\Gamma_4(a) = a \rightharpoonup a$, the set of partial functions from a to itself. The Γ_4-forms are the partial functions from some set X of urelements into itself. For any set a, $\emptyset \in \Gamma_4(a)$. So $\emptyset$ will belong to the least fixed point, as will $\{\langle \emptyset, \emptyset \rangle\}$. In fact, we get functions f_α by recursion on the ordinals so that

$$f_\alpha \quad = \quad \{f_\beta \mid \beta < \alpha\} \times \{\emptyset\}.$$

Each of these belongs to the domain of the previous ones, so since they are all wellfounded, they are different. Further, each belongs to the least fixed point. This shows that the least fixed point of Γ_4 is a proper class.

Assuming *AFA*, the greatest fixed point also contains functions like $\Omega = \{\langle \Omega, \Omega \rangle\}$.

Exercise 15.3 Carry out an analysis of the forms and fixed points of Γ_5, similar to the analysis we have just given of Γ_1 and Γ_3.

15.2 Least fixed points

We now proceed to prove a number of results illustrating parallels between least and greatest fixed points. We start with those having to do with least fixed points and then turn to the parallel results for greatest fixed points.

Theorem 15.2 (Least fixed point theorem) *Every monotone operator Γ has a least fixed point which we denote by Γ_*. This least fixed point may be characterized in either of the following ways:*[2]

1. $\Gamma_* = \bigcup_\alpha \Gamma_\alpha$, *where Γ_α is defined by recursion on the ordinals by*

$$\Gamma_\alpha = \bigcup_{\beta < \alpha} \Gamma(\Gamma_\beta).$$

2. Γ_* *is the least Γ-closed class.*

If you are not familiar with such definitions you should note that this definition of Γ_α is equivalent to the following definition by cases:

$$\begin{aligned}
\Gamma_0 &= \emptyset \\
\Gamma_{\alpha+1} &= \Gamma(\Gamma_\alpha) \\
\Gamma_\lambda &= \bigcup_{\beta < \lambda} \Gamma_\beta \quad (\lambda \text{ a limit})
\end{aligned}$$

Proof Let Γ_* be defined by (1). We will show that it satisfies (2) and is a fixed point, from which it follows that it is the least fixed point. To prove (2), we first show that Γ_* is Γ-closed: $\Gamma(\Gamma_*) \subseteq \Gamma_*$. Let $a \subseteq \Gamma_*$ be a set; we must show that $\Gamma(a) \subseteq \Gamma_*$. By Replacement, there is some α such that $a \subseteq \Gamma_\alpha$. But then by monotonicity, $\Gamma(a) \subseteq \Gamma_{\alpha+1} \subseteq \Gamma_*$.

Next, suppose G is any Γ-closed class: $\Gamma(G) \subseteq G$. Then an easy induction on α shows that $\Gamma_\alpha \subseteq G$ for all α. Hence $\Gamma_* \subseteq G$.

It only remains to show that Γ_* is a fixed point of Γ. We already know that $\Gamma(\Gamma_*) \subseteq \Gamma_*$. By monotonicity, $\Gamma(\Gamma(\Gamma_*)) \subseteq \Gamma(\Gamma_*)$. So $\Gamma(\Gamma_*)$ is Γ-closed. Therefore $\Gamma_* \subseteq \Gamma(\Gamma_*)$. So equality holds. ⊣

This theorem is pretty abstract, but it corresponds very closely to what was going on in the examples given earlier. It gives us the following method of proof for sets defined inductively:

Method of Proof [Preliminary Induction Principle for Γ_*] To prove that $\Gamma_* \subseteq G$, show that G is Γ-closed ($\Gamma(G) \subseteq G$).

This principle is justified by the second characterization of the least fixed point given in the theorem. Often it is convenient to have a stronger principle, though.

Method of Proof [Induction Principle for Γ_*] To prove that $\Gamma_* \subseteq G$, show that $\Gamma(G \cap \Gamma_*) \subseteq G$.

---

[2]This form of definition is an instance of transfinite recursion on the ordinals (see page 20). If this style of definition is new to you, then you should take it on faith that in set theory we can prove the existence of a class Γ_* that is the least fixed point of Γ. You should also know that ordinals are used in the solutions to Exercises 15.8 and 15.11 below.

Exercise 15.4 Justify this principle. If you have trouble, the solutions contain a hint.

15.3 Greatest fixed points

With least fixed points under our belts, we turn to the parallel results for greatest fixed points. The first of these is the existence of greatest fixed points. As it happens, the proof is rather different from that of least fixed points.

Readers who are familiar with the theory of inductive definitions on the subsets of some fixed *set* may find this lack of symmetry puzzling. The difficulty comes from the fact that the theory ZFA, like ZFC, has a rather impoverished theory of classes. For example, there is no way to state, let alone justify, recursive definitions like the following:

$$\Gamma^\alpha \ = \ \bigcap_{\beta < \alpha} \Gamma(\Gamma^\beta).$$

We will return to discuss this lack of duality between greatest and least fixed points on page 219.

Theorem 15.3 (Greatest fixed point theorem) *Every monotone operator Γ has a greatest fixed point, which we denote by Γ^*. This greatest fixed point may be characterized in either of the following ways:*

1. $\Gamma^* = \bigcup\{a \mid a$ *is a Γ-correct set*$\}$.
2. Γ^* *is the greatest Γ-correct class.*

We need the following lemma in the proof of this theorem.

Lemma 15.4 *Let Γ be monotone and let $\Gamma^* = \bigcup\{a \mid a$ is Γ-correct$\}$.*

1. *Let $a \subseteq \Gamma^*$. Then there is some Γ-correct set b such that $a \subseteq b$.*
2. *Let $a \subseteq G$ where G is Γ-correct. Then there is some set $b \subseteq G$ such that $a \subseteq \Gamma(b)$.*

Proof (1) For each $c \in a$, $c \in \Gamma^*$. So there is some b_c such that $c \in b_c \subseteq \Gamma(b_c)$. Let $b = \bigcup_{c \in a} b_c$. Then $a \subseteq b$, and by monotonicity,

$$b \ = \ \bigcup_{c \in a} b_c \ \subseteq \ \bigcup_{c \in a} \Gamma(b_c) \ \subseteq \ \Gamma(b).$$

Part (2) is proved similarly. ⊣

Proof of Theorem 15.3 Let Γ^* be defined by (1). We will show Γ^* satisfies (2) in the statement of the theorem, and that Γ^* is a fixed point. It follows that Γ^* is the greatest fixed point. To prove (2) we first show that Γ^* is Γ-correct. Let $b \in \Gamma^*$. Then $b \in a$ for some a such that $a \subseteq \Gamma(a)$. By definition, $a \subseteq \Gamma^*$. So $\Gamma(a) \subseteq \Gamma(\Gamma^*)$ by monotonicity. But

$$b \ \in \ a \ \subseteq \ \Gamma(a) \ \subseteq \ \Gamma(\Gamma^*),$$

so $b \in \Gamma(\Gamma^*)$. This proves that $\Gamma^* \subseteq \Gamma(\Gamma^*)$, i.e., that Γ^* is Γ-correct.

Next let G be any Γ-correct class: $G \subseteq \Gamma(G)$. We want to show that $G \subseteq \Gamma^*$. Let $b \in G$. Let $a_0 = \{b\}$. By part (2) of Lemma 15.4, let $a_1 \subseteq G$ be such that $a_0 \subseteq \Gamma(a_1)$. Continuing, we get $a_2 \subseteq G$ so $a_1 \subseteq \Gamma(a_2)$. Find $a_3, a_4, \ldots$ similarly. For each n, $a_n \subseteq \Gamma(a_{n+1})$. Let $a = \bigcup_n a_n$ Then $a \subseteq \Gamma(a)$ because if $d \in a$, then for some n,

$$d \quad \in \quad a_n \quad \subseteq \quad \Gamma(a_{n+1}) \quad \subseteq \quad \Gamma(a).$$

The point is that $a \subseteq \Gamma^*$. Since $b \in a_0 \subseteq a$, $b \in \Gamma^*$. This proves that $G \subseteq \Gamma^*$.

Finally, to show that Γ^* is a fixed point of Γ, we repeat the analogous argument about Γ_*, reversing all the inclusions. ⊣

Incidentally, the characterization of Γ^* as the union of the Γ-correct sets is often much less useful than the analogous result characterizing Γ_* in terms of ordinals. In Chapter 16, we'll see a parallel result (Theorem 16.6) that represents the greatest fixed point of Γ as the class of sets occurring in the solution sets of certain systems of equations. But that result is fairly involved, and it does not hold for all monotone operators.

Exercise 15.5 (1) Let $\Gamma_3(c) = A \times c$ be the operator whose greatest fixed point is the set of streams over A. Let $s \in A^\infty$, and let $a = \{s\}$. Following the proofs of Lemma 15.4 and Theorem 15.3, find a Γ_3-correct set b such that $a \subseteq b$. (Of course, A^∞ is such a set, but if you follow the proof, you can find the least such b.)

(2) For the operator $\Gamma_1(a) = \mathcal{P}(a)$, let b be an arbitrary element of $V_{afa}[\mathcal{U}]$. Find the least Γ-correct c such that $\{b\} \subseteq c$.

Exercise 15.6 A monotone operator Γ has the property that for all b and c, $\Gamma(b \cap c) \subseteq \Gamma(b) \cap \Gamma(c)$. Γ *commutes with binary intersections* if for all b and c, $\Gamma(b \cap c) = \Gamma(b) \cap \Gamma(c)$. Γ *commutes with all intersections* if for all b, $\Gamma(\bigcap b) = \bigcap_{c \in b} \Gamma(c)$.

(1) Of the monotone operators $\Gamma_1, \ldots, \Gamma_5$ at the beginning of this section, which commute with all intersections?

(2) Give an example of a monotone operator that does not commute with binary intersections.

(3) Prove that if Γ commutes with all intersections, then the intersection of a family of Γ-correct sets is itself Γ-correct.

(4) Prove that if Γ commutes with all intersections, and $a \subseteq \Gamma^*$, then there is a $\subseteq$-minimal Γ-correct b such that $a \subseteq b$. We call b the Γ-*closure of* a.

(5) Does your example in part (2) have the property that every $a \subseteq \Gamma^*$ has a Γ-closure?

As a consequence of Theorem 15.3, we have some general coinduction principles.

Method of Proof [Preliminary Coinduction Principle for Γ^*] To prove that $G \subseteq \Gamma^*$, show that G is Γ-correct ($G \subseteq \Gamma(G)$).

This principle is justified by the second characterization of the greatest fixed point given in the theorem.

Method of Proof [Coinduction Principle for Γ^*] To prove that $G \subseteq \Gamma^*$, show that $G \subseteq \Gamma(G \cup \Gamma^*)$.

Proof Assume $G \subseteq \Gamma(G \cup \Gamma^*)$. Since Γ^* is a fixed point, $\Gamma(\Gamma^*) = \Gamma^*$. Then by monotonicity, $\Gamma(\Gamma^*) \subseteq \Gamma(G \cup \Gamma^*)$. So $\Gamma^* \subseteq \Gamma(G \cup \Gamma^*)$ Hence, by our assumption, $(G \cup \Gamma^*) \subseteq \Gamma(G \cup \Gamma^*)$. But then by the second characterization of Γ^*, $(G \cup \Gamma^*) \subseteq \Gamma^*$, from which it follows that $G \subseteq \Gamma^*$. ⊣

Relation to induction on natural numbers Consider the operator

$$\Gamma(a) = \{\emptyset\} \cup \{b \cup \{b\} \mid b \in a \text{ is a set}\}.$$

The goal of a certain amount of every elementary set theory book is the study of the properties of the least fixed point of this operator, including the verification that it is a set, not a proper class. Indeed, the least fixed point of this operator is the set ω (or N) of the natural numbers. The induction principle can be restated for this example as follows. (This is the standard formulation of induction for natural numbers.)

Method of Proof [Induction Principle for the Natural Numbers] To prove that every natural number belongs to some set S, show that $0 \in S$ and that if n is a natural number in S, then $n + 1 = n \cup \{n\}$ is in S also.

Turning to the greatest fixed point, we claim first that if G is any fixed point of Γ, then $G \cap WF \subseteq \omega$. One shows that if $a \in G \cap WF$ then $a \in \omega$ using $\in$-induction. We'll omit this argument. It follows that under *FA*, the greatest fixed point is ω. Under *AFA*, though, the greatest fixed point is $\Omega \cup \omega$. Clearly this is a fixed point. To see that it is the greatest, let G be any fixed point, and let $a_0 \in G$ be non-wellfounded. Then for some non-wellfounded $a_1 \in G$, $a_0 = a_1 \cup \{a_1\}$. Continuing, we get non-wellfounded a_2, a_3, etc., such that $a_n = a_{n+1} \cup \{a_{n+1}\}$. Then $TC(a_0) = \bigcup_n a_n$ is a set of pure, non-wellfounded sets. Ω is the only set with this property.

Exercise 15.7 Let Γ be the operator taking a to the set of non-empty subsets of a.

1. Compute Γ_* and Γ^*.
2. Let A be a fixed set of urelements, and let $\Delta(a)$ be the non-empty subsets of $A \cup a$. What is the greatest fixed point of Δ? Prove that if $A \neq \emptyset$, then Δ_* is a proper class.

Exercise 15.8 Recall that a set is reflexive if it is a member of itself. Let A be a fixed set of urelements and let $\Delta(a)$ be the set of reflexive subsets of $A \cup a$. Identify the least and greatest fixed points of Δ. Show in particular that the former is a set while the latter is a set iff $A = \emptyset$.

Exercise 15.9 Suppose that Γ and Δ are monotone operators with the property that for all sets a, $\Gamma(a) \subseteq \Delta(a)$. Prove that $\Gamma_* \subseteq \Delta_*$ and $\Gamma^* \subseteq \Delta^*$. [Hint: Under the assumption, $\Gamma(C) \subseteq \Delta(C)$ will also hold for classes C.]

Exercise 15.10 Suppose that Γ and Δ are monotone operators with the property that for all sets a, $\Gamma(a) \subseteq \Delta^*$. Prove that $\Gamma^* \subseteq \Delta^*$.

Exercise 15.11 Let Γ be a monotone operator, let Γ^* be its greatest fixed point, and let C be a fixed class. Let $\Delta_{\Gamma,C}$ and $\Phi_{\Gamma,C}$ be the operators defined by

$$
\begin{aligned}
\Delta_{\Gamma,C}(a) &= \Gamma(C \cup a) \\
\Phi_{\Gamma,C}(a) &= \Gamma(C \cup \Gamma^* \cup a)
\end{aligned}
$$

1. Prove that $(\Phi_{\Gamma,C})_* \subseteq (\Delta_{\Gamma,C})^*$.
2. In some cases, the sets in the conclusion of this proposition are identical. This happens with $\Gamma(a) = A \times a$, for example. Prove this. [Hint: see Exercise 14.2 on page 200.]
3. On the other hand, in cases like $\Gamma(a) = \mathcal{P}(a)$ and $C = \{x\}$, $(\Phi_{\Gamma,C})_*$ is a proper subset of $(\Delta_{\Gamma,C})^*$. Prove this.

Duals of operators One aspect of the classical theory of monotone operators is that operators have *duals* that are sometimes useful to study. We briefly review the appropriate definitions at this point. For each class $C \subseteq V_{afa}[\mathcal{U}]$, let

$$
-C = \{a \in V_{afa}[\mathcal{U}] \mid a \notin C\}
$$

be the *complementary class to* C. The first things to note about this are that $-(-C) = C$, and also that if $C \subseteq D$, then $-D \subseteq -C$.

For an operator Γ, let $\hat{\Gamma}$ be defined by

$$
\hat{\Gamma}(a) = -\Gamma(-a).
$$

As we have defined it, $\hat{\Gamma}$ takes sets to classes, and thus it is not strictly speaking an operator at all. However, we'll see that except for this defect, $\hat{\Gamma}$ has most of the nice properties of Γ.

Exercise 15.12 Let $\Gamma(a) = \mathcal{P}(a)$. Prove that $\hat{\Gamma}(\emptyset)$ is a proper class.

Exercise 15.13 Assume that Γ is monotone.

1. Prove that as an operator on classes, $\hat{\Gamma}$ is monotone. That is, if $C \subseteq D$, then $\hat{\Gamma}(C) \subseteq \hat{\Gamma}(D)$.
2. Prove that if Γ is monotone and commutes with all intersections (of

classes), then

$$\hat{\Gamma}(C) \quad = \quad \bigcup \{\hat{\Gamma}(a) \mid a \subseteq C \text{ is a set}\}.$$

3. Give an example of a monotone Γ where (2) is true for Γ but false for $\hat{\Gamma}$.
4. Assume that Γ is monotone, so that Γ^* and Γ_* exist. Prove that $-\Gamma^*$ is the least fixed point of $\hat{\Gamma}$ and that $-\Gamma_*$ is the greatest fixed point.

In our definition of an operator, we demanded that $\Gamma(a)$ be a set whenever a is a set. This property was used in proving that both Γ_* and Γ^* exist. However, the dual $\hat{\Gamma}$ need not take sets to sets, as Exercise 15.12 shows. Nevertheless, the fixed points $\hat{\Gamma}_*$ and $\hat{\Gamma}^*$ will exist.

The point of these exercises is that someone might prefer to *define* the greatest fixed point of Γ by passing to the dual $\hat{\Gamma}$, taking the least fixed point, and then taking the complement. However, this approach does not quite work out: if we do not know that Γ^* exists, then there does not seem to be any reason to think that $\hat{\Gamma}_*$ exists. Indeed, the existence of Γ_* used (2) for Γ, and as we see in Exercise 15.13.3, this may fail for $\hat{\Gamma}$.

15.4 Games and fixed points

Let Γ be a monotone operator. There is a very intuitive characterization of the least and greatest fixed points of Γ in game-theoretic terms. We associate two games, $\mathcal{G}_*(\Gamma)$ and $\mathcal{G}^*(\Gamma)$ with Γ. In both games, I and II play as follows. Player I initiates the play by playing some set a_1. Given a play a_n by I, player II is required to play some set b_n such that $a_n \in \Gamma(b_n)$. Given a play b_n by II, player I is required to play some set $a_{n+1} \in b_n$. This can be depicted as follows:

I	II
a_1	
	b_1 with $a_1 \in \Gamma(b_1)$
$a_2 \in b_1$	
	b_2 with $a_2 \in \Gamma(b_2)$
$\vdots$	$\vdots$

Notice that if II is ever able to play $b_n = \emptyset$, then II wins, since there is no way for I to make a legal next move. In the game $\mathcal{G}_*(\Gamma)$, this is the only way for II to win. In other words, the set of winning plays for II in $\mathcal{G}_*(\Gamma)$ is the set of those plays that eventually result in II playing the empty set; otherwise I wins. In the game $\mathcal{G}^*(\Gamma)$ we make things much easier for II: a sequence of legal plays constitutes a win for II unless I makes a move that II cannot answer.

Proposition 15.5 *Let Γ be a monotone operator. The game $\mathcal{G}^*(\Gamma)$ is an open game, while $\mathcal{G}_*(\Gamma)$ is closed. Hence both games are determined.*

Proof This is immediate from the definitions of open and closed games, and the Gale-Stewart Theorem. ⊣

Theorem 15.6 *Let* Γ *be a monotone operator.*

1. $a_1 \in \Gamma_*$ *iff II has a winning strategy in* $\mathcal{G}_*(\Gamma)$ *if I begins with* a_1.
2. $a_1 \in \Gamma^*$ *iff II has a winning strategy in* $\mathcal{G}^*(\Gamma)$ *if I begins with* a_1.

Proof Let G_* and G^* be the sets of losing moves for I in $\mathcal{G}_*(\Gamma)$ and $\mathcal{G}^*(\Gamma)$ respectively. Let us first show that G_* is Γ-closed, and hence contains Γ_*. To show this, let $a \in \Gamma(G_*)$. Then $a \in \Gamma(b)$ for some $b \subseteq G_*$. If I begins play with a, then let II respond with this b. Then no matter how I continues, II has a winning response, since $b \subseteq G_*$. Thus $a \in G_*$, so G_* is Γ-correct. On the other hand, an easy proof by induction shows that each $a \in \Gamma_\alpha$ is in G_*, by Theorem 15.2.

Next, we show that G^* is Γ-correct, and hence is contained in Γ^*. To see this, let $a \in G^*$. We need to see that $a \in \Gamma(G^*)$. Let b be a winning move for II in response to a in $\mathcal{G}^*(\Gamma)$. Then $a \in \Gamma(b)$. But clearly $b \subseteq \Gamma^*$. Hence, by monotonicity, $a \in \Gamma(G^*)$. Finally, we need to show that if $a \in \Gamma^*$, then II has a winning move. But this is an immediate consequence of Lemma 15.4. Indeed, this lemma shows us that, given $a \in \Gamma^*$, II can find a single b such that he can respond with b repeatedly. ⊣

Historical Remarks

The theory of least fixed points and their connection with induction is a standard part of set theory. The parallel development for greatest fixed points is due to Aczel (1988).

The basic connection between least fixed points and open games was discovered by Moschovakis (1974). There is a general duality between least and greatest fixed points (see Exercise 15.13.4) that makes it natural to look for the kind of result obtained in Theorem 15.6.

16

Uniform operators

But in fact the opposition of instinct and reason is mainly illusory. Instinct, intuition, or insight is what first leads to the beliefs which subsequent reason confirms or confutes; but the confirmation, where it is possible, consists, in the last analysis, of agreement with other beliefs no less instinctive. ... Even in the most purely logical realm, it is insight that first arrives at what is new.

Russell, "Mysticism and Logic"

One of the most powerful ideas in mathematics is *generalization*. A generalization of an idea is a more abstract, but more useful form of it. In the best cases, a generalization allows one to see a familiar mathematical idea as one instance of a repeated pattern. Then one uses the experience with the concrete case to gain intuition about the general situation. Of course, there is a price to pay: one must be willing to work with abstract ideas, and also the details of a generalization are often subtly different than the original.

This chapter presents generalizations of several ideas from earlier in the book. First, we generalize the notion of a flat system of equations. The basic idea is that flat systems using a set A of atoms are related to the operator $\Gamma(a) = \mathcal{P}(A \cup a)$. When we generalize this, we get the analogue of flat systems for other operators.

The second generalization has to do with solving systems. In our discussion of streams, we saw that solving stream equations always gave us streams, and that this gave us powerful ways of defining operations on streams. In this chapter we generalize the method from streams to a wide class of monotone operators. Working out the details of this generalization will occupy us for most of the chapter.

16.1 Systems of equations as coalgebras

A flat system of equations $\mathcal{E} = \langle X, A, e \rangle$ is basically just a set X together with a map $e : X \to \mathcal{P}(X \cup A)$, where X is the set of indeterminates of $\mathcal{E}$ and A is the set of atoms. The Flat Solution Lemma formulation of *AFA* says that every such system has a solution. In Chapter 8, we used *AFA* to get solutions to more systems: we were able to solve systems of the form $\langle X, A, e \rangle$, where e might map into the class $V_{afa}[X \cup A]$. This is a common way of doing things in mathematics. We started with a simple form of our axiom, then motivated stronger forms, and proved that the stronger one is actually derivable given the original one. The key idea is that $V_{afa}[X \cup A]$ is the greatest fixed point of a certain operator, as we see in the next example.

Example 16.1 Let Γ be the operator $\Gamma(a) = \mathcal{P}(a \cup A)$. Consider also the operator Γ_X given by

$$\Gamma_X(a) \quad = \quad \Gamma(X \cup a),$$

so that $\Gamma_X(a) = \mathcal{P}(X \cup A \cup a)$. It is not difficult to see that $V_{afa}[X \cup A]$ is the greatest fixed point of Γ_X. The point of this example is that a *general* systems of equations is given by a set X of urelements and a map e from X into Γ_X^* and that all of these general systems have unique solutions, the solutions being drawn from Γ^*.

Exercise 16.1 Verify that $V_{afa}[X \cup A]$ is the greatest fixed point of Γ_X.

As a step on the road to generalization, we see the same kind of thing happen with streams.

Example 16.2 Fix some set A and consider the operator $\Gamma(a) = A \times a$. A flat system of stream equations over A is a general system $\langle X, e \rangle$, where $e : X \to \Gamma(X)$. All such systems have solutions, and the solution sets belong to $\Gamma^* = A^\infty$. A more general system allows parametric streams over X as the right sides of equations. The parametric streams over some set X of urelements form the largest collection H such that $H = A \times (X \cup H)$. Here is how we can view this in terms of fixed points. Consider the operator Γ_X given by the same formula we saw just before: $\Gamma_X(a) = \Gamma(X \cup a)$. The set of parametric streams is the greatest fixed point Γ_X^*. $\Gamma_X(a) = \Gamma(X \cup a)$. We saw in the Chapter 14 that every general system of stream equations has a unique solution, and that the solution set is included in A^∞. This "Solution Lemma for Streams" is what allowed us to define functions on streams by corecursion.

The work of this section generalizes these two examples. We first generalize the notion of flat system of equations and flat system of stream equations to monotone operators Γ. We then strengthen the results on flat systems to arbitrary systems in the manner suggested by the above results.

We begin by generalizing the notion of a flat system of equations. The name "coalgebra" used below comes from category theory. We want to explain this terminology. Although this discussion is not strictly needed for our later work, it should help make some of the definitions become less mysterious.

Background on algebras and homomorphisms Let's consider an "algebra." To be more concrete, let's consider an algebraic system $\mathfrak{A}$ making use of the symbols $+$ and $\times$. $\mathfrak{A}$ is *usually* given by a set A together with functions $+_{\mathfrak{A}}$ and $\times_{\mathfrak{A}}$ from A^2 to A. (As always, one drops the subscripts whenever possible.) But this is just one way to formalize the concept of an algebra. We want to have another look, from a more abstract perspective.

For any set A, let $T(A)$ be the set of triples consisting of an operation symbol and two elements of A:

$$T(A) \quad = \quad \{+, \times\} \times A \times A.$$

Once again, $+$ and $\times$ in the first set are simply objects being used as algebraic symbols.

So if $A = \{0, 5\}$, then $T(A)$ has elements like $\langle \times, 5, 0 \rangle$. When we do this, we are thinking of the elements of $T(A)$ (and indeed also those of A) as purely syntactic objects. As yet, we have no way of "cashing in" the expressions from $T(A)$ back into elements of A. For example, we can't say that $0 + 0$ is the same as 0. Every full-bodied algebraic system $\mathfrak{A}$ will give us a way to evaluate all the complex expressions from $T(A)$. And conversely, every way to interpret expressions (even just the binary ones like $0 + 5$) will give a map from $T(A)$ to A. Accordingly, we get a different way to formalize the concept of an algebra by *defining* an algebra to be a function $f : T(A) \to A$.

One of the main concepts of algebra is that of a *homomorphism*, or *structure-preserving* map. Continuing our discussion of algebraic systems using the symbols $+$ and $\times$, let's recall the concept of a homomorphism $h : \mathfrak{A} \to \mathfrak{B}$ between two such systems. It would be a map h from A to B with the property that if we take two elements a_1 and a_2 from A, then it makes no difference whether we (1) add (or multiply) them in A and then transfer via h to get an element of B, or (2) transfer them over to $h(a_1)$ and $h(a_2)$, and then add (or multiply) those in B. In symbols: for all $a_1, a_2 \in S, h(a_1 + a_2) = h(a_1) + h(a_2)$ and $h(a_1 \times a_2) = h(a_1) \times h(a_2)$.

This definition of homomorphism is suited to our first formalization of the concept. How would the definition go when we treat algebras as functions? Given algebras $f : T(A) \to A$ and $g : T(B) \to B$, and a map $h : A \to B$, we get a map from $T(A)$ to $T(B)$ by just applying h inwardly. For example, suppose again that $A = \{0, 5\}$, and now let $B = \{4, 23, \Omega\}$. Let $h(0) = 4$, and let $h(5) = \Omega$. Then our map on $T(A)$ would take $\langle \times, 5, 0 \rangle$ from $T(A)$ to

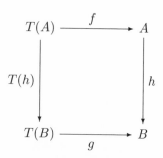

FIGURE 8 A homomorphism of algebras.

the element $\langle \times, \Omega, 4 \rangle$. In this way, given *any* function $h : A \to B$, we get a unique function from $T(A)$ to $T(B)$; we call this function $T(h)$.

The way to cast the notion of a homomorphism when we treat algebras as functions is easiest to explain using the diagram shown in Figure 8.

Asserting that a function $h : A \to B$ is a homomorphism of the algebras f and g is the same thing as saying that the diagram *commutes*. This means that given any expression $e \in T(A)$, it makes no difference whether we (1) evaluate it in A using f, and then use h to transfer to B; or (2) first use $T(h)$ to transfer the expression to $T(B)$, and then evaluate the result in B using g. More pictorially, if we take any $e \in T(A)$ and follow the arrows around the square to get an element of B, then it doesn't matter which of the two routes we take.

Flat coalgebras With this background, we return to our discussion at the beginning of this section. We considered two examples of operators Γ, and we noted that a flat system for each of these operators is given by a pair $\langle X, e \rangle$ consisting of a set X and a map $e : X \to \Gamma(X)$. This looks a little bit like the formalization of an algebra, except that the arrow is turned around, going from X to $\Gamma(X)$ instead of from $\Gamma(X)$ to X. In category theory, it is common to use the prefix "co" whenever one notion is obtained from another by reversing arrows. Thus we see that the notion of a system of equations is basically the "co" of the notion of an algebra, when viewed abstractly. This reversal is the source of the "*co*" in the definition of *coalgebra* below. The reader should think of coalgebras as a generalization of our notion of a system of equations to an arbitrary monotone operator Γ.[1]

Definition Let Γ be a monotone operator. A Γ-*coalgebra* is a pair $\mathcal{E} = \langle b, e \rangle$, where b is a set or class (possibly not a set or class of urelements), and a function $e : b \to \Gamma(b)$.

---

[1] Indeed, the reader is encouraged to read "system of Γ-equations" for "Γ-coalgebra" until the latter terminology becomes familiar.

Now we would like to take a *flat* coalgebra to be one where the set b is a subset of $\mathcal{U}$. As we have seen in the stream case, we must insure that the urelements involved have nothing to do with the innards of Γ.

Definition Let Γ be a monotone operator. A set $X \subseteq \mathcal{U}$ is *new for* Γ if for all substitutions t with domain X, and all sets a, $\Gamma(a[t]) = \Gamma(a)[t]$.

The idea is that Γ should commute with the action of all substitutions based on X.

Example 16.3 Let's look back at $\Gamma_1, \ldots, \Gamma_5$ from Section 15.1, and figure out which sets of urelements are new for these.

First, consider $\Gamma_1(a) = \mathcal{P}(a)$. We claim that every set $Y \subseteq \mathcal{U}$ is new for Γ_1. Fix a set a, and let t be a substitution with domain Y. We need to check that $\mathcal{P}(a[t]) = \mathcal{P}(a)[t]$. Let $b \in \mathcal{P}(a[t])$. Then $b \subseteq a[t]$. Let $b' = \{c' \in a \mid c'[t] \in b\}$. Then for every $c \in b$ there is some $c' \in a$ such that $c'[t] = c$. This means that $b'[t] = b$. Since $b' \in \mathcal{P}(a)$, $b \in \mathcal{P}(a)[t]$.

Conversely, let $b \in \mathcal{P}(a)[t]$. Then there is some $c \in \mathcal{P}(a)$ such that $b = c[t]$. Since $c \subseteq a$, $c[t] \subseteq a[t]$. Thus $b \in \mathcal{P}(a[t])$. We conclude that every set of urelements is new for Γ_1. Γ_2 works the same way.

For Γ_3, we showed in Proposition 14.3 that X is new for Γ_3 iff $X \cap support(A) = \emptyset$. (Actually, we defined new for A by the latter condition and then showed it was equivalent to the condition involving substitution.)

Recall that $\Gamma_4(a) = a \rightharpoonup a$, the set of partial functions from a to a. Γ_4 is monotone. We claim that no set of three distinct urelements, say $X = \{x, y, z\}$ can be new for Γ. To see this, let $a = \{x, y, z\}$. Let s be defined on $\{x, y, z\}$ by $s(x) = y$, $s(y) = y$, $s(z) = z$. Then $\Gamma_4(a)[s]$ contains sets like

$$\{\langle x, y \rangle, \langle y, z \rangle\}[s] \quad = \quad \{\langle y, y \rangle, \langle y, z \rangle\},$$

and the second set above isn't a function. So $\Gamma_4(a)[s] \neq \Gamma_4(a[s])$.

Incidentally, this example allows us to clear up something that might be confusing you about the definition just above. For a given operator Γ, we speak new *sets* of urelements not new urelements themselves. As Γ_4 shows, the newness is a property of sets. However, in most cases a set $X \subseteq \mathcal{U}$ is new for Γ if and only if each $\{x\} \subseteq X$ is new for Γ. This is true for all the operators that play the biggest roles in the rest of the book.

Exercise 16.2 What sets of urelements are new for Γ_5?

With this notion at our disposal, we can define the notion of a flat coalgebra.

Definition Let Γ be a monotone operator. A *flat* Γ-*coalgebra* is a coalgebra $\mathcal{E} = \langle X, e \rangle$ where $X \subseteq \mathcal{U}$ is a set of urelements that is new for Γ.

In other words, a flat Γ-coalgebra consists of a new set X of urelements and a function e that assigns to each $x \in X$ some Γ-form e_x over X. In particular, a flat $\mathcal{P}$-coalgebra is just a flat system of set equations since all sets of urelements are new. And a flat Γ_3-coalgebra (for the operator $\Gamma_3(a) = A \times a$) is just a flat system of stream equations over the set A, using the right kind of urelements. One of our main goals is to develop a theory of Γ-coalgebras in such a way as to generalize our earlier results on systems of equations.

16.2 Morphisms

The concept of homomorphism plays such an important role in algebra that it would be surprising if no corresponding concept came up when the arrows are turned around. Thus, since we have the notion of a Γ-coalgebra, it behooves us to ask about the definition of a Γ-morphism between such coalgebras.

Let's first think about the flat case. In general, homomorphisms make things simpler by identifying things that are distinct. So a morphism from a flat coalgebra $\mathcal{E} = \langle X, e \rangle$ to a flat coalgebra $\mathcal{E}' = \langle Y, e' \rangle$ should somehow show that $\mathcal{E}'$ is a simpler version of $\mathcal{E}$; the morphism should give us a simplification. Naturally, we want to get a definition by turning the arrows around in the figure concerning morphisms of algebras.

A Γ-morphism should be given by a function from $r : X \to Y$ with a certain commutativity property. In order to state this property, we need to know what map to put between $\Gamma(X)$ and $\Gamma(Y)$. Actually, there is only one natural choice. We know that r is a substitution. Hence the operation $[r]$ is defined on all of $V_{afa}[\mathcal{U}]$. So we take a restriction of $[r]$ to $\Gamma(X)$.

Definition For any substitution r and any set b, let $[r]_b$ be the restriction of the substitution operation $[r]$ to b, that is:

$$[r]_b = \{ \langle a, a[r] \rangle \mid a \in b \}.$$

Returning to our particular $r : X \to Y$, consider the restriction of $[r]$ to $\Gamma(X)$. The fact that X is a set of new urelements insures that $\Gamma(X)[r] = \Gamma(X[r])$. The definition of $[r]$ implies that $X[r] \subseteq Y$. So by monotonicity, $\Gamma(X)[r] = \Gamma(X[r]) \subseteq \Gamma(Y)$. Thus, $[r]_{\Gamma(X)}$ maps $\Gamma(X)$ into $\Gamma(Y)$.

This matches the situation of algebras and T, so we seem to be on the right track. (If $[r]$ didn't map $\Gamma(X)$ into $\Gamma(Y)$, then there would not be much of a reason to call $\mathcal{E}'$ a simplification of $\mathcal{E}$.) It is convenient to draw a picture at this point (and you should do this every time you have a morphism of coalgebras): see Figure 9. We should remark that the r on the arrow from X to Y could just as well have been $[r]_X$, since formally the substitution r is identical to $[r]_X$. Now it is clear that the definition should require that the diagram commute. This means that for all $x \in X$, $e_x[r]_{\Gamma(X)} = e'_{r(x)}$.

This definition looks reasonable. But having made it, we see that we have not used the fact that the set Y is new for Γ, or even that $\mathcal{E}'$ is flat. Accordingly,

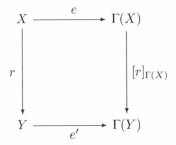

FIGURE 9 A Morphism of Flat Γ-coalgebras.

$\mathcal{E}'$ can be *any* Γ-coalgebra. What is more, we do not even really need that $X \subseteq \mathcal{U}$, since for any substitution r, the restriction $[r]_X$ of $[r]$ to X is a function defined on X.

We can summarize this discussion by means of the following official definition:

Definition Let Γ be a monotone operator, let $\mathcal{E} = \langle a, e \rangle$ and $\mathcal{E}' = \langle b, e' \rangle$ be Γ-coalgebras.

1. A Γ-*morphism* from $\mathcal{E}$ into $\mathcal{E}'$ is a substitution r such that
$$e' \circ [r]_a = [r]_{\Gamma(a)} \circ e.$$
   (That is, these are functions defined on a, and they are the same.) We write
$$r : \mathcal{E} \to \mathcal{E}'$$
   to indicate that r is a morphism.[2]

2. A *reorganization of* $\mathcal{E}_1$ *onto* $\mathcal{E}_2$ is a morphism r with the property that $a[r] = b$. Equivalently, a reorganization is a morphism that is surjective on the first component.

Here is an example and an exercise to illustrate these definitions.

Example 16.4 Let $A = \{p, q\} \subseteq \mathcal{U}$, and let $\Gamma(c) = \mathcal{P}(c \cup A)$. Consider two flat Γ-coalgebras $\mathcal{E}$ and $\mathcal{E}'$ given by

$$
\begin{aligned}
x &= \{p, y, z\} & v &= \{q, w\} \\
y &= \{q, z\} & w &= \{p, v, w\} \\
z &= \{p, x, y\} &
\end{aligned}
$$

Let r be the substitution that maps x and z to w and y to v. Then r is a reorganization of $\mathcal{E}$ onto $\mathcal{E}'$. (You should check this to make sure you understand the above definition.)

---

[2]Frequently we'll drop the "Γ" when it is fixed in a discussion.

Exercise 16.3 Suppose that r reorganizes $\mathcal{E} = \langle X, e \rangle$ onto $\mathcal{E}' = \langle b, e' \rangle$. Recall that $[r]$ is defined on all sets, in particular on the set $\mathcal{E}$ itself. What is $\mathcal{E}[r]$?

Exercise 16.4 Suppose that $\mathcal{E}$ and $\mathcal{E}'$ are flat Γ-coalgebras, as in Figure 9. Let $r : \mathcal{E} \to \mathcal{E}'$ be a Γ-morphism with domain X. Find an equation that holds between r, e and e', considered as substitutions. (Use the $\star$ operation of Section 8.4.)

16.3   Solving coalgebras

A flat Γ-coalgebra $\mathcal{E} = \langle X, e \rangle$ basically a system of equations, so we would like a notion of a solution. Assume first that each e_x is a set rather than an urelement. Then $\mathcal{E}$ is a general system of equations (see page 92), so $\mathcal{E}$ will indeed have a unique solution. In order for each e_x to be a set, it must be the case that $e : X \to V_{afa}[\mathcal{U}]$; i.e., that $\Gamma(X) \subseteq V_{afa}[\mathcal{U}]$. This point is so important that we turn it into a definition.

Definition An operator Γ is *proper* if for all sets a, $\Gamma(a) \subseteq V_{afa}[\mathcal{U}]$.

All of the example operators $\Gamma_1, \ldots, \Gamma_5$ are proper. On the other hand, the identity operator is not proper. (If a is a non-empty set of urelements, then $a \not\subseteq V_{afa}[\mathcal{U}]$.) Since $\mathcal{P}(V_{afa}[\mathcal{U}]) \subseteq V_{afa}[\mathcal{U}]$ (see Exercise 15.2), the proper operators are closed under composition.

Thus, for a proper operator Γ, we have a notion of solution: we say that a substitution s is a solution to $\mathcal{E}$ if $s_x = e_x[s]$ for all $x \in X$. This definition is fine, but it does not, at first sight, tell us anything about those solutions. In particular, it does not insure that the solutions take values in Γ^*. Luckily, everything works out as one would hope using the notion of Γ-morphism, as we shall see. First a definition, generalizing our notion of a canonical system of equations.

Definition Let Γ be a monotone operator. A Γ-coalgebra $\langle b, e \rangle$ is a *canonical coalgebra* if for all $c \in b$, $e_c = c$.

For example, if i is the identity on $\Gamma^* = \Gamma(\Gamma^*)$, then $\langle \Gamma^*, i \rangle$ is a canonical Γ-coalgebra. The definition clearly generalizes our notion of a canonical flat system of equations.

Lemma 16.1 *Let Γ be a monotone, proper operator, and let $\mathcal{E} = \langle X, e \rangle$ be a flat Γ-coalgebra. For any substitution s the following are equivalent:*

(1) s is the solution to $\mathcal{E}$: $dom(s) = X$, and for all $x \in X$, $s_x = e_x[s]$.
(2) $s \star e = s$.
(3) $dom(s) = X$, and $\mathcal{E}[s]$ is a canonical Γ-coalgebra.
(4) s reorganizes $\mathcal{E}$ onto some canonical Γ-coalgebra.

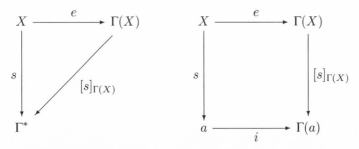

FIGURE 10 Equivalent views of solutions: parts (1) and (4).

(5) s is a Γ-morphism of $\mathcal{E}$ into $\langle \Gamma^*, i \rangle$.

Proof The overall plan of this proof is to show $(1) \implies (2) \implies (3) \implies (4)$ $\implies (5) \implies (1)$.

The proof that $(1) \implies (2)$ is contained in the proof of Theorem 8.5. We show $(2) \implies (3)$. First, $s \star e$ has the same domain as e, and this is X. Note next that $\mathcal{E}[s] = \langle X, e \rangle[s] = \langle X[s], e[s] \rangle$. For each $x \in X$,

$$\langle x, e_x \rangle[s] \quad = \quad \langle s_x, (s \star e)_x \rangle \quad = \quad \langle s_x, s_x \rangle.$$

Thus $e[s] = \{ \langle s_x, s_x \rangle \mid x \in X \}$ is identity map on $X[s]$. To finish the proof, we show that $\mathcal{E}[s]$ is a Γ-coalgebra; we need to see that $X[s] \subseteq \Gamma(X)[s]$. To see this, recall that $e : X \to \Gamma(X)$. So $X[e] \subseteq \Gamma(X)$. Here is where we use the newness of X for Γ:

$$X[s] \quad = \quad X[e][s] \quad \subseteq \quad \Gamma(X)[s] \quad = \quad \Gamma(X[s]).$$

Continuing, assume (3). Let s be such that $\langle X, e \rangle[s]$ is a canonical Γ-coalgebra. That is, $\langle X[s], e[s] \rangle$ consists of a set $X[s]$ together with the identity map on that set. We repeat: $e[s]$ is the identity map on $X[s]$. We need to check that s is a reorganization. This means that $[s]_{\Gamma(X)} \circ e = e[s] \circ [s]_X$. (Draw a picture like Figure 9.)

We need to do some preliminary work on this. What does it mean to say that $e[s]$ is the identity map on $X[s]$? Each element of e is of the form $\langle x, e_x \rangle$. So for each $x \in X$, $\langle x, e_x \rangle[s]$ belongs to the identity map on $X[s]$. Therefore $s_x = e_x[s]$ for all $x \in X$.

Now we can show that $[s]_{\Gamma(X)} \circ e = e[s] \circ [s]_X$. On the left we have a function with domain X. For each $x \in X$, $e_x \in \Gamma(X)$. So applying the left side to x gives $e_x[s] = s_x$. On the right, we have the function $e[s]$ applied to $x[s]_X = s_x$. But as we know, $e[s]$ is the identity map. So we get s_x for both sides. This concludes the proof of (4).

Next, we show that $(4) \implies (5)$. Assume (4). Then b is the solution set of $\mathcal{E}$. By Theorem 15.3, $b \subseteq \Gamma^*$. The inclusion map gives a Γ-morphism of $\mathcal{E}'$

into Γ^*, and the composition of s with this gives a Γ-morphism into Γ^*. Going back, if $s : \mathcal{E} \to \Gamma^*$, then the image $X[s]$ gives a canonical coalgebra, say $\mathcal{E}'$; s reorganizes $\mathcal{E}$ onto $\mathcal{E}'$.

Finally, (5) $\Longrightarrow$ (1) is a simple consequence of the definitions. ⊣

Alternatives (1) and (4) of the Lemma are pictured in Figure 10. (Alternative (1) is on the left, and (5) has been incorporated to stress the fact that solutions take their values in Γ^*.) On the other hand, sometimes an appeal to version (2) or (3) leads to a short proof of some fact. We'll see this several times in the rest of the book. If you prefer, you can always draw a commutative diagram to get the ideas behind a more algebraic-looking proof.

Here's one immediate consequence of Lemma 16.1 that's worth noting.

Proposition 16.2 (Solution Lemma for Operators) *Let Γ be a proper, monotone operator, and let $\mathcal{E} = \langle X, e \rangle$ be a flat Γ-coalgebra. Then $\mathcal{E}$ has a unique solution and solution-set$(\mathcal{E}) \subseteq \Gamma^*$.*

The next result shows a further connection between the concepts of reorganization and solution.

Proposition 16.3 *Let Γ be a proper operator, let $\mathcal{E} = \langle X, e \rangle$ and $\mathcal{E}' = \langle Y, e' \rangle$ be flat Γ-coalgebras, and let $r : \mathcal{E} \to \mathcal{E}'$ be a Γ-morphism with domain X. Let s be the solution to $\mathcal{E}'$. Then $s \circ r$ is the solution to $\mathcal{E}$. Hence if r is a reorganization, then $\mathcal{E}$ and $\mathcal{E}'$ have the same solution sets.*

Proof We use the notation and results of Section 8.4. By Exercise 16.4, $r \star e = e' \star r$. Now since s is the solution to $\mathcal{E}'$, $s \star e' = s$. Note that e and r both have domain X. Also, s and e' have Y as their domain. Thus by Exercise 8.8,

$$
\begin{aligned}
(s \star r) \star e &= s \star (r \star e) \\
&= s \star (e' \star r) \\
&= (s \star e') \star r \\
&= s \star r.
\end{aligned}
$$

This just says that $s \star r = s \circ r$ is the solution to $\mathcal{E}$.

For the second statement, assume r is surjective. Then the solution set of $\mathcal{E}$ is

$$
X[s \star r] = \{r_x[s] \mid x \in X\} = \{y[s] \mid y \in Y\},
$$

and this is the solution set of $\mathcal{E}'$. ⊣

Exercise 16.5 One of our earlier worries concerned the operator $\Gamma(b) = A \times b$, when A is some fixed subset of $\mathcal{U}$. As it turns out, we have avoided a potential problem by the use of a delicate point in our definitions. To understand this, consider the following argument:

Let $a \in A$, and consider the flat Γ-coalgebra $a = \langle a, a \rangle$. (Technically, this

would be
$$\mathcal{E} = \langle \{a\}, \{\langle a, \langle a, a \rangle \rangle \} \rangle,$$
but of course we seldom write such coalgebras out explicitly.) Now the solution to $\mathcal{E}$ is clearly Ω. By Lemma 16.2, the solution must be a *stream* s over A. But this means that there is a stream s with the property that $s = \langle s, s \rangle$.

Find the error in this reasoning.

16.4   Representing the greatest fixed point

We can now relate the material from Chapter 15 to the previous two sections. The main result shows that the greatest fixed point Γ^* is in fact the set of solutions to flat Γ-systems. This is stated more precisely as Theorem 16.6 below. In order to prove this result, though, we'll need to establish some preliminary facts first, and we'll need to restrict to "uniform" operators, operators satisfying one additional natural condition.

Definition An operator Γ *commutes with almost all substitutions* if almost all sets of urelements are new for Γ, that is, if that there is a *set* $X_\Gamma \subseteq \mathcal{U}$ such that for all sets $Y \subseteq \mathcal{U}$ with $Y \cap X_\Gamma = \emptyset$, the set Y is new for $\mathcal{U}$. We call X_Γ *an avoidance set for* Γ in this case.[3] Γ is *uniform* if

(1) Γ is monotone

(2) Γ is proper

(3) Γ commutes with almost all substitutions.

The thrust of this definition is that a uniform operator should commute with substitutions "nearly all" of the time. The exceptions should form a set. All of our examples are uniform, except for Γ_4.

Example 16.5 The operators Γ_1, Γ_2, and Γ_3 are all uniform. In Example 16.3 we determined which sets were new for each of these. Γ_5 is also uniform. This follows from Exercise 16.2. Our work on page 227 shows that Γ_4 is not uniform.

Proposition 16.4 (Closure conditions on uniform operators)

1. *If a is a set of sets, then the constant operation $\Gamma(b) = a$ is uniform.*

2. *If Γ is monotone and commutes with almost all substitutions and Δ is uniform, then $\Delta \circ \Gamma$ is uniform. So if Γ and Δ are both uniform operators, then so is $\Delta \circ \Gamma$.*

---

[3]It does not follow from this definition that if Γ has an avoidance set then it has a smallest avoidance set. However, in all cases of interest in this book, Γ will have a smallest avoidance set. For example, if Γ commutes with all unions, then Γ will have a smallest avoidance set.

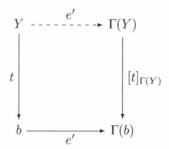

FIGURE 11 Obtaining one coalgebra from another using a reorganization.

3. *If Γ and Δ are uniform operators, then so is $\Gamma \times \Delta$, where*

$$(\Gamma \times \Delta)(b) \quad = \quad \Gamma(b) \times \Delta(b).$$

4. *If Γ and Δ are uniform operators, then so is $\Gamma \cup \Delta$, where*

$$(\Gamma \cup \Delta)(b) \quad = \quad \Gamma(b) \cup \Delta(b).$$

Exercise 16.6 Prove this proposition.

Exercise 16.7 Consider the operator $\Gamma(a) = a \times a$. What are the fixed points Γ_* and Γ^*? Is Γ uniform?

Exercise 16.8 Prove that the operation $\Gamma(a) = \{a\}$ commutes with all substitutions but is not monotone, and hence not uniform.

Exercise 16.9 Let Γ be uniform, say with $X_\Gamma \subseteq \mathcal{U}$ as an avoidance set. Let a be any set. Prove that

$$support(\Gamma(a)) - X_\Gamma \quad \subseteq \quad support(a) - X_\Gamma.$$

Specializing to the case of a set $X \subseteq \mathcal{U}$, we see that $support(\Gamma(X)) \subseteq X \cup X_\Gamma$. (While we only use this result several more times in the book, we have found it to be a key observation and so recommend it for your consideration.)

The import of our definitions of newness and uniformity come out in the next two results. Of particular importance is the Representation Theorem 16.6 below.

Proposition 16.5 *Let $\mathcal{E} = \langle b, e \rangle$ be a Γ-coalgebra, let Y be a set of new urelements for Γ, and let $t : Y \to b$ be a surjective map. Then there is a map $e' : Y \to \Gamma(Y)$ such that t reorganizes $\langle Y, e' \rangle$ onto $\mathcal{E}$.*

Proof See Figure 11. First, since t is surjective, $Y[t] = b$. That is, the left-hand arrow is surjective. We claim that the right-hand arrow too is surjective.

This is because Y is new for Γ. That is, the image is $\Gamma(Y)[t]$ and this is $\Gamma(Y[t]) = \Gamma(b)$.

Now we turn to the definition of e'. For each $y \in Y$, $e_{t(y)}$ is some element of $\Gamma(b)$. By surjectivity, there is some $a \in \Gamma(Y)$ such that $a[t] = e_{t(y)}$. This a might depend on y, of course, and there might well be more than one of them.[4] Using the Axiom of Choice, we get a function $e' : Y \to \Gamma(Y)$ such that for all $y \in Y$, $e'_y[t] = e_{t(y)}$. But this is exactly the statement that t is a Γ-morphism. Since $t : Y \to b$ is surjective, t is a reorganization. $\dashv$

Theorem 16.6 (Representation Theorem for Γ^*) *If Γ is uniform then*

$$\Gamma^* = \bigcup \{\textit{solution-set}(\mathcal{E}) \mid \mathcal{E} \textit{ a flat } \Gamma\textit{-coalgebra}\}.$$

Proof Let G be the union of all solution sets of flat Γ-coalgebras $\mathcal{E} = \langle X, e \rangle$. We want to prove that $\Gamma^* = G$. The inclusion $G \subseteq \Gamma^*$ is an immediate consequence of the Solution Lemma for Operators (Lemma 16.2). To prove that $\Gamma^* \subseteq G$, we use the characterization of Γ^* as the union of the family of Γ-correct sets (Theorem 15.3). Let b be a set so $b \subseteq \Gamma(b)$; we show that $b \subseteq G$. Since almost all sets of urelements are new for Γ, there is some such set $X \subseteq \mathcal{U}$ that is new for Γ and in bijective correspondence with b. Fix a bijection $t : X \to b$.

Let i be the inclusion of b in $\Gamma(b)$, and consider the canonical coalgebra $\mathcal{E} = \langle b, i \rangle$. By Proposition 16.5, there is some $e' : X \to \Gamma(X)$ such that t reorganizes $\mathcal{E}' = \langle X, e' \rangle$ onto $\mathcal{E}$. Since t is surjective, $\textit{solution-set}(\mathcal{E}') = b$. So $b \subseteq G$, as desired. $\dashv$

Exercise 16.10 Prove that the Representation Theorem is false for the operator $\Gamma_4(a) = a \rightharpoonup a$. This shows that monotonicity is not enough for the Representation Theorem; some condition like uniformity is needed.

16.5 The Solution Lemma Lemma

Just as we generalized the Flat Solution Lemma to the General Solution Lemma, we can generalize the solution lemma for flat coalgebras to general coalgebras. But first, we need to define the notion of a "general" Γ-coalgebra.

Definition Let Γ be a uniform operator and let $X \subseteq \mathcal{U}$.

1. The operator Γ_X is defined by

$$\Gamma_X(a) = \Gamma(X \cup a).$$

---

[4]This fact, that a is not unique, leads to our consideration of "smooth" operators in the next chapter.

We let $\Gamma^*[X]$ be the greatest fixed point of Γ_X; its members are called *parametric Γ-objects over X*.

2. A *general Γ-coalgebra* is a pair $\langle X, e \rangle$ consisting of a set X of urelements that is new for Γ, and a map $e : X \to \Gamma^*[X]$.[5]

Example 16.6 Two examples have appeared in earlier chapters. Consider first the operator $\Gamma_1(a) = \mathcal{P}(a)$. $\Gamma_1^*[X] = V_{afa}[X]$ (see page 224). So a general Γ_1-coalgebra is a pair $\langle X, e \rangle$, and $e : X \to V_{afa}[X]$. This is exactly a general system of equations in the sense of Chapter 8, but where the set A of atoms is empty.

Next consider $\Gamma_3(a) = A \times a$. We studied $\Gamma_3^*[X]$ in Chapter 14; it is the set of parametric A-streams over X. A general Γ_3-coalgebra is a general system of equations whose set of indeterminates X is disjoint from $support(A)$, and whose right-hand sides are parametric A-streams over X.

A general Γ-coalgebra $\mathcal{E}$ is yet another general system of equations of a special kind. As such, we already know what it means to be a solution to $\mathcal{E}$, and moreover that $\mathcal{E}$ has a unique solution. The main point is to show that *solution-set*$(\mathcal{E})$ must be included in Γ^*.

Theorem 16.7 (The Solution Lemma Lemma) *If Γ is a uniform operator, then every general Γ-coalgebra $\mathcal{E}$ has a unique solution and*

$$\textit{solution-set}(\mathcal{E}) \quad \subseteq \quad \Gamma^*.$$

This is the main result of this section. We'll prove it by way of another lemma. That preliminary result deals with decomposing one big system into several smaller ones. For example, consider the following system $\mathcal{E}$

$$
(1) \quad
\begin{aligned}
x_1 &= \{1, x_1, x_2, y_1\} \\
x_2 &= \{2, x_1, y_1, y_2\} \\
y_1 &= \{3, x_2, y_1, y_2\} \\
y_2 &= \{4, y_1\}
\end{aligned}
$$

Suppose we take the subsystem consisting of the bottom two equations, and solve it as it is. We would get a substitution, say t, defined on $\{y_1, y_2\}$ such that

$$
(2) \quad
\begin{aligned}
t(y_1) &= \{3, x_2, t(y_1), t(y_2)\} \\
t(y_2) &= \{4, t(y_1)\}
\end{aligned}
$$

Note that x_2 is in the support of both $t(y_1)$ and $t(y_2)$. Now let's go back and apply $[t]$ to the right sides of the top two lines of (1). We get a new system that

---

[5]This terminology is a bit misleading since a general Γ-coalgebra is not in general a Γ-coalgebra. It might be better to think of a general*ized* coalgebra.

we'll call $\mathcal{E}_1$:

$$x_1 \quad = \quad \{1, x_1, x_2, t(y_1)\}$$
$$x_2 \quad = \quad \{2, x_1, t(y_1), t(y_2)\}$$

In doing the applications, we used the fact that each x_i is fixed by $[t]$. Let s' be the solution to $\mathcal{E}_1$. So s' is defined on $\{x_1, x_2\}$ and

$$s'(x_1) \quad = \quad \{1, s'(x_1), s'(x_2), t(y_1)[s']\}$$
$$s'(x_2) \quad = \quad \{2, s'(x_1), t(y_1)[s'], t(y_2)[s']\}$$

Notice that we cannot say $t(y_1)[s'] = t(y_1)$; the fact that $x_2 \in support(t(y_1))$ means that this would be false. We really are interested in using t and s' to get the solution to the original system $\mathcal{E}$ given in (1). Call this solution s. We claim that s is s' on $\{y_1, y_2\}$, and $s(y_i) = t(y_i)[s]$. To verify this means to show that

$$s'(x_1) \quad = \quad \{1, s'(x_1), s'(x_2), t(y_1)[s']\}$$
$$s'(x_2) \quad = \quad \{2, s'(x_1), t(y_1)[s'], t(y_2)[s']\}$$
$$t(y_1)[s'] \quad = \quad \{3, s'(x_2), t(y_1)[s'], t(y_2)[s']\}$$
$$t(y_2)[s'] \quad = \quad \{4, t(y_1)[s']\}$$

We have already seen the first two lines. We get the last two by taking (2) and applying $[s']$ to both sides.

Here is a general result based on this example.

Lemma 16.8 *Let Γ be a uniform operator. Let X and Y be disjoint sets of urelements that are new for Γ. $\mathcal{E} = \langle X \cup Y, e \rangle$ be a Γ-coalgebra, and let s be its solution. Let $\mathcal{E}_0 = \langle Y, e \upharpoonright Y \rangle$ be the general system of equations obtained by restricting e to Y, and let t be the solution to $\mathcal{E}_0$. Let $\mathcal{E}_1 = \langle X, e_1 \rangle$, where $e_1(x) = e(x)[t]$. Let s' be the solution to $\mathcal{E}_1$. Then s is given by*

$$s(z) \quad = \quad \begin{cases} s'(z) & \text{if } z \in X \\ t(z)[s'] & \text{if } z \in Y \end{cases}$$

In particular, the solution to $\mathcal{E}_1$ is the restriction to X of the solution to $\mathcal{E}$.

Proof For all $y \in Y$ we have $t_y = e_y[t]$. Also, for $x \in X$ we have $s'_x = e(x)[t][s']$. We first claim that for all $z \in X \cup Y$, $s_z = z[t][s']$. This is clear for $z \in Y$. For $z \in X$ it follows because t fixed the elements of X.

Let Z be an avoidance set for Γ. For all $z \in Z$, $s_z = z[t][s']$ as well. By Exercise 8.3, the two substitution-like operations $[s]$ and $[t][s']$ agree on $V_{afa}[X \cup Y \cup Z]$. Now $\Gamma(X \cup Y) \subseteq V_{afa}[X \cup Y \cup Z]$ by Exercise 16.9. So for all $z \in X \cup Y$, $e_z[s] = e_z[t][s']$. For $x \in X$ we have $s_x = x[t][s'] = s'_x = e_x[s]$. And for $y \in Y$ we have $s_y = y[t][s'] = e_y[t][s'] = e_y[s]$. This proves that s is the solution of $\mathcal{E}$. $\dashv$

Proof of the Solution Lemma Lemma We are going to use the result of Lemma 16.8, *except in the reverse order* to that which we have seen. We begin with a general Γ-coalgebra $\mathcal{E}_1 = \langle X, e_1 \rangle$. Recall that this means that

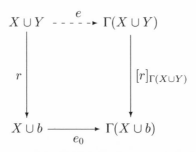

FIGURE 12 Proof of the Solution Lemma Lemma

$e_1 : X \to \Gamma^*[X]$. By Lemma 15.4, we can find a set $b \subseteq \Gamma^*[X]$ such that $e[X] \subseteq \Gamma_X(b) = \Gamma(X \cup b)$ and $b \subseteq \Gamma(X \cup b)$. Let Y be a set of urelements such that $Y \cap support(b) = \emptyset$, $Y \cap X = \emptyset$, Y is new for Γ, and Y is in correspondence with b. Fix a bijection $t : Y \to b$. Let $r : X \cup Y \to X \cup b$ be given by

$$r(z) \;=\; \begin{cases} z & \text{if } z \in X \\ t(z) & \text{if } z \in Y \end{cases}$$

Then $r : X \cup Y \to X \cup b$ is a bijection. Consider the following map $e_0 : X \cup b \to \Gamma(X \cup b)$:

$$e_0(z) \;=\; \begin{cases} e_z & \text{if } z \in X \\ z & \text{if } z \in b \end{cases}$$

Notice that e_0 really does map into $\Gamma(X \cup b)$; this is because e maps into $\Gamma(X) \subseteq \Gamma(X \cup b)$, and $b \subseteq \Gamma(X \cup b)$ by its definition. This gives a Γ-coalgebra $\mathcal{F} = \langle X \cup b, e_0 \rangle$. Applying Proposition 16.5 to $\mathcal{F}$ and r we obtain a Γ-coalgebra $\mathcal{E} = \langle X \cup Y, e \rangle$ such that r reorganizes $\mathcal{E}$ onto $\mathcal{F}$ (see Figure 12).

At this point we have the big coalgebra $\mathcal{E}$ using $X \cup Y$ as its indeterminates. Further, we have the restriction $\mathcal{E}_0$ based on Y. We claim that t is the solution to $\mathcal{E}_0$. Let $y \in Y$. Then $e_0(y) = y$ and $y[t] = y[r]$. So $y[t] = y[r] = e_0(y)[r] = e(y)[r]$. The reorganization condition tells us that $e(y)[r] = e(y[r])$. This last is $e(y[t])$. This means that t is indeed the solution to $\mathcal{E}_0$.

Recall that in Lemma 16.8 we also consider a coalgebra $\mathcal{E}_1$ defined using X as a set of indeterminates and defined by $e_1(x) = e(x)[t]$. The coalgebra $\mathcal{E}_1$ is exactly the one we started with: if $x \in X$, $e(x)[t] = e_0(x) = e(x)$. This is what we meant by working with Lemma 16.8 in reverse. In any case the Lemma tells us that the solution to $\mathcal{E}_1$ is the restriction to X of the solution s of $\mathcal{E}$. But $\mathcal{E}$ is flat coalgebra, so $\mathcal{E}$ maps into Γ^* by Theorem 16.6. This completes the proof.

We close this section with an exercise that applies the results of this section.

Although interesting, it is not used in the rest of the book. Furthermore, the results of the following section are also not used.

Exercise 16.11 Let Γ be a uniform operator, and let Δ be given by $\Delta(b) = \Gamma(\Gamma(b))$.

1. Check that Δ is uniform.
2. Prove that Γ and Δ have the same least and greatest fixed points.
3. Give an example of a uniform Γ, and a fixed point C of the associated Δ, such that C is not a fixed point of Γ. (This may not be easy to come by.)

16.6 Allowing operations in equations

Cantor's Theorem tells us that the powerset $\mathcal{P}(c)$ of any set c has larger cardinality than the set c itself. This implies, of course, that there can be no set c that is a solution to the equation $c = \mathcal{P}(c)$. Somewhat surprisingly, closely-related equations can be solved. For example, we can show that there is a set $c = \{\mathcal{P}(c)\}$.

This equation does not immediately fall under Theorem 8.2, however, because of the occurrence of the powerset operation symbol $\mathcal{P}$ on the right-hand side. But this can be circumvented. To see how, suppose there were such a set c. Let $b = \mathcal{P}(c)$, so that $c = \{\mathcal{P}(c)\} = \{b\}$. But then it follows that $b = \mathcal{P}(\{b\}) = \{\emptyset, \{b\}\}$. The Solution Lemma shows that there is indeed a set b satisfying *this* equation. But then its singleton $c = \{b\}$ satisfies the original equation (check this if it is not obvious).

Exercise 16.12 Generalize the preceding construction by showing that for any set a there is a set $c = \{\mathcal{P}(a \cup c)\}$.

We can generalize this idea to show that we can allow operations in systems of equations, as long as two restrictions are met: (1) all the operations must be "guarded" by some set brackets, and (2) the operations involved must commute with almost all substitutions. To see a condition is necessary, try the next exercise.

Exercise 16.13 Let $f(a) = \bigcup a$.

1. Show that F does not commute with almost all substitutions.
2. Show that $x = \{f(x)\}$ has a proper class of solutions.

Now we turn to the main work of this section.

Definition Let $\mathcal{F}$ be some set of set-theoretic operations, and assume that each $f \in \mathcal{F}$ comes with some natural number $arity(f)$. Each $f \in \mathcal{F}$ is an operation on sets, and we also assume that we have an urelement f^* associated with it.

Let $f \in \mathcal{F}$. An $\mathcal{F}$-*term* is a tuple $\langle f^*, x_1, \ldots, x_n \rangle$, where $f \in \mathcal{F}$, $n = arity(f)$ and $x_1, \ldots, x_n$ are any sets or urelements.

We want to define an operation $[s]_{\mathcal{F}}$ of $\mathcal{F}$-substitution. The basic idea is that $\mathcal{F}$-substitution should behave like the substitution $[s]$ that we have seen, except when we encounter an $\mathcal{F}$-term $\langle f^*, x_1, \ldots, x_n \rangle$. In this case we should not use $[s]$ but rather compute $x_1[s]_{\mathcal{F}}$, ..., $x_n[s]_{\mathcal{F}}$ and then apply the operator f to these values. (If we used the usual substitution $[s]$ on $\langle f^*, x_1, \ldots, x_n \rangle$, we would get $\langle f^*[s], x_1[s], \ldots, x_n[s] \rangle$ that is not at all what we want.) This is justified by the following result.

Proposition 16.9 *Let s be a substitution. Then there is a unique operation $[s]_{\mathcal{F}}$ such that $[s]_{\mathcal{F}}$ agrees with s on urelements, and for all sets a,*

$$a[s]_{\mathcal{F}} = \{ f(x_1[s]_{\mathcal{F}}, \ldots, x_n[s]_{\mathcal{F}}) \mid \langle f^*, x_1, \ldots, x_n \rangle \text{ is an } \mathcal{F}\text{-term in } a \}$$
$$\cup \{ b[s]_{\mathcal{F}} \mid b \in a \text{ is not an } \mathcal{F}\text{-term} \}$$

Proof This is very much like the proof of the existence and uniqueness of sub that we saw in Chapter 8. ⊣

We need the following lemma in order to define the a uniform operator $\Gamma_{\mathcal{F}}$ to go along with $\mathcal{F}$:

Lemma 16.10 *If each $f \in \mathcal{F}$ commutes with almost all substitutions, then almost all sets $X \subseteq \mathcal{U}$ are new for each $f \in \mathcal{F}$.*

Proof Since $\mathcal{F}$ is a set, we can find a set Y so that for all X disjoint from Y, all substitutions defined on X, all sets a, and all $f \in \mathcal{F}$, $f(a)[s] = f(a[s])$. ⊣

We are now in a position to define an operator $\Gamma_{\mathcal{F}}$.

Definition Let $\mathcal{F}$ be a set of operators, and let $\mathcal{F}^*$ be a corresponding set of urelements new for each $f \in \mathcal{F}$. The operator $\Gamma_{\mathcal{F}}$ is defined as follows:

$$\Gamma_{\mathcal{F}}(a) = \{ \langle f^*, b_1, \ldots, b_n \rangle : f \in \mathcal{F} \ \& \ arity(f) = n \ \& \ b_1, \ldots, b_n \in a \}.$$

A *flat $\Gamma_{\mathcal{F}}$-coalgebra* is a pair $\mathcal{E} = \langle X, e \rangle$, where $e : X \to \mathcal{P}(\Gamma_{\mathcal{F}}(X))$. An *$\mathcal{F}$-solution* to $\mathcal{E}$ is a substitution s defined on X so that for all $x \in X$, $s_x = e_x[s]_{\mathcal{F}}$.

To see that we are heading in the right direction, let's consider an example.

Example 16.7 Suppose we want sets a and b that satisfy the following equations:

$$\begin{aligned} a &= \{ \langle c, b \rangle \} \\ b &= \{ \mathcal{P}(a) \} \end{aligned}$$
(3)

Here c is some fixed set or urelement. We take the operators f and $\mathcal{P}$, where $f(a') = \langle c, a' \rangle$ for all sets a'. These operators commute with almost all substitutions. Let $X = \{ x, y \}$ be a set of urelements that is new for f and $\mathcal{P}$ operators. (This just means that neither of these belongs to $support(c)$.) Let

$\mathcal{F}^* = \{f^*, \mathcal{P}^*\}$ be a set of new urelements. Then $\Gamma_{\mathcal{F}}(X)$ contains the $\mathcal{F}$-terms $\langle f^*, y \rangle$ and also $\langle \mathcal{P}^*, x \rangle$. It follows that the following is a flat $\Gamma_{\mathcal{F}}$-coalgebra:

$$
\begin{aligned}
x &= \{\langle f^*, y \rangle\} \\
y &= \{\langle \mathcal{P}^*, x \rangle\}
\end{aligned}
\tag{4}
$$

Call this flat $\Gamma_{\mathcal{F}}$-coalgebra $\mathcal{E}$. An $\mathcal{F}$-solution for $\mathcal{E}$ will give us exactly what we are after.

Theorem 16.11 ($\mathcal{F}$-Solution Lemma) *Suppose that every $f \in \mathcal{F}$ commutes with almost all substitutions. Then every flat $\Gamma_{\mathcal{F}}$-coalgebra has a unique $\mathcal{F}$-solution.*

Proof We illustrate the proof by showing how to solve the $\Gamma_{\mathcal{F}}$-coalgebra described in (4) above. Let v and w be new urelements. The idea is that x stands for the a in (3) and y for the b. Then v stands for $f(b)$ and w stands for $\mathcal{P}(a)$. So we consider the following general system $\mathcal{E}'$:

$$
\begin{aligned}
x &= \{v\} \\
y &= \{w\} \\
v &= f(e'_y) = \langle c, e'_y \rangle = \langle c, \{w\} \rangle \\
w &= \mathcal{P}(e'_x) = \mathcal{P}(\{v\}) = \{\emptyset, \{v\}\}
\end{aligned}
$$

That is, we get e'_x and e'_y directly, and we use these together with the functions involved to get e'_v and e'_w.

This system $\mathcal{E}'$ has a solution, call it t. Let s be the restriction of t to X. We claim that s is an $\mathcal{F}$-solution to the original $\mathcal{E}$. This is where we use the fact that we used indeterminates that are new for our operators. Indeed, the newness implies that $f(e'_y)[t] = f(e_y[t]) = f(t_y)$ and also that $\mathcal{P}(e'_x)[t] = \mathcal{P}(e'_x[t]) = \mathcal{P}(t_x)$. So t gives us sets with the following properties:

$$
\begin{aligned}
t_x &= \{t_v\} \\
t_y &= \{t_w\} \\
t_v &= f(t_y) \\
t_w &= \mathcal{P}(t_x)
\end{aligned}
$$

So we have $t_x = \{f(t_y)\}$ and $t_y = \{\mathcal{P}(t_x)\}$. This means that t_x and t_y satisfy the equations for a and b in (3).

Turning to s, we see that

$$
s_x = \{f(s_y)\} = \{\langle f^*, y \rangle [s]_{\mathcal{F}}\} = \{\langle f^*, y \rangle\}[s]_{\mathcal{F}} = e_x[s]_{\mathcal{F}}.
$$

Similarly, $s_y = e_y[s]_{\mathcal{F}}$. These two facts tell us that s is an $\mathcal{F}$-solution to $\mathcal{E}$. This concludes our sketch of the proof of the $\mathcal{F}$-Solution Lemma. ⊣

The Flat $\mathcal{F}$-Solution Lemma gives us solutions to many interesting equations. For example, suppose we want to find a and b to solve the following modification of (3):

$$(5) \qquad \begin{aligned} a &= \{\langle c, b \rangle, \Omega\} \\ b &= \{\mathcal{P}(\mathcal{P}(a) \cup \{c\})\} \end{aligned}$$

At first glance, we seem to be out of luck. One problem is that this system involves Ω and this is not like any term we have seen. However, we can take a 0-ary function f_Ω whose value is Ω. So for all substitutions s, $\langle f_\Omega^* \rangle [s]_{\mathcal{F}} = \Omega$. To contend with $\mathcal{P}(\mathcal{P}(a) \cup \{c\})$, we note that as a function of a, it does commute with almost all substitutions. Call this operator g. We take as operators f (from the proof above), f_Ω, and g. Let x and y be new for these operators. Then (5) corresponds to the Flat $\mathcal{F}$-coalgebra

$$\begin{aligned} x &= \{\langle f^*, y \rangle, \langle f_\Omega^* \rangle\} \\ y &= \{\langle g^*, x \rangle\} \end{aligned}$$

Exercise 16.14 Solve the following:

1. $x = \{\mathcal{P}(\mathcal{P}(x))\}$. Is the solution to this the same as the solution to $x = \{\mathcal{P}(x)\}$?

2. $x = \{x \times \{p, q\}\}$, where p and q are distinct urelements. How many elements are in $TC(\{x\})$?

Historical Remarks

A number of people have formulated principles of coinduction in recent years. Some sources are Fiore (1993), Pitts (1993), and Rutten (1993). Also, Rutten and Turi (1994) have a development of a "coalgebraic" view very similar to the one we are advancing in this part of the book.

There are results similar to the Solution Lemma Lemma in Aczel (1988) and Devlin (1993), but with somewhat different formulations. Our proof was influenced by conversations with Glen Whitney.

The $\mathcal{F}$-Solution Lemma of Section 16.6 has not been published before. It emerged a few years ago in a discussion among Peter Aczel and us.

17

Corecursion

Unsheathe your dagger definitions.

James Joyce, *Ulysses*

Recursive definitions allow one to define a mapping from the least fixed point Γ_* of some monotone operator into some set or class C. Thanks to the work of many logicians during the past century such definitions are now well understood. Corecursive definitions turn things on their head. The goal of a corecursive definition is to define a mapping from some set or class C into the greatest fixed point Γ^* of Γ. Such definitions are far less well understood. The goal of the chapter is to investigate tools that permit one to define functions by corecursion.

In Chapter 14, we investigated functions defined by corecursion into the set A^∞ of streams over A. Given any functions $G : C \to A$ and $H : C \to C$ we saw that there is a unique function $\varphi : C \to A^\infty$ satisfying the following for all $c \in C$:

(1)
$$\varphi(c) \quad = \quad \langle G(c), \varphi(H(c)) \rangle$$

A prime example of this is the iteration function $iter_f$. Given any function $f : A \to A$, we saw that there is a unique $iter_f : A \to A^\infty$ satisfying

$$iter_f(a) \quad = \quad \langle f(a), iter_f(f(a)) \rangle.$$

In order to generalize corecursion from streams to other greatest fixed points, it is useful to observe that we can pack these two functions G and H into a single function $\pi : C \to A \times C$; let $\pi(c) = \langle G(c), H(c) \rangle$. Notice that this function is basically a coalgebra for the operator in question. The

Corecursion Principle for Streams tells us there is a unique $\varphi : C \to A^\infty$ satisfying the equation

$$\varphi(c) \;\;=\;\; \langle 1^{st}\pi(c), \varphi(2^{nd}\pi(c))\rangle.$$

As we know, the collection A^∞ is the greatest fixed point of the uniform operator $\Gamma(a) = A \times a$. In this chapter we want to show how to extend the corecursion method to define functions into the greatest fixed points of other operators. For an example, intuitively we should be able to define a function $\varphi : A^\infty \to Can_A$ that assigns to each stream s a canonical state over A given as follows:

$$\varphi(s) \;\;=\;\; \{\langle 1^{st}s, \varphi(2^{nd}2^{nd}s)\rangle, \langle 1^{st}2^{nd}s, \varphi(2^{nd}2^{nd}s)\rangle\}$$

For example, if $s = \langle a, \langle b, s_1\rangle\rangle$ then $\varphi(s)$ is the state that can perform one of two actions, a or b, going into the state $\varphi(s_1)$ in either case. How does one justify such a function? What are the properties of the monotone operator Γ that are needed in the justification of such definitions?

As with the parallel subject of defining functions by recursion on wellorder or on wellfounded sets, there are various ways to formulate the results. There is a trade-off between generality, on the one hand, and intelligibility on the other. In line with our policy, we do not formulate the most general results possible. Rather, we present a simple Corecursion Theorem in Section 17.2. In Section 17.3 we extend this result to wider classes of corecursive definitions. The last section of this chapter gets a generalization of bisimulation; the theory there uses results from earlier in the chapter.

On sets and classes in this chapter

We formulated the Strong Axiom of Plenitude in Chapter 2. This strong form will be used in this chapter. Specifically, we will use flat coalgebras based on proper classes of urelements, not just on sets. This necessitates a departure from our earlier treatments. Prior to this, we defined substitution operations $[s]$ whenever s was a substitution defined on a *set* of urelements. Now we need to allow s to be defined on a proper class. There is no problem in extending our work in Section 8.2 to define $[s]$ when s is such a *large* substitution. Our earlier work gives a definable, two-place operation sub taking small substitutions s and sets a as arguments; $sub(s, a) = a[s]$. When we extend to large substitutions, we do not have a uniform definition of sub. But we still can prove that every s extends to an operation $[s]$ that is defined on all sets a.

17.1   Smooth operators

We saw above the function $\varphi : A^\infty \to Can_A$ defined by

$$\varphi(s) \;\;=\;\; \{\langle 1^{st}s, \varphi(2^{nd}2^{nd}s)\rangle, \langle 1^{st}2^{nd}s, \varphi(2^{nd}2^{nd}s)\rangle\}$$

Notice that we make implicit use of some very particular facts about the domain and range of φ. For one, we use the function $2^{nd} : A^\infty \to A^\infty$ that allows us to get hold of the next element to which we want to apply φ. More crucially, we use the fact that if $a, b \in A$, then the Γ_5-form

$$\{\langle a, x \rangle, \langle b, x \rangle\}$$

gives rise to a canonical state when x is replaced by a state, say a canonical state s, or even by $\varphi(2^{nd}s)$. The main obstacle to giving a general treatment of corecursion is finding a way to state this sort of thing. To give ourselves the needed machinery, we introduce the notion of a "notation scheme." However, before we give this definition, we need to introduce an additional assumption on our operators.

In Chapter 16, the main condition on operators was uniformity. This key assumption is satisfied by nearly all of the interesting operators, and it allows for results like the Representation Theorem for Γ^* and the Solution Lemma Lemma. The theory of corecursion needs a stronger condition.

Definition Let Γ be an operator.

1. A set or class $X \subseteq \mathcal{U}$ is *very new* for Γ if for all substitutions s with domain X and all sets a,

   a. $\Gamma(a)[s] = \Gamma(a[s])$
   b. If $[s]_a : a \to a[s]$ is a bijection, then $[s]_{\Gamma(a)} : \Gamma(a) \to \Gamma(a[s])$ is also a bijection.

   Note that in (b), these maps are automatically surjective, so the real content of this definition is that if $[s]_a$ is injective, so is $[s]_{\Gamma(a)}$.

2. Γ is *smooth* iff Γ is monotone, proper, and if almost all sets of urelements are very new for Γ; that is, there is a set $X \subseteq \mathcal{U}$ so that the complement $\mathcal{U} - X$ is very new for Γ. We call X an *avoidance set* for Γ.

Exercise 17.1 This exercise is on the properties of smooth operators.

1. Prove that if Γ is smooth, then Γ is uniform.
2. Show that the smooth operators are closed under composition.
3. Show that our example uniform operators $\Gamma_1, \Gamma_2, \Gamma_3$, and Γ_5 are smooth.

Exercise 17.2 Show that the operator $\Gamma(a) = \mathcal{P}(a) \cup \mathcal{P}(\mathcal{P}(a))$ is uniform but not smooth.

We now define the notion of a Γ-notation scheme.

Definition Let C be a set or class, and let Γ be an operator. A Γ-*notation* *scheme for C* is a bijection *den* : $X \to C$ whose domain X is a set or class of urelements that is very new for Γ.

Proposition 17.1 *For every set or class C and every smooth operator Γ, there is a Γ-notation scheme for C, say den : $X \to C$, such that $X[den] = C$.*

Proof Let $Y \subseteq \mathcal{U}$ be an avoidance set for Γ. Let $f : C \to \mathcal{U}$ be given by $f(c) = \mathsf{new}(c, Y)$. Then f is injective. Let $X = f[C]$, and let *den* be the inverse of f. $\dashv$

Given a Γ-notation scheme *den* : $X \to C$, if $c = den(x)$ we say that c is the *denotation* of x and that x is the *name* of c. We write $\ulcorner c \urcorner = x$ interchangeably with $c = den(x)$. Notice that *den* is a substitution, so by uniformity $[den]_{\Gamma(X)} : \Gamma(X) \to \Gamma(C)$. Indeed, since $X[den] = C$, we have $\Gamma(X)[den] = \Gamma(X[den]) = \Gamma(C)$. Thus, every element of $\Gamma(C)$ can be thought of as being given by a form in $\Gamma(X)$.

The next result is the main reason why we need smooth operators in our study of corecursion. The same result does not hold for all uniform operators (see Exercise 17.8).

Proposition 17.2 *Let Γ be a smooth operator and let den : $X \to C$ be a Γ-notation scheme. For any Γ-coalgebra $\mathcal{E} = \langle C, e \rangle$ there is a unique map $\bar{e} : X \to \Gamma(X)$ such that $e \circ [den]_X = [den]_{\Gamma(X)} \circ \bar{e}$. That is, $\bar{e}$ is the unique map such that den : $\langle X, \bar{e} \rangle \to \mathcal{E}$ is a Γ-morphism. The map $\bar{e}$ is given by*

$$(2) \qquad\qquad \bar{e} = [den]_{\Gamma(X)}^{-1} \circ e \circ [den]_X.$$

Proof The proof is straightforward, given our definitions. Since $[den]_{\Gamma(X)}$ is injective, it has an inverse. This means that (2) defines a function $\bar{e}$. It is immediate that $e \circ [den]_X = [den]_{\Gamma(X)} \circ \bar{e}$. The uniqueness of $\bar{e}$ is due to the injectivity of $[den]_{\Gamma(X)}$; and this is precisely where the smoothness of Γ is important. $\dashv$

Using the notation of Proposition 17.2, we say that $\bar{e}$ is *the Γ-lift of e by den*. The intuitive idea is that $\bar{e}$ acts on names of things in X in a way completely parallel to the way e acts on the corresponding elements of C.

Suppose that $C = \Gamma^*$, and let *den* : $X \to C$ be a Γ-notation scheme. The inverse of *den* gives us a name $\ulcorner a \urcorner \in X$ for each $a \in \Gamma^*$. The Γ-lift $c = \overline{id}$ of the identity *id* : $\Gamma^* \to \Gamma^*$ is a bijection, called the *canonical map* $c : X \to \Gamma(X)$. It is depicted in Figure 13. The figure also explains the terminology of lifting. Note also that we use a double-arrow for Γ-notation schemes *den* and for the associated map $[den]_{\Gamma(X)}$. We do this to remind ourselves that these are bijections, and also because we think of the two sides of these arrows as essentially the same.

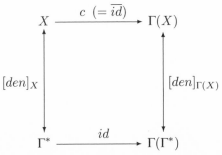

FIGURE 13 The canonical map $c : X \to \Gamma X$ is the Γ-lift of the identity.

The set $c(\ulcorner a \urcorner)$ is a Γ-form over X; it is called the *canonical Γ-form* of a (relative to the Γ-notation scheme den). Notice that for all a

$$(3) \qquad c(\ulcorner a \urcorner)[den]_{\Gamma(X)} = a$$

In words, you get a from its canonical Γ-form $c(\ulcorner a \urcorner)$ by replacing the names in the form by what they denote. To check (3), we consider the square of Figure 13 and we follow $\ulcorner a \urcorner$ as it goes from the upper left to the lower right corners by the two different paths. One path gives us the left side of the equation, the other path the right side. Since the square commutes, the two paths give the same result.

Example 17.1 In the cases of interest, we usually have a simple formula for computing canonical forms at our finger tips. For example, consider the case where Γ is $\mathcal{P}$. Then $\Gamma^* = V_{afa}$, the class of all pure sets. So X is in bijective correspondence with Γ^*. The formula for c is

$$c(\ulcorner a \urcorner) \quad = \quad \{\ulcorner b \urcorner \mid b \in a\}$$

for all $a \in V_{afa}$. Note that every element of a pure set a is itself pure; so for each $b \in a$ we have $\ulcorner b \urcorner \in X$. In the example of streams,

$$c(\ulcorner s \urcorner) \quad = \quad \langle 1^{st}s, \ulcorner 2^{nd}(s) \urcorner \rangle$$

for all streams s. Notice that $1^{st}s$ is an element of A whereas $\ulcorner 2^{nd}s \urcorner$ is an urelement naming $2^{nd}s$.

Before turning to the Corecursion Theorem, we need one more technical result on lifting, the Lifting Lemma. First, it is convenient to establish a general property concerning the composition of Γ-morphisms.

Lemma 17.3 *Let Γ be uniform, and suppose that $\mathcal{E}_i = \langle X_i, e_i \rangle$ are Γ-coalgebras for $i = 1, 2, 3$, with $\mathcal{E}_1$ and $\mathcal{E}_2$ flat. Let $r_1 : \mathcal{E}_1 \to \mathcal{E}_2$ and $r_2 : \mathcal{E}_2 \to \mathcal{E}_3$ be Γ-morphisms. Then $r_2 \circ r_1 : \mathcal{E}_1 \to \mathcal{E}_3$ is again a Γ-morphism.*

Proof For $x \in X_1$ we know that $e_1(x)[r_1] = e_2(r_1(x))$. Since $r_1(x) \in X_2$, we also have $e_2(r_1(x))[r_2] = e_3(r_2(r_1(x)))$. Putting these together shows that

$$e_1(x)[r_1][r_2] \quad = \quad e_3(r_2 \circ r_1(x)).$$

We need to see that $e_1(x)[r_1][r_2] = e_1(x)[r_1 \circ r_2]$.

For this, we need to recall some facts from the past. The operations $[r_1] \circ [r_2]$ and $[r_1 \circ r_2]$ are substitution-like, by Exercise 8.2. Let $Y \subseteq \mathcal{U}$ be an avoidance set for Γ. By Exercise 16.9, $support(\Gamma(X_1)) \subseteq X_1 \cup Y$. Clearly, $[r_1] \circ [r_2]$ and $[r_1 \circ r_2]$ agree on X_1. By newness, $X_1 \cap Y = \emptyset$ and $X_2 \cap Y = \emptyset$. So $[r_1] \circ [r_2]$ and $[r_1 \circ r_2]$ also agree on Y. Therefore, they agree on $V_{afa}[X_1 \cup Y]$, by Exercise 8.3. Since $e_1(x) \in V_{afa}[X_1 \cup Y]$, we are done.   ⊣

Lemma 17.4 (Lifting Lemma) *Let $\langle C, e \rangle$ be a Γ-coalgebra, and consider two Γ-notation schemes for C, say den $: X \to C$ and $den' : Y \to C$. Let*

$$\epsilon \quad = \quad den^{-1} \circ den',$$

so $\epsilon : Y \to X$ is a Γ-notation scheme for X. Let $\bar{e}$ be the Γ-lift of e by den, and let $\bar{e}'$ be the Γ-lift of e by den'. Then $\bar{e}'$ is also the Γ-lift of $\bar{e}$ by ϵ.

Proof In terms of coalgebra morphisms, we have den $: \langle X, \bar{e} \rangle \to \langle C, e \rangle$ and $\epsilon : \langle Y, \bar{e}' \rangle \to \langle X, \bar{e} \rangle$. By Lemma 17.3, $den \circ \epsilon = den'$ is a morphism

$$den' : \langle Y, \bar{e}' \rangle \to \langle C, e \rangle.$$

Now we are done by the uniqueness part of Proposition 17.2.   ⊣

17.2 The Corecursion Theorem

We now have our general machinery all set up. To define a function $\varphi :$ $C \to \Gamma^*$, we need something to keep the corecursion going. That is, we need something to tell us how to operate on a complex Γ-object. This is provided by a coalgebra $\pi : C \to \Gamma(C)$, called the *pump* of the corecursive definition.

Example 17.2 Let's see how we can look at the corecursion for streams in these terms. Given any function $\pi : C \to A \times C$ we saw that there was a unique $\varphi : C \to A^\infty$ satisfying the equation

$$\varphi(c) \quad = \quad \langle 1^{st}\pi(c), \varphi(2^{nd}\pi(c)) \rangle.$$

As we noted earlier, the function π is just a Γ_3-coalgebra. If we lift it, we obtain a function $\bar{\pi}$ taking us from names to Γ_3-forms in those names:

$$\bar{\pi}(\ulcorner c \urcorner) \quad = \quad \langle 1^{st}\pi(c), \ulcorner 2^{nd}\pi(c) \urcorner \rangle.$$

Notice that $\varphi \circ den$ is a substitution defined on X. Since X is new for Γ_3, $[\varphi \circ den]$ leaves unaffected the elements of the set A. Notice also that $\bar{\pi}(\ulcorner c \urcorner)[\varphi \circ den] = \varphi(c)$. This is because den and $\ulcorner \quad \urcorner$ are inverses of one another. In fact, this is the whole point: we define the action of φ by first passing from C to X. Then we go back at the end and replace each $\ulcorner c \urcorner$ by the corresponding c. Thus, substituting with $[\varphi \circ den]$ gives us an equation satisfied by our original function φ:

$$\varphi(c) \quad = \quad \pi(\ulcorner c \urcorner)[\varphi \circ den]$$

Our Corecursion Theorem will insure us that these sorts of equations always uniquely determine a function φ, and it gives us a way to obtain φ. Since the φ which we constructed earlier satisfies the equation, it follows that the method of corecursion explored in this chapter generalizes our earlier work.

Now let's try the same thing on a non-stream example.

Example 17.3 To define the map $\varphi : A^\infty \to Can_A$ into the set of canonical states by the equation

(4) $$\varphi(s) \quad = \quad \{\langle 1^{st}s, \varphi(2^{nd}2^{nd}s)\rangle, \langle 1^{st}2^{nd}s, \varphi(2^{nd}2^{nd}s)\rangle\}$$

we will take as a pump the function $\pi : A^\infty \to \mathcal{P}(A \times A^\infty)$ defined by

$$\pi(s) \quad = \quad \{\langle 1^{st}s, 2^{nd}2^{nd}s\rangle, \langle 1^{st}2^{nd}s, 2^{nd}2^{nd}s\rangle\}$$

This is easier to read if we rewrite it a bit as

$$\pi(\langle a, \langle b, s\rangle\rangle) \quad = \quad \{\langle a, s\rangle, \langle b, s\rangle\}.$$

The two forms are equivalent, since every stream t can be written uniquely as $\langle a, \langle b, s\rangle\rangle$ for some a, b, and s. This gives a Γ_5-coalgebra $\langle A^\infty, \pi\rangle$. Lifting this pump we get a flat Γ_5-coalgebra $\langle X, \overline{\pi}\rangle$. The formula for $\overline{\pi}$ is

$$\overline{\pi}(\ulcorner \langle a, \langle b, s\rangle\rangle \urcorner) \quad = \quad \{\langle a, \ulcorner s \urcorner\rangle, \langle b, \ulcorner s \urcorner\rangle\}$$

Let's assume that we had a function φ satisfying (4). Again, using $[\varphi \circ den]$, the denotation function den composed with φ, considered as a substitution, gives us a function φ that satisfies

$$\varphi(s) \quad = \quad \overline{\pi}(\ulcorner s \urcorner)[\varphi \circ den].$$

This function satisfies our defining equation. Of course this is circular, since it presupposes that φ exists, but that is what we will show in the Corecursion Theorem.

We have seen the same equation for φ several times already, and it is worthwhile to make a definition.

Definition Let Γ be a smooth operator, let $\pi : C \to \Gamma(C)$ be a Γ-coalgebra, and let $den : X \to C$ be a Γ-notation scheme. A map $\varphi : C \to \Gamma^*$ *satisfies* Γ-*corecursion for π relative to den* if for all $c \in C$,

(5) $$\varphi(c) \quad = \quad \pi(\ulcorner c \urcorner)[\varphi \circ den]$$

Here $\overline{\pi}$ is the Γ-lift of π by den.

As the name indicates, pumps are what keeps corecursive definitions going. The following shows that this process always works, and that the result is independent of the Γ-notation scheme employed. (It is for this latter fact that we need the assumption that Γ is smooth, not just uniform.)

Theorem 17.5 (Corecursion Theorem) *Let Γ be a smooth operator, and let $\pi : C \to \Gamma(C)$. There is a unique map*

$$\varphi : C \to \Gamma^*$$

such that φ satisfies Γ-corecursion for π relative to some Γ-notation scheme for C. Furthermore, φ satisfies Γ-corecursion for π relative to any *Γ-notation scheme for C.*

In words, this tells us that there is a unique function φ that we can compute in the following way:

1. Pick a Γ-notation scheme $den : X \to C$.

2. For each $c \in C$, find its associated name $\ulcorner c \urcorner$.

3. Apply $\overline{\pi}$ to get $\overline{\pi}(\ulcorner c \urcorner)$. In general, this will contain the names of some elements $c' \in C$, possibly including c.

4. For each of these names $\ulcorner c' \urcorner$, find $\varphi(c')$ and substitute this value for $\ulcorner c' \urcorner$ in $\overline{\pi}(\ulcorner c \urcorner)$.

We call Theorem 17.5 a Corecursion Theorem because φ itself is involved in part (4) of this procedure.

Note that the Corecursion Theorem is a generalization of the Solution Lemma for Operators (see page 232). In fact, if C happens to be a set of urelements and den is the identity map, then $\overline{\pi} = \pi$, and (5) just says that φ is the solution of π.

Note also that the conclusion of the theorem speaks of Γ-corecursion, and not of corecursion alone. Similarly, we must speak of the Γ-lift of a coalgebra, and not the lift *simpliciter*. This is because it is possible to have two different operators, say Γ and Δ, a set C so that $\Gamma(C) = \Delta(C)$ and a single pump π so that $\pi : C \to \Gamma(C)$ and $\pi : C \to \Delta(C)$. In this case, the corecursive function φ would not be determined by π alone. It would depend on which operator we used (See Exercise 17.9).

We now turn to the proof of the Corecursion Theorem.

Proof We begin with a diagrammatic proof of the existence half of the proof. We are given C and π as follows:

$$C \xrightarrow{\quad \pi \quad} \Gamma(C)$$

Let $\langle X, den \rangle$ be a Γ-notation scheme. Any Γ-notation scheme will do. As you can see from our first diagram, we abbreviate $[den]_{\Gamma(X)}$ to $[den]$ in the arrows.

Let $\overline{\pi}$ be the Γ-lift of π by *den*,

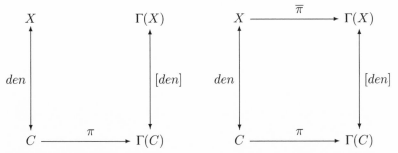

as in the second diagram. Since X is new for Γ, $\mathcal{E} = \langle X, \overline{\pi} \rangle$ is a flat Γ-coalgebra. Let s be its solution. By Lemma 16.2, s maps X into Γ^*. We have indicated s below, along with $[s]_{\Gamma(X)}$ which we've also shortened to $[s]$. Using the fact that *den* is a bijection, define φ by $\varphi(c) = s(\ulcorner c \urcorner)$. In other words, $\varphi = s \circ den^{-1}$. This gives us our last diagram. In it, the outer square commutes, as do the two inner triangles heading inwards to Γ^*.

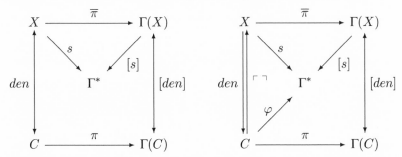

Since s is a solution to $\mathcal{E}$, we see that for all $a \in \Gamma^*$, $s(\ulcorner a \urcorner) = \overline{\pi}(\ulcorner a \urcorner)[s]$. Thus we get

$$
\begin{aligned}
\varphi(c) &= s(\ulcorner c \urcorner) \\
&= \overline{\pi}(\ulcorner c \urcorner)[s]_{\Gamma(X)} \\
&= \overline{\pi}(\ulcorner c \urcorner)[\varphi \circ den]
\end{aligned}
$$

as desired.

We need to prove that φ is unique. Suppose we started with an arbitrary Γ-notation scheme, say $den' : Y \to C$. We'll write $den'(y) = c$ by 'c' $= y$. Let $\overline{\pi}'$ be the Γ-lift of π by den'. Suppose that we happened to come across a function φ' that satisfies Γ-corecursion for π relative to den'. This means that for all $c \in C$,

$$
\varphi'(c) = \overline{\pi}'(\text{'}c\text{'})[\varphi' \circ den'].
$$

We must prove that $\varphi = \varphi'$. Since each $c \in C$ is $den'(\text{'}c\text{'})$, we have

$$
\varphi' \circ den'(\text{'}c\text{'}) = \overline{\pi}'(\text{'}c\text{'})[\varphi' \circ den'].
$$

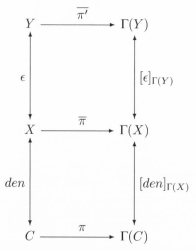

FIGURE 14 Proof of the Corecursion Theorem: Uniqueness.

Now every $y \in Y$ is of the form 'c' for some unique $c \in C$. Thus, for all $y \in Y$,

$$\varphi' \circ den'(y) \quad = \quad \overline{\pi}'(y)[\varphi' \circ den'].$$

Therefore, $\varphi' \circ den'$ is the solution to the coalgebra $\langle Y, \overline{\pi}' \rangle$.

Let $\epsilon = den^{-1} \circ den'$, so $\epsilon : Y \to X$ is a bijection. Therefore $\langle Y, \epsilon \rangle$ is a Γ-notation scheme. By the Lifting Lemma, $\overline{\pi}'$ is the Γ-lift of $\overline{\pi}$ by ϵ. In particular, the top square of Figure 14 is a Γ-morphism. Proposition 16.3 tells us that the solutions to the two flat coalgebras $\langle Y, \overline{\pi}' \rangle$ and $\langle X, \overline{\pi} \rangle$ are related by ϵ. More precisely, $\varphi' \circ den' = s \circ \epsilon$. Therefore,

$$
\begin{aligned}
\varphi' \quad &= \quad (s \circ \epsilon) \circ (den')^{-1} \\
&= \quad s \circ (den^{-1} \circ den') \circ (den')^{-1} \\
&= \quad s \circ den^{-1} \\
&= \quad \varphi
\end{aligned}
$$

This confirms that $\varphi' = \varphi$.

So at this point, we know that φ is the unique map that satisfies corecursion for π relative to *some* Γ-notation scheme. Since the Γ-notation scheme den used in getting φ is arbitrary, this shows that φ satisfies corecursion relative to *any* Γ-notation scheme $den' : Y \twoheadrightarrow C$. ⊣

Example 17.4 Here is another application of corecursion. Let A be a set of urelements. Suppose we want an operation $\varphi : V_{afa}[\mathcal{U}] \to V_{afa}[A]$ that takes a set a, and applies itself corecursively to the *sets* in a but keeps only those urelements that happen to belong to A. To get this, we use the operator

$\Gamma(a) = \mathcal{P}(A \cup a)$. This operator is not smooth, but it is 'nearly smooth', and the Corecursion Theorem will apply (see page 263 for details). We use

$$\pi : V_{afa}[\mathcal{U}] \to \mathcal{P}(A \cup V_{afa}[\mathcal{U}])$$

given by $\pi(a) = a \cap (A \cup V_{afa}[\mathcal{U}])$. Let $den : X \to V_{afa}[\mathcal{U}]$ be a Γ-notation scheme for Γ. Then the Γ-lift of π by den is

$$\overline{\pi} : X \to \mathcal{P}(A \cup X),$$

given by

$$\overline{\pi}(\ulcorner a \urcorner) \quad = \quad \{\ulcorner b \urcorner : b \in a \text{ is a set}\} \cup (A \cap a).$$

(We stress that b ranges over the *sets* in a.) Then by the Corecursion Theorem there is a unique map $\varphi : V_{afa}[\mathcal{U}] \to V_{afa}[A]$ so that for all $a \in V_{afa}[\mathcal{U}]$,

$$\begin{aligned} \varphi(a) \quad &= \quad \{\ulcorner b \urcorner : b \in a\} \cup (A \cap a)[\varphi \circ den] \\ &= \quad \{\varphi(b) : b \in a\} \cup (A \cap a) \end{aligned}$$

We are using the fact that X is a set that is new for Γ, so that the elements of A are fixed by $[\varphi \circ den]$. This is the operation we want.

We close this section with a number of exercises. Exercises 17.3–17.6 give some applications of the method. Exercise 17.7 gives an extension of the results which we shall pursue in the next section. The final exercises have to do with the Corecursion Theorem itself. In Exercise 17.8, we show that the uniqueness part of the theorem does not hold for all uniform operators; thus, smoothness or some similar condition is necessary. Second, Exercise 17.9 amplifies a point we made earlier about the need for Γ in the statement of the Corecursion Theorem. Finally, Exercise 17.10 shows that the uniqueness part of the Corecursion Theorem cannot hold for functions going *out of* a greatest fixed point, even when the operator is smooth.

Exercise 17.3 Recall map_f from our earlier work on streams. If A is a set and $f : A \to A$, then $map_f : A^{\infty} \to A^{\infty}$ is given by

$$map_f(s) \quad = \quad \langle f(1^{st}s), map_f(2^{nd}s) \rangle.$$

Show how to use the Corecursion Theorem to define map_f. This means: state what C, Γ, and π are, and check that the φ you get satisfies the equation above.

Exercise 17.4 In the setting of the Corecursion Theorem, let $\Gamma(a) = A \times a$, and let $C = \Gamma^*$. Let π be given by $\pi(s) = \langle 1^{st}(s), 2^{nd}(2^{nd}(s)) \rangle$. What is the formula for the function φ? What is the most appropriate name for it?

Exercise 17.5 Let A be a set, and consider the operator $\Delta(a) = a \times A$. We call the elements of Δ^* *smaerts* over A. Use corecursion to get a natural bijection between streams and smaerts over A.

The set or class C in corecursion can be a product $C = A \times B$. In such a case, it is natural to write $\varphi(a, b)$ rather than $\varphi(\langle a, b \rangle)$. Our machinery applies directly to give functions from C to some greatest fixed point.

Exercise 17.6 Let A and B be sets. Define an operation

$$\mu : A^\infty \times B^\infty \to (A \times B)^\infty$$

that merges streams in the obvious way:

$$\mu(s, t) \quad = \quad \langle\langle 1^{st}(s), 1^{st}(t)\rangle, \mu(2^{nd}(s), 2^{nd}(t))\rangle\rangle.$$

Exercise 17.7 Let A and B be sets, and let $f, g : A \to B$ be two functions from A to B. Let $\Gamma(a) = B \times a$ be the stream-forming operator. Suppose we want a map $d : A \to B^\infty$ satisfying

$$d(a) \quad = \quad \langle f(a), \langle g(a), d(2^{nd}(a)) \rangle\rangle.$$

So d is a "doubled" embedding of A into B^∞. The natural idea is to try to employ the function $\pi(a) = \langle f(a), \langle g(a), a \rangle\rangle$ in corecursion. In this case, π maps A not to $\Gamma(A)$ but to $\Gamma(\Gamma(A))$. So our theory needs to be extended a bit to handle this kind of example. Let $den : X \to A$ be a Γ-notation scheme.

1. What should the Γ-lift of π be?
2. Decide how, in general, a map $\pi : C \to \Gamma(\Gamma(C))$ lifts by den to a map $\overline{\pi} : X \to \Gamma(\Gamma(X))$. Check that your general definition specializes to your answer to part (1).
3. Continuing the general discussion, let s be the solution of the coalgebra $\langle X, \overline{\pi} \rangle$. Show that s maps into Γ^*. The point is that corecursion into Γ^* can be driven by functions $\pi : C \to \Gamma(\Gamma(C))$.
4. Returning back to the concrete case, show that the function φ obtained by corecursion satisfies the defining condition for d.

Exercise 17.8 Let $\Gamma(a) = \mathcal{P}(a) \cup \mathcal{P}(\mathcal{P}((a))$ be the operator from Exercise 17.2. Find a coalgebra $\pi : C \to \Gamma(C)$ and a Γ-notation scheme $den : X \to C$ such that π has two different Γ-lifts $\overline{\pi}$ and $\overline{\pi}'$. Then check that this gives two different functions $g : C \to \Gamma^*$, each of which satisfies (5) for one of the lifts.

Exercise 17.9 Give an example of two different operators, say Γ and Δ, a set C so that $\Gamma(C) = \Delta(C)$ and a single pump π so that $\pi : C \to \Gamma(C)$ and $\pi : C \to \Delta(C)$, and (crucially) so that the corecursive functions φ and ψ determined by these operators are different.

Exercise 17.10 The method of corecursion is designed for functions going *into* a greatest fixed point. It is tempting to try to get a result for functions going *out of* a greatest fixed point, too. This exercise shows that this is not always possible.

Let $A = \{0, 1\}$, and let min : $A \to A$ be the minimum function on A. Suppose that we want to define the minimum value reached on a stream. We therefore might try to get a unique $\varphi : A^\infty \to A$ such that for all $s \in A^\infty$,

(6) $$\varphi(s) \quad = \quad \min(1^{st}s, \varphi(2^{nd}s)).$$

Prove that there are at least two maps φ that satisfy (6). The point is that to define the minimum value, we must use some method other than corecursion.

When defining functions by recursion, once one understands the basic process, one rarely if ever goes back and explicitly puts a recursive definition in the form required by some particular theorem justifying definitions by recursion. So too with definitions by corecursion. Once we understand the basic process, it is seldom necessary to resort to the Corecursion Theorem for justification. Still, it is nice to know that the result is there if we need it. In the remainder of this chapter we sketch some generalizations of it.

17.3   Simultaneous corecursion

No treatment of recursion can be complete without simultaneous recursion. We'll work out the theory of simultaneous corecursion in a simple case, defining two functions by simultaneous corecursion. The extension to more than two functions is routine.

Here is an example. Suppose we have two sets B and C, and also functions $f : B \to C$ and $g : C \to B$. Suppose we also have a set A and that we consider the operator $\Gamma(a) = A \times a \times a$. The greatest fixed point Γ^* satisfies $\Gamma^* = A \times \Gamma^* \times \Gamma^*$. This set is called the set of *infinite binary trees over A*, and we study it further in Section 18.2. Suppose that we want to define functions from B and C into Γ^*. For example, let us see how to define $\varphi : B \to \Gamma^*$ and $\psi : C \to \Gamma^*$ so that

(7) $$\begin{aligned} \varphi(b) &= \langle a_0, \psi(f(b)), \varphi(b) \rangle \\ \psi(c) &= \langle a_1, \varphi(g(c)), \varphi(g(f(g(c)))) \rangle \end{aligned}$$

Here a_0 and a_1 are two fixed elements of A. We want to generalize our earlier work to cover this kind of corecursion.

Our first guess is to use pumps $\pi_1 : B \to B \cup C$ and $\pi_2 : C \to B \cup C$. We try to take

$$\begin{aligned} \pi_1(b) &= \langle a_0, f(b), b \rangle \\ \pi_2(c) &= \langle a_1, g(c), g(f(g(c))) \rangle \end{aligned}$$

And then we would try to use our earlier technology of lifting.

However, there is a problem with the approach. It may well be that $B \cap C \neq \emptyset$. In fact, in many similar cases of interest, $B = C$. In such a situation, if we lifted these pumps and found some $x \in X$ in the right hand side, we would not know whether to apply φ or ψ in order to continue. In other words, the right hand sides above do not contain enough information. For the elements $f(b)$, b, $g(c)$, and $g(f(g(c)))$, the expressions above fail to specify which function, φ or ψ, is supposed to apply to which arguments.

The main new idea for simultaneous corecursion is to consider the *disjoint union* $B + C$. This set is given by

$$B + C \quad = \quad (\{0\} \times B) \cup (\{1\} \times C)$$

(see page 13). With this idea in place, we can specify the pumps for our simultaneous corecursion. They are given by

$$\pi_1(b) \quad = \quad \langle a_0, \langle 1, f(b) \rangle, \langle 0, b \rangle \rangle$$
$$\pi_2(c) \quad = \quad \langle a_1, \langle 0, g(c) \rangle, \langle 0, g(f(g(c))) \rangle \rangle$$

Once again, the point is that by taking maps into $\Gamma(B+C)$ rather than $\Gamma(B \cup C)$, we have some extra information. Given a pair $\langle i, b \rangle$, we can see whether φ or ψ is supposed to apply next.

Here is a series of definitions for binary corecursion.

Definition Let Γ be an operator.

1. A *binary Γ-notation scheme* is a pair $d = \langle den_1, den_2 \rangle$ of Γ-notation schemes with disjoint domains. In other, words,

$$den_1 : X \to B$$
$$den_2 : Y \to C$$

   with X and Y very new for Γ, and $X \cap Y = \emptyset$.

2. Concerning the inverses of these denotation maps, we'll write $\ulcorner b \urcorner = x$ to mean that $den_1(x) = b$, and 'c' $= x$ to mean that $den_2(x) = c$.

3. Along with d comes a Γ-notation scheme

$$den : X \cup Y \to B + C$$

   given by

$$den(z) \quad = \quad \begin{cases} \langle 0, den_1(z) \rangle & \text{if} \quad z \in X \\ \langle 1, den_2(z) \rangle & \text{if} \quad z \in Y \end{cases}$$

4. In the binary setting, corecursion will be run by two pumps,

$$\pi_1 : B \to \Gamma(B + C)$$
$$\pi_2 : C \to \Gamma(B + C)$$

5. These pumps can be put together in a natural way to get a coalgebra $\pi : B + C \to \Gamma(B + C)$.

6. π can be lifted to $\overline{\pi} : X \cup Y \to \Gamma(X \cup Y)$. This is a flat Γ-coalgebra.

7. Let $\varphi : B \to \Gamma^*$ and $\psi : C \to \Gamma^*$. The pair $\langle \varphi, \psi \rangle$ *satisfies corecursion for* $\langle \pi_1, \pi_2 \rangle$ *relative to* $d = \langle den_1, den_2 \rangle$ *if for all* $b \in B$ *and* $c \in C$,

$$
(8) \qquad
\begin{aligned}
\varphi(b) &= \overline{\pi}(\ulcorner b \urcorner) \quad [(\varphi \circ den_1) \cup (\psi \circ den_2)] \\
\psi(c) &= \overline{\pi}(`c`) \quad [(\varphi \circ den_1) \cup (\psi \circ den_2)]
\end{aligned}
$$

Theorem 17.6 (Simultaneous Corecursion Theorem) *Let* Γ *be a smooth operator, and let* $\pi_1 : B \to \Gamma(B + C)$ *and* $\pi_2 : C \to \Gamma(B + C)$ *be* Γ-*coalgebras. There is a unique pair of maps* $\langle \varphi, \psi \rangle$,

$$
\varphi : B \to \Gamma^*
$$
$$
\psi : C \to \Gamma^*
$$

so that for some binary Γ-*notation scheme*

$$
d = \langle den_1 : X \to B, \; den_2 : Y \to C \rangle,
$$

$\langle \varphi, \psi \rangle$ *satisfies corecursion for* $\langle \pi_1, \pi_2 \rangle$ *relative to d. Moreover,* $\langle \varphi, \psi \rangle$ *satisfies corecursion relative to such d.*

As part of the proof of this result, we need to know that binary Γ-notation schemes always exist. So the following exercise generalizes Proposition 17.1.

Exercise 17.11 Let B and C be sets or classes, and let Γ be smooth. Prove that there is a binary Γ-notation scheme $\langle den_1, den_2 \rangle$ such that the range of den_1 is B and the range of den_2 is C.

Exercise 17.12 Sketch a proof of the existence half of the Simultaneous Corecursion Theorem.

Corecursion in parameters We now want to consider a variant of the functions defined by simultaneous corecursion in equation (7) back on page 255. Suppose we want to define functions φ and ψ that satisfy

$$
\begin{aligned}
\varphi(b) &= \langle a_0, \psi(f(b)), \varphi(b) \rangle \\
\psi(c) &= \langle a_1, t_0, \varphi(g(f(g(c)))) \rangle
\end{aligned}
$$

Here a_0 and a_1 are two fixed elements of A, and t_0 is a fixed element of Γ^*.

We need to generalize our earlier work to account for this kind of definition. We sketch a way to do it, building on our earlier work on simultaneous corecursion.

Along with our binary notation scheme $\langle den_1, den_2 \rangle$, we also take a canonical map $m : P \to \Gamma(P)$. Here P is a class or urelements which is very new for Γ, and also disjoint from X and from Y. We also have a bijection $d : P \to \Gamma^*$. Notice that d is a substitution; in fact it is the solution to m. In this case the pumps would be allowed to map into a bigger class: for example, $\pi_1 : B \to \Gamma(B + C + P)$. Then we also get

$$
\overline{\pi} : X \cup Y \cup P \to \Gamma(X \cup Y \cup P).
$$

This lift is obtained from π_1, π_2 and our canonical map m. Finally, the definition of satisfying corecursion changes a bit. The pair $\langle \varphi, \psi \rangle$ *satisfies corecursion for* $\langle \pi_1, \pi_2 \rangle$ *relative to* $\langle den_1, den_2 \rangle$, m, *and* d if for all $b \in B$ and $c \in C$,

$$\varphi(b) = \overline{\pi}(\ulcorner b \urcorner) \quad [(\varphi \circ den_1) \cup (\psi \circ den_2) \cup d]$$
$$\psi(c) = \overline{\pi}(`c') \quad [(\varphi \circ den_1) \cup (\psi \circ den_2) \cup d]$$

This concludes our discussion of corecursion.

17.4 Bisimulation generalized

One of the fundamental concepts of this book is bisimulation. We devoted a chapter to it early on, Chapter 7. When we turned to streams in Chapter 14, we had a different notion, that of *stream bisimulations*. The facts concerning the two notions were entirely parallel, but the definitions were rather different. Whenever this happens, the mathematician's urge is to develop a common generalization. The purpose of this section is to provide such a development in this particular case. We begin with some definitions.

Definition Let Γ be a smooth operator, say with W as an avoidance set. Let $X \subseteq \mathcal{U}$. Then we define

$$X \otimes X = \{ \mathsf{new}(\langle y, z \rangle, W) \mid \langle y, z \rangle \in X \times X \}.$$

$X \otimes X$ is in one-to-one correspondence with $X \times X$, the only difference being that the former is a set of urelements. For every $R \subseteq X \times X$ we have a corresponding set $R^o \subseteq X \otimes X$. Finally, we have two substitutions $\pi_1, \pi_2 : X \otimes X \to X$ given by

$$\pi_1(\mathsf{new}(\langle y, z \rangle, W) = y$$
$$\pi_2(\mathsf{new}(\langle y, z \rangle, W) = z$$

Whenever possible, we identify $X \otimes X$ and $X \times X$. This will help us suppress a certain amount of notation involving the operation new. For example, we'll write $\pi_1(\langle x, y \rangle) = x$ and $\pi_2(\langle x, y \rangle) = y$ instead of the formulas above. In analogy with similar concepts in other areas, π_1 and π_2 are called *projections*.

We next have the fundamental definition of this section:

Definition Let Γ be a smooth operator, and let $\mathcal{E} = \langle X, e \rangle$ be a flat coalgebra. A relation $R \subseteq X \times X$ is a Γ-*bisimulation on* $\mathcal{E}$ if there is a map $e' : R^o \to \Gamma(R^o)$ giving a flat coalgebra $\mathcal{E}' = \langle R^o, e' \rangle$ such that π_1 and π_2 are Γ-morphisms from $\mathcal{E}'$ to $\mathcal{E}$.

Example 17.5 Let $A = \{p, q\} \subseteq \mathcal{U}$, and consider the operator $\Gamma(a) = \mathcal{P}(A \cup a)$. Let $X = \{x, y, z\}$ be a set of urelements disjoint from A and hence new for Γ. Consider the Γ-coalgebra $\mathcal{E} = \langle X, e \rangle$ given by

$$
\begin{aligned}
e_x &= \{p, y\} \\
e_y &= \{p, y\} \\
e_z &= \{p, q, z\}
\end{aligned}
$$

We claim that $R \subseteq X \times X$ is a Γ-bisimulation on $\mathcal{E}$, where $R = \{\langle x, y \rangle, \langle y, y \rangle\}$. To see this, we need to find a map $e' : R^o \to \Gamma(R^o)$ and then verify that the projections π_1 and π_2 are morphisms. Recall that we are identifying R^o and R, so we'll find $e' : R \to \Gamma(R)$. We define e' as follows:

$$
\begin{aligned}
e'_{\langle x, y \rangle} &= \{p, \langle y, y \rangle\} \\
e'_{\langle y, y \rangle} &= \{p, \langle y, y \rangle\}
\end{aligned}
$$

To check that R is a bisimulation, note that

$$
\begin{aligned}
e'_{\langle x, y \rangle}[\pi_1] &= \{p, y\} &= e_x \\
e'_{\langle y, y \rangle}[\pi_1] &= \{p, y\} &= e_y \\
e'_{\langle x, y \rangle}[\pi_2] &= \{p, y\} &= e_x \\
e'_{\langle x, y \rangle}[\pi_2] &= \{p, y\} &= e_y
\end{aligned}
$$

This concludes the verification.

This relation R is not the only Γ-bisimulation on $\mathcal{E}$. The most important is the largest bisimulation relation R_{max} which is given by $u \, R_{max} \, v$ iff $s_u = s_v$, for s the solution to $\mathcal{E}$. Explicitly,

$$
R_{max} = R \cup \{\langle x, x \rangle, \langle y, x \rangle, \langle z, z \rangle\}.
$$

The relation R above is not an equivalence relation, but R_{max} always is an equivalence relation.

Now that we have seen an example, here is what we are going to do with this notion of a Γ-bisimulation. First, we want to see how it is that our new notion generalizes what we have seen before. That is, the bisimulations on flat systems should be exactly the Γ-bisimulations for the operator $\Gamma(a) = pow(A \cup a)$. Likewise, the stream bisimulations should be the Γ-bisimulations for the operator $\Gamma(a) = A \times a$. Second, we'll see how to extend the notion from flat Γ-coalgebras (as in the definition) to all Γ-coalgebras. Third, we'll generalize an important fact from our earlier theory on bisimulations to the new context. We'll also see an example of a fact that does not generalize.

Proposition 17.7 *Let Γ be the operator $\Gamma(a) = \mathcal{P}(A \cup a)$. Let $\langle X, e \rangle$ be a flat Γ-coalgebra, and let $R \subseteq X \times X$. Then R is a Γ-bisimulation iff the following condition holds: if xRy, then $e_x \cap A = e_y \cap A$, and*

1. *For all $x' \in e_x \cap X$ there is some $y' \in e_y \cap X$ such that $x' \, R \, y'$.*
2. *For all $y' \in e_y \cap X$ there is some $x' \in e_x \cap X$ such that $x' \, R \, y'$.*

Proposition 17.8 *Let Γ be the operator $\Gamma(a) = A \times a$, and let $\langle X, e \rangle$ and R be as in Proposition 17.7. R is a Γ-bisimulation iff the following condition holds: if $x\,R\,y$, then $1^{st}(e_x) = 1^{st}(e_y)$ and $2^{nd}(e_x)\,R\,2^{nd}(e_y)$.*

We will not prove Propositions 17.7 and 17.8. Instead, we invite you to work the following exercise. The solution contains all of the details needed in the propositions above.

Exercise 17.13 Recall that our example operator Γ_5 is given by $\Gamma_5(a) = \mathcal{P}(Act \times a)$ for some fixed set Act. In the notation above, prove that R is a Γ_5-bisimulation iff the following condition holds: if xRy and $a \in Act$, then

1. For all x' such that $\langle a, x' \rangle \in e_x$ there is some y' such that $\langle a, y' \rangle \in e_y$ and $x'\,R\,y'$.
2. For all x' such that $\langle a, y' \rangle \in e_y$ there is some y' such that $\langle a, x' \rangle \in e_x$ and $x'\,R\,y'$.

At this point, we want to digress to explain how to get a notion of generalized bisimulation for all Γ-coalgebras, not just the flat ones. The basic idea is to use a notation system to obtain a flat coalgebra from an arbitrary one.

Definition Let $\mathcal{E} = \langle a, e \rangle$ be a Γ-coalgebra and let $R \subseteq a \times a$. We say that R is a Γ-*bisimulation on* $\mathcal{E}$ if there is a Γ-notation scheme $den : X \to a$ and a Γ-bisimulation $S \subseteq X \times X$ on the lifted coalgebra $\langle X, \bar{e} \rangle$ such that $S[den] = R$.

$\mathcal{E}$ is *strongly extensional* if every Γ-bisimulation on $\mathcal{E}$ is a subrelation of the identity. That is, whenever R is a Γ-bisimulation on $\mathcal{E}$ and $b\,R\,b'$, then $b = b'$.

It is worth checking that whether R is a Γ-bisimulation or not is independent of the choice of notation scheme.

Exercise 17.14 Let $\mathcal{E} = \langle a, e \rangle$ be a Γ-coalgebra. Suppose that $den : X \to a$ and $den' : Y \to a$ are two Γ-notation schemes for a. Let $\bar{e}$ and $\bar{e}'$ be the lifts of e by den and den', respectively. Suppose that $S \subseteq X \times X$ is a Γ-bisimulation on $\langle X, \bar{e} \rangle$. Prove that there is a Γ-bisimulation T on $\langle Y, \bar{e}' \rangle$ such that $T[den'] = S[den]$. [Hint: Use Lemma 17.3 and the Lifting Lemma.]

As Propositions 17.7–17.8 and Exercise 17.13 show, the notion of Γ-bisimulation specializes to notions we have seen before. This is the main point of it. We also have results for the new notion, generalizing some facts that we have seen earlier in the book.

Proposition 17.9 *Let $\mathcal{E} = \langle X, e \rangle$ be a flat Γ-coalgebra, and let s be its solution. Suppose that R is a bisimulation on $\mathcal{E}$. Then $x\,R\,y$ implies $s_x = s_y$.*

Proof The definition of a Γ-bisimulation gives us a coalgebra $\langle R^o, e' \rangle$ which we write as $\mathcal{E}'$. We'll also write s' for the solution of $\mathcal{E}'$. We also get two

Γ-morphisms $\pi_1, \pi_2 : \mathcal{E}' \to \mathcal{E}$. Suppose that $x \, R \, y$ so that $\langle x, y \rangle \in R^o$. Then by Proposition 16.3, $s' = s \circ \pi_1 = s \circ \pi_2$. In particular, for all $\langle x, y \rangle \in R^o$,

$$ s_x \quad = \quad (s \circ \pi_1)(\langle x, y \rangle) \quad = \quad (s \circ \pi_2)(\langle x, y \rangle) \quad = \quad s_y. $$

This completes the proof. $\dashv$

Theorem 17.10 *For every smooth operator Γ, the greatest fixed point coalgebra $\langle \Gamma^*, i \rangle$ is strongly extensional.*

Proof Suppose that R is a Γ-bisimulation on Γ^* and $a \, R \, b$. Let $den : X \to \Gamma^*$ be a Γ-notation scheme, and let $c : X \to \Gamma(X)$ be the canonical coalgebra. Then den is the solution to c. We therefore have a Γ-bisimulation S on $\langle X, c \rangle$ so that $\ulcorner a \urcorner S \ulcorner b \urcorner$. This means that we have a Γ-coalgebra $\mathcal{E}' = \langle S^o, e' \rangle$ and two Γ-morphisms π_1 and π_2 from $\mathcal{E}'$ to $\mathcal{E}$; finally, $\langle \ulcorner a \urcorner, \ulcorner b \urcorner \rangle \in S^o$. By Proposition 17.9, $den(\ulcorner a \urcorner) = den(\ulcorner b \urcorner)$. This means that $a = b$. $\dashv$

We conclude this chapter with a discussion of a result on bisimulations of flat systems from earlier in the book which does not generalize. Recall that flat systems are the coalgebras for the operator $\Gamma(a) = \mathcal{P}(A \cup a)$. We know from Chapter 7 that if $\mathcal{E} = \langle X, e \rangle$ is a Γ-coalgebra, then the following relation R_{max} would be a Γ-bisimulation on $\mathcal{E}$:

$$ x \, R_{max} \, y \quad \text{iff} \quad s_x = s_y. $$

This holds for Γ and also for all of our examples of smooth operators. However, it is not true for *all* smooth operators whatsoever. In fact, there is an even weaker property that also does not hold for all smooth operators: if s is a constant function on X, then $R_{max} = X \times X$ is a Γ-bisimulation on $\mathcal{E}$. We conclude this section with a discussion of a counterexample to this property.

The *Aczel-Mendler* operator Γ_{AM} is given by

$$ \Gamma_{AM}(a) \quad = \quad \{ \langle b, c, d \rangle \in a \times a \times a \mid |\{b, c, d\}| \leq 2 \}. $$

For example, if $X = \{x, y\}$ then $\Gamma(X) = X \times X \times X$. This holds whether $x = y$ or not. But if a is a set with at least three elements, then $\Gamma_{AM}(a)$ is a proper subset of $a \times a \times a$. It is not hard to show that Γ_{AM} is a smooth operator. Furthermore, its greatest fixed point is Ω.

Exercise 17.15 Consider the following coalgebra $\mathcal{E}$ for the operator Γ_{AM}: $\mathcal{E} = \langle X, e \rangle$, where $X = \{x, y\}$ and

$$ e_x \quad = \quad \langle x, x, y \rangle $$
$$ e_y \quad = \quad \langle x, y, y \rangle $$

Note that $s_x = s_y = \Omega$, so that $R_{max} = X \times X$. Prove that $X \times X$ is *not* a Γ_{AM}-bisimulation on $\mathcal{E}$. [Hint: Consider $e'_{\langle x, y \rangle}$.]

The point again is that in order that $X \times X$ be a Γ-bisimulation, we would need to assume a condition on our operators that is stronger than smoothness.

There is such a condition in the literature called *preservation of weak pullbacks*. All of the smooth operators in this book have this property, except of course for Γ_{AM}. Thus it should be important for future studies of operators. You can read about this property in several sources, including Aczel (1988), Aczel and Mendler (1989), and Rutten and Turi (1993). We'll study this condition (actually an even weaker form of it), adapted to our setting.

Definition Let a be any set, and let t be a substitution. A *cover* for $\langle a, t \rangle$ is a triple $\langle b, u_1, u_2 \rangle$ such that:

1. $[t]_a \circ [u_1]_b = [t]_a \circ [u_2]_b$.

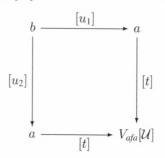

2. If $a_1, a_2 \in a$ are such that $a_1[t] = a_2[t]$, then there is some $b' \in b$ such that $b'[u_1] = a_1$ and $b'[u_2] = a_2$.

Example 17.6 The main example of a cover is obtained by taking a set a and a substitution t, and then considering the relation R on a given by

$$a_1 \, R \, a_2 \quad \text{iff} \quad a_1[t] = a_2[t],$$

and then taking $\langle b, u_1, u_2 \rangle = \langle R^o, \pi_1, \pi_2 \rangle$. The reason this is a cover is that if $a_1 \, R \, a_2$, we need only take $b = \langle a_1, a_2 \rangle \in R^o$. This property also explains the terminology of covering; the idea is that every pair a_1, a_2 such that $a_1[t] = a_2[t]$ is "covered" by some $b' \in b$.

Not every cover of $\langle a, t \rangle$ is of this form. The reason is that we do not insist that there be a *unique* b' for each suitable a_1 and a_2. So given a cover $\langle b, u_1, u_2 \rangle$ we could basically add on a disjoint copy of b to get a set $b + b$. Then by adjusting the substitutions accordingly, we could get another cover.

Definition An operator on sets Γ is *cover preserving* if there is a set $X \subseteq \mathcal{U}$ such that the complement $\mathcal{U} - X$ is very new for Γ, and also such that whenever t, u_1, and u_2 are substitutions with domains disjoint from X, if $\langle b, u_1, u_2 \rangle$ is a cover of $\langle a, t \rangle$, then also $\langle \Gamma(b), u_1, u_2 \rangle$ is a cover of $\langle \Gamma(a), t \rangle$.

Example 17.7 The main smooth operators of this book all preserve covers. For example, we consider $\Gamma(a) = \mathcal{P}(a)$. We let $X = \emptyset$ in the definition. (This is also the avoidance set for Γ that shows it to be a smooth operator.)

Suppose that $\langle b, u_1, u_2 \rangle$ is a cover of $\langle a, t \rangle$ and that $a_1, a_2 \in \Gamma(a) = \mathcal{P}(a)$ are such that $a_1[t] = a_2[t]$. Let $c_1 \in a_1$. Then as $c_1[t] \in a_1[t] = a_2[t]$, there is some $c_2 \in a_2$ such that $c_2[t] = c_1[t]$. For this pair $\langle c_1, c_2 \rangle$ there is some $b^* = b^*(c_1, c_2) \in b$ such that $b^*[u_1] = c_1$, and $b^*[u_2] = c_2$. This is for an arbitrary $c_1 \in a_1$; of course we could have started with some $c_2 \in a_2$. Let

$$ b' = \{b^*(c_1, c_2) \mid c_1 \in a_1, c_2 \in a_2, \text{ and } c_1[t] = c_2[t]\}. $$

Then $b' \in \mathcal{P}(b)$, and also $b'[u_1] = a_1$ and $b'[u_2] = a_2$. This verifies that $\langle \Gamma(b), u_1, u_2 \rangle$ is a cover of $\langle \Gamma(a), t \rangle$.

The cover preserving property is easy to check for our other operators. It is also easy to see that the composition of cover preserving operators is again cover preserving. An example of a smooth operator that is not cover preserving is the Aczel-Mendler operator Γ_{AM}.

The following result explains our interest in operators that preserve covers.

Theorem 17.11 *Suppose that Γ preserves covers. Let $\mathcal{E} = \langle X, e \rangle$ be a flat Γ-coalgebra, and let s be the solution to $\mathcal{E}$. Let t be a substitution defined on X with the following property:*

$$ \textit{If } t_x = t_y, \textit{ then } e_x[t] = e_y[t]. $$

Then t determines a Γ-bisimulation $R \subseteq X \times X$ by

$$ x \, R \, y \quad \textit{iff} \quad t_x = t_y. $$

Consequently, $t_x = t_y$ implies $s_x = s_y$.

Proof As we know, $\langle R^\circ, \pi_1, \pi_2 \rangle$ is a cover of $\langle X, t \rangle$. So since Γ preserves covers, we also know that $\langle \Gamma(R^\circ), \pi_1, \pi_2 \rangle$ is a cover of $\langle \Gamma(X), t \rangle$.

For each $\langle x, y \rangle \in R$, e_x and e_y are members of $\Gamma(X)$ with the property that $e_x[t] = e_y[t]$. By the definition of a cover, there is some $e'_{\langle x,y \rangle} \in \Gamma(R^\circ)$ such that $e'_{\langle x,y \rangle}[\pi_1] = e_x$ and $e'_{\langle x,y \rangle}[\pi_2] = e_y$. So using the Axiom of Choice we get a map $e' : R^\circ \to \Gamma(R^\circ)$. This gives us a flat Γ-coalgebra. It has the property that whenever $x \, R \, y$,

$$ e'_{\langle x,y \rangle}[\pi_1] = e_x = e(\langle x, y \rangle [\pi_1]). $$

The same equation holds for π_2 in place of π_1. It follows that R is a Γ-bisimulation on $\mathcal{E}$. So by Proposition 17.9, if $t_x = t_y$, then $s_x = s_y$. This is what we are trying to prove. $\dashv$

On the operator $\Gamma(a) = \mathcal{P}(A \cup a)$. This operator is a key one for some of our work, owing to its connection to flat systems of equations (see Chapter 6) and modal logic (Chapter 11). It is unfortunate that this operator is not smooth.

To see what goes wrong, suppose that x is any urelement, $a \in A$, $b = \{x\}$, and s is defined on b by $s(x) = a$. Then s is trivially one-to-one on b. However, $[s]$ is not one-to-one on $\Gamma(b)$ because $\{x\}[s] = \{a\} = \{x, a\}[s]$. But mapping into A is the only source of obstacles: for all substitutions $s : Y \to B$ *such that* $B \cap A = \emptyset$, $[s]$ is one-to-one on $\Gamma(Y)$.

What happens to the main theorems of this chapter for this operator? In the Corecursion Theorem, we start with $\pi : C \to \Gamma(C)$. Provided that $C \cap A = \emptyset$, we can prove that there is a unique φ as in the Corecursion Theorem. This operator Γ is also not cover preserving. But once again, it fails to have this property in a way that only minimally affects our results. For example, consider Theorem 17.11. The version of this result would be one that restricted t be be a substitution which does not take as a value any element of A.

There is an operator closely related to Γ which does have all the nice properties studied in this book: the operator $\Delta(a) = \mathcal{P}(A) \times \mathcal{P}(a)$. The connection between Γ and Δ is that if $a \cap A = \emptyset$, then $\Gamma(a)$ and $\Delta(a)$ are in bijective correspondence. It is easy to put down a mapping $j : \Gamma^* \to \Delta^*$ explicitly and to prove that it is a bijection. The point of this is that instead of extending our theory to cover corecursions into Γ^*, we could instead map into Δ^* and transfer by j. In applications, we could reformulate those applications that use Γ by using Δ instead.

Historical Remarks

The Corecursion Theorem is our version of a result in Aczel (1988) called the Special Final Coalgebra Theorem. As the name indicates, the statement of this result requires familiarity with category-theoretic ideas. While we have come part of the way in introducing the motivation for these ideas, we wanted a purely set-theoretic treatment since the material works without the extra category-theoretic overhead and so should be accessible to everyone with a basic knowledge of set theory. For this reason, we thought about corecursion theorems from scratch and were lead to smooth operators and the work of this section. The relation between our work and the original ideas is explored in Moss and Danner (to appear).

The work on generalized bisimulations in Section 17.4 are from Aczel and Mendler (1989), as is Exercise 17.15 and Theorem 17.11.

Part VI

Further Applications

18

Some Important Greatest Fixed Points

In this chapter we put the machinery developed in Part V to work by studying the greatest fixed points of some important monotone operators.

operator	Γ^*	name for Γ^*
$\Delta_1(a) = \mathcal{P}_{fin}(A \cup a)$	$HF^1[A]$	hereditarily finite sets over A
$\Delta_2(a) = A \times a \times a$	A^{trees}	infinite binary trees over A
$\Delta_3(a) = \mathcal{P}(Act \times a)$	$Can(Act)$	canonical labeled transition systems over Act
$\Delta_4(a) = 2 \times (Act \to a)$	$Aut(Act)$	canonical deterministic automata over Act
$\Delta_5(a) = A \times \mathcal{P}(a)$	$LSets(A)$	A-labeled sets

These operators are some of the simplest possible operators beyond $\Gamma(a) = \mathcal{P}(a)$ and $\Gamma(b) = A \times b$, so we would be interested in them even if their greatest fixed points did not have some independent motivation. However, each of them does have a motivation and the greatest fixed points are important objects of study.

Each of these operators is uniform, and all but the first are smooth. (These facts follow from closure properties of the uniform and smooth operators.) Therefore, our general results on induction, coinduction, and corecursion apply to give us powerful tools for studying these operators. For example, we get two principles right away, just knowing that Δ_1 is uniform:

Method of Proof [Coinduction Principle for $HF^1[A]$] To show that $Z \subseteq HF^1[A]$, show that $Z \subseteq \mathcal{P}_{fin}(HF^1[A] \cup Z)$.

Theorem 18.1 (Solution Lemma for $HF^1[A]$) *Let* $\mathcal{E} = \langle X, e \rangle$ *be a (possibly infinite) general* Δ_1*-coalgebra. (That is,* $X \cap support(A) = \emptyset$, *and each set* e_x *belongs to* $HF^1[X \cup A]$.) *Then the solution set of* $\mathcal{E}$ *is a subset of* $HF^1[A]$.

We get similar results for the other operators, but we won't bother to write them down explicitly.

Another thing we can do is to define a function from some set A into some greatest fixed point Δ_i^* by corecursion; by the Corecursion Theorem, it suffices to give a coalgebra $\pi : A \to \Delta_i(A)$. We'll see examples of this in Sections 18.2–18.4.

The last set of general results concerns bisimulations on coalgebras. In Section 17.4, we saw a general definition of a bisimulation on a coalgebra. In Sections 18.2–18.4, we'll specialize this very general work to get appropriate notions of bisimulation for Δ_i-coalgebras. Along with this will come strong extensionality principles.

18.1 Hereditarily finite sets

Suppose we are interested in finite sets, sets of finite sets, and so forth: those objects which are in some sense "hereditarily" finite. So, for example, every natural number is clearly hereditarily finite.

What about the set Ω? Is it hereditarily finite? Intuitively, it seems it should be, since it has only one member, namely itself. But there are three rather different ways of defining the hereditarily finite sets. On two of these definitions, Ω is hereditarily finite, but on one it isn't. The three ways of defining the hereditarily finite sets agree when it comes to wellfounded sets, but diverge on the non-wellfounded.

Definition Let A be a set. Let $\Delta_1(b) = \mathcal{P}_{fin}(b \cup A)$. Then Δ_1 is a monotone operator. We define $HF^0[A]$ to be its least fixed point, $(\Delta_1)_*$. Similarly, we take $HF^1[A]$ to be its greatest fixed point Δ_1^*. Finally, $HF^{1/2}[A]$ is the set of all sets $b \in HF^1[A]$ whose transitive closures are finite.[1] As usual, we write, for example, HF^0 for $HF^0[\emptyset]$, and similarly for $HF^{1/2}$ and HF^1.

In addition to the coinduction principle for $HF^1[A]$ from the last section, we also need the following induction principle.

Method of Proof [Induction Principle for $HF^0[A]$] To show $HF^0[A] \subseteq Z$,

---

[1] The definition of $HF^{1/2}[A]$ is rather specific to cardinality, and does not generalize to arbitrary monotone operators.

show that $\mathcal{P}_{fin}(Z \cap HF^0[A]) \subseteq Z$.

Of course this holds precisely because $HF^0[A]$ is the least fixed point of a monotone operator.

Another observation is that $HF^{1/2}[A]$, too, is a fixed point of Δ_1. This is a consequence of the following general formula:

$$TC(b) \;=\; \bigcup\{TC(c) \mid c \in b \text{ is a set}\} \cup (b \cap \mathcal{U}).$$

We have two results about all three fixed points, beginning with a description of the most basic relations between them.

Proposition 18.2 *Consider the fixed points of* Δ_1.

1. *If A is transitive, then $HF^0[A]$, $HF^{1/2}[A]$, and $HF^1[A]$ are all transitive.*
2. $HF^0[A] \subseteq HF^{1/2}[A] \subseteq HF^1[A]$.
3. $b \in HF^0[A]$ *iff b is wellfounded and $b \in HF^1[A]$.*
4. *Every natural number n belongs to HF^0.*
5. $\Omega \in HF^{1/2}$ *(and hence in HF^1) but not in HF^0. The infinite stream*

$$s_0 \;=\; \langle 0, \langle 1, \langle 2, \ldots \rangle\rangle\rangle$$

is in HF^1 but not in $HF^{1/2}$.

Proof In part (1) the transitivity of $HF^0[A]$ and $HF^1[A]$ is a consequence of more general observation: if A is transitive and D is a fixed point of Δ_1, then D is transitive. As for $HF^{1/2}[A]$, note that if $a \in b$, then $TC(a) \subseteq TC(b)$. Thus if the transitive closure of a set is finite, so is the transitive closure of each of its elements.

As for (2), for any operator Γ whatsoever, $\Gamma_* \subseteq \Gamma^*$. Therefore $HF^0[A] \subseteq HF^1[A]$. To see that $HF^0[A] \subseteq HF^{1/2}[A]$, we use the induction principle for $HF^0[A]$. We show that

$$\mathcal{P}_{fin}(A \cup (HF^{1/2}[A] \cap HF^0[A])) \;\subseteq\; HF^{1/2}[A].$$

To see this, let b be a finite subset of $(A \cup (HF^{1/2}[A] \cap HF^0[A]))$. Then the transitive closure of b is finite: $TC(b) = b \cup \bigcup_{a \in b} TC(a)$ is a finite union of finite sets.

The left-to-right direction of (3) follows from (2) and the fact that every element of HF^0 is wellfounded. To prove this, we again use the induction principle for HF^0. Let WF be the class of wellfounded sets. Then

$$\mathcal{P}_{fin}((WF \cap HF^0) \cup A) \;\subseteq\; WF$$

because every set whose elements are either wellfounded or urelements is wellfounded. By induction, $HF^0 \subseteq WF$.

To prove the converse direction of (3), we use the principle of $\in$-induction for wellfounded sets. Assume that b is wellfounded and belongs to $HF^1[A]$, and that every element c of b with both of these properties belongs to $HF^0[A]$.

Now b is a finite subset of $HF^1[A] \cup A$. The wellfounded sets and $HF^1[A]$ are both transitive collections, and our induction hypothesis implies that b is a finite subset of $HF^0[A] \cup A$. Thus $b \in HF^0[A]$.

Part (4) is proved by induction on the natural numbers. Suppose that every element of n belongs to HF^0. Since n is finite, n is a finite subset of HF^0, and hence an element of HF^0. Therefore every n belongs to HF^0.

The first claim in (5) is clear, since the transitive closure of Ω has only one element. The fact that the stream s_0 mentioned in (5) belongs to HF^1 follows from the Solution Lemma for HF^1; see Exercise 18.4. Also, $s_0 \notin HF^{1/2}$ since $TC(s_0)$ contains each natural number. $\dashv$

One very natural question that should come to mind is why exactly $HF^1[A]$ is always a *set*. For that matter, why is $HF^0[A]$ always a set? It should be noted, for example, that it is possible to have similar definitions whose least or greatest fixed points are proper classes. For example, if we strike the word "finite" from the conditions above, and took $A = \emptyset$, then the least fixed point would be the wellfounded sets (not itself a set, even with *AFA*). The greatest fixed point would be the class V_{afa} of pure sets, if we assume *AFA*. The point we want to make is that in the case of Δ_1, the greatest fixed point (and hence the least) is a *set*. The key to this is the second part of the next theorem.

Theorem 18.3 *Consider the fixed points of Δ_1.*

1. *$HF^0[A] = (\Delta_1)_\omega[A]$. That is, $HF^0[A] = \bigcup_n V_n[A]$, where $V_0[A] = \emptyset$, and $V_{n+1}[A] = \Delta_1(V_n \cup A)$.*

2. *Let $X = \{x_1, x_2, \dots, x_n, \dots\}$ be a fixed infinite set disjoint from A, and let C be the collection of flat Δ_1-coalgebras whose indeterminates are subsets of X. Then C is a set, and*

   (1) $\qquad HF^1[A] \quad = \quad \bigcup \{\text{solution-set}(\mathcal{E}) : \mathcal{E} \in C\}.$

   In particular, $HF^1[A]$ is a set.

Proof Let $C = \bigcup_n V_n[A]$. For the first part, we first check that C is a fixed point of Δ_1. If a is a finite subset of $A \cup C$, then there is some natural number n such that $a \cap C \subseteq V_n[A]$. Then $a \in V_{n+1}[A]$. This shows that C is a fixed point of Δ_1. To see that it is the smallest such collection, let C' be any fixed point. We show by induction on n that $V_n[A] \subseteq C'$. Clearly this is true for $n = 0$. Assuming it for n, note that

$$V_{n+1}[A] \quad = \quad \Delta_1(V_n[A]) \quad \subseteq \quad \Delta_1(C') \quad = \quad C'.$$

We turn to the second part. We know from Theorem 16.6 that every $a \in HF^1[A]$ belongs to the solution set of some flat Δ_1-coalgebra, say $\mathcal{E} = \langle Y, e \rangle$ and $a = s_x$ for some $x \in Y$. We may assume that Y is a countable set. That is, we can replace Y with the set of $y \in Y$ *accessible from x*.

These are the (finitely many) $y \in Z_0 = e_x \cap Y$ together with the (finitely many) $y \in Z_1 = \bigcup\{e_x \cap Y : z \in Z_0\}$, etc. Replacing Y with the subset accessible from x does not change the solutions to the relevant urelements. (See Exercise 9.1 for a similar argument.) In particular, x is still in the solution set.

The problem at this point is that there is no reason at this point to think that Y is the set X that we declared in this part of the theorem. Although Y might be either finite or countably infinite, we'll assume that it is infinite. Let $den : X \to Y$ be a bijection. Since Y is new for Δ_1, we can lift e by den to get $\bar{e} : X \to \Delta_1(X)$, giving a coalgebra $\mathcal{E}' = \langle X, \bar{e} \rangle$; the point is that $\mathcal{E}'$ belongs to the class $\mathcal{C}$. Also, t reorganizes $\mathcal{E}'$ onto $\mathcal{E}$. By Proposition 16.3, $\mathcal{E}$ and $\mathcal{E}'$ have the same solution set. In particular $a \in$ solution-set$(\mathcal{E}')$. Since a is arbitrary, this establishes (1).

$\mathcal{C}$ is a set because we have taken X to be a fixed set, and A, too, is a fixed set.[2] Since $\mathcal{C}$ is a set, the Replacement Scheme tells us that $HF^1[A]$ is a set. $\dashv$

We close this section with a series of results presented as exercises.

Exercise 18.1 Suppose $A \subseteq B$. Prove that $HF^0[A] \subseteq HF^0[B]$ and $HF^1[A] \subseteq HF^1[B]$.

Exercise 18.2 Let $Y \subseteq \mathcal{U}$ and suppose that $A \subseteq HF^1[Y]$. Prove that $A^\infty \subseteq HF^1[Y]$. This gives a proof that A^∞ is a set, without using any details of what A^∞ "looks like."

Exercise 18.3 Let C_n be the class of all sets a such that

$$a \cup \bigcup a \cup \bigcup\bigcup a \cup \cdots \cup \bigcup^n a$$

is finite. In Exercise 2.6 on page 16, we showed that each C_n is a proper class. Prove that $HF^1 = \bigcap_n C_n$.

Exercise 18.4 The Solution Lemma for HF^1 shows an important regard in which HF^1 is better behaved than HF^0. Consider, for example, the equation $x = \langle x, x \rangle$. The right-hand side of this equation is in $HF^0[x]$. Its solution is not in HF^0, but it is in HF^1, by the Solution Lemma for HF^1. Generalize this to show that if $a_0, a_1, \ldots$ are in HF^1 then so is $\langle a_1, \langle a_2, \langle a_3, \ldots \rangle \rangle \rangle$

Exercise 18.5 For those who know a little set theory, prove that HF^1 is uncountable.

Exercise 18.6 (A Solution Lemma for $HF^{1/2}$) Suppose that $\mathcal{E} = \langle X, A, e \rangle$ is a finite system which belongs to $HF^{1/2}[A]$. That is, $\mathcal{E}$ is a finite system of equations, and that each set e_x is in $HF^{1/2}[X \cup A]$. Prove that solution-set$(\mathcal{E}) \subseteq HF^{1/2}[A]$.

---

[2] It is here that we use the assumption that A is a set. However, the definitions of $HF^1[A], HF^1[A]$ etc., make sense even when A is a proper class of urelements, and all of the other results of this section would still hold.

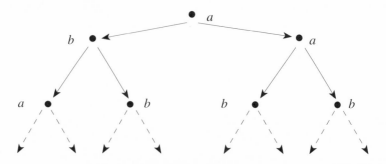

FIGURE 15 A binary tree.

Exercise 18.7 (For those who know a bit more set theory.) Prove that $HF^{1/2}$ is a model of all the axioms of *ZFA* (including *AFA*), except for the Axiom of Infinity. [The proof that *AFA* holds is essentially the Solution Lemma. The fact that $HF^{1/2}$ is transitive and closed in various ways makes most of the other axioms easy to verify.]

Exercise 18.8 What happens to the results in this section if we change "finite" to "countable?" To "of size less than κ" for some infinite regular cardinal κ?

18.2 Infinite binary trees

Intuitively, an *infinite binary tree over* A is a structure of the kind depicted in Figure 15. Notice that the nodes of the tree are labeled with elements of A. Let t be such a tree. We naturally think of the two children of t, say t_1 and t_2, as being trees in their own right. This suggests defining a binary tree over A corecursively to be anything of the form $\langle a, t_1, t_2 \rangle$ such that t_1, t_2 are binary trees over A. This can be put into our framework by means of the following definition.

Definition Let A be any set, and let Δ_2 be the operator $\Delta_2(b) = A \times b \times b$. Δ_2 is monotone, and indeed smooth. Let A^{trees} be the greatest fixed point of Δ_2. The elements of A^{trees} are called *(infinite binary) trees over* A. If t is such a tree, we write $1^{st}(t)$, $2^{nd}(t)$, and $3^{rd}(t)$ for the components of t. Note that $1^{st}(t)$ is an element of A (think of a label on a node of the infinite binary tree); $2^{nd}(t)$ and $3^{rd}(t)$ are again trees over A.

Recall that with streams we had a mental crutch: A^∞ is in correspondence with the function space $N \to A$. Using our work in Part V, we get results about

A^{trees} just by knowing that it is the greatest fixed point of a smooth operator, without introducing any representational apparatus.

Our general theory tells us that A^{trees} is a set, just as in Exercise 18.2: Suppose that $A \subseteq HF^1[Y]$. First, since $\Delta_2(X) \subseteq HF^1[Y \cup X]$ for all $X \subseteq \mathcal{U}$, we see that $A^{trees} = \Delta_2^* \subseteq HF^1[Y]$. In particular, A^{trees} is a set.[3]

Here is an exercise which also uses the method of Exercise 18.2:

Exercise 18.9 Let $A \subseteq \mathcal{U}$. Borrowing a term from formal language theory, call a tree t over A *rational* if $TC(t)$ is a finite set. This is equivalent to having $t \in HF^{1/2}[A]$. Prove that the set of rational trees is a fixed point of Δ_2.

More generally, let Γ be a uniform operator. Assume that $A \subseteq \mathcal{U}$ has the property that for all sets $X \subseteq \mathcal{U}$, $\Gamma(X) \subseteq HF^{1/2}[A \cup X]$. Let $\Gamma^r = \Gamma^* \cap HF^{1/2}[A]$.

1. Prove that for all sets a, $a \in \Gamma^r$ iff a belongs to the solution set of a finite Γ-coalgebra.

2. Prove that Γ^r is a fixed point of Γ.

We also get a principle of proof by coinduction for trees from our general theory, and a Solution Lemma for trees. These are straightforward instances of our earlier work. Since we won't need the exact details, we omit the statements altogether. However, we will use Δ_2-bisimulations, which we call *tree bisimulations*.

Proposition 18.4 *Let $\langle b, e \rangle$ be a Δ_2-coalgebra. A relation R on $b \times b$ is a tree bisimulation if tRu implies that $1^{st}(e_t) = 1^{st}(e_u)$, $2^{nd}(e_t) R 2^{nd}(e_u)$, and $3^{rd}(e_t) R 3^{rd}(e_u)$.*

The proof is an easy generalization of Proposition 17.8.

STRONG EXTENSIONALITY PRINCIPLE FOR A^{trees} Let R be a tree bisimulation on A^{trees}. If t and u belong to A^{trees} and tRu, then $t = u$.

This result is an immediate consequence of Theorem 17.10.

We close this section with some further examples of corecursion and coinduction pertaining to trees. For example, consider $lb : A^{trees} \to A^{\infty}$ (for *left branch*) given by

(2) $$lb(t) = \langle 1^{st}(t), lb(2^{nd}(t)) \rangle$$

It is not hard to use the Corecursion Theorem from Chapter 17 to prove that there is a unique function lb as in (2). We take $\Gamma(a) = A \times a$, $C = A^{trees}$ and the pump $\pi : C \to \Gamma(C)$ to be given by $\pi(\langle a, t_1, t_2 \rangle) = \langle a, t_1 \rangle$. When we work out a formula for the function φ given in the Theorem, we have (2).

---

[3]Notice that this implies that each $t \in A^{trees}$ is a finite set. The reason why we call t an *infinite* binary tree is that the tree in Figure 15 is infinite. Put another way, the word *infinite* refers to the depth of the set involved, not the breadth.

In the other direction, we can define a map $copy : A^\infty \to A^{trees}$ by

$$copy(s) \quad = \quad \langle 1^{st}(s), copy(2^{nd}(s)), copy(2^{nd}(s)) \rangle.$$

For example, if $s = \langle 0, \langle 1, \langle 2, \ldots \rangle \rangle \rangle$, then $copy(s)$ is a tree whose top node is 0, whose next level is two 1's, and whose next level after that is four 2's, etc. Another use of the Corecursion Theorem proves the existence and uniqueness of $copy$.

Define $invert : A^{trees} \to A^{trees}$ by

$$invert(t) \quad = \quad \langle 1^{st}(t), invert(3^{rd}(t)), invert(2^{nd}(t)) \rangle.$$

We might like to prove an identity such as $invert(invert(t)) = t$. To do this, consider the relation R on trees over A such that for all s,

$$invert(invert(s)) \quad R \quad s.$$

We claim that R is a tree bisimulation. Suppose that $invert(invert(t))\ R\ t$. Note that

$$1^{st}(invert(invert(t))) \quad = \quad 1^{st}(invert(t)) \quad = \quad 1^{st}(t).$$

Furthermore,

$$
\begin{aligned}
2^{nd}(invert(invert(t))) \quad &= \quad invert(3^{rd}(invert(t))) \\
&= \quad invert(invert(2^{nd}(t))).
\end{aligned}
$$

So indeed $2^{nd}(invert(invert(t)))\ R\ 2^{nd}(t)$; this is part of what we need to show to know that R is a tree bisimulation. The same argument works for 3rd, finishing the verification.

Exercise 18.10 Define $rb : A^{trees} \to A^\infty$ by

$$rb(t) \quad = \quad \langle 1^{st}(t), rb(3^{rd}(t)) \rangle.$$

Prove that for all t, $lb(t) = rb(invert(t))$.

Exercise 18.11 The last exercise can be viewed as establishing an equational law: $lb = rb \circ invert$. Similarly, the discussion before it shows that $lb \circ copy = i_{str}$, where i_{str} is the identity function on streams. Consider all possible equations between compositions involving rb, lb, $copy$, $invert$, i_{str}, and i_{tr}. Of course, we are not concerned with ill-typed compositions like $rb \circ lb$ but only with ones that actually make sense. Write down a complete set of laws for these kinds of equations. ("Complete" here means that all other laws can be derived from these using equational logic. If you do not know about equational logic, you should try to write down all the additional laws you can think of, and read the answer for a short discussion.)

18.3   Canonical labeled transition systems

Labeled transition systems (lts's) were introduced in Section 3.2. You probably should review that material now, because our discussion here picks up where that left off.

Definition Let Act be some fixed alphabet. (That is, Act is an arbitrary set that we think of as a set of atomic actions.) Let Δ_3 be the operator $\Delta_3(b) = \mathcal{P}(Act \times b)$. A Δ_3-coalgebra $\mathcal{T} = \langle S, \delta \rangle$ is called a *labeled transition system* (over Act). We call S the *state set* of $\mathcal{T}$. A canonical coalgebra is then a *canonical lts*. The elements of Δ_3^* are called *canonical (lts) states*, and we denote the set of these by $Can(Act)$.

Example 18.1 We revisit the clock example discussed in Section 3.2. Let $Act = \{tick, break\}$. We show that there are unique canonical states *clock* and *stuckclock* in $Can(Act)$ such that $stuckclock = \emptyset$ and

$$clock \quad = \quad \{\langle tick, clock \rangle\} \cup \{\langle break, stuckclock \rangle\}$$

In more conventional notation,

$$clock \quad = \quad tick.clock + break.stuckclock$$

To do this, we simply take two urelements x and y which do not belong to Act, and we consider the Δ_3-coalgebra

$$
\begin{aligned}
x &= \{\langle tick, x \rangle, \langle break, y \rangle\} \\
y &= \emptyset
\end{aligned}
$$

The Solution Lemma for Operators (Lemma 16.2) tells us that the solution set of this system will indeed be a subset of $Can(Act)$.

As we know, for every monotone operator Γ, the greatest fixed point Γ^* gives us a coalgebra $\langle \Gamma^*, i \rangle$, where i is just the identity function on Γ^*. We denoted this coalgebra by Γ^*, too. Specializing to the case of Δ_3, we have a lts $Can(Aut) = \langle Can(Aut), i \rangle$. (The fact that $Can(Aut)$ is a proper class is not an issue here, since we don't require coalgebras to be based on sets.)

At this point we need a definition of *lts bisimulation*. As with trees, we look back at Section 17.4, and specialize the general definition to Δ_3. Fortunately, this is done for us in Exercise 17.13. We reformulate that result a bit so that it matches the standard notion of a bisimulation of lts's.

Proposition 18.5 Let $\mathcal{T} = \langle S, \delta \rangle$ be an lts. A relation R on S is an lts bisimulation *if sRt implies that for all $a \in Act$,*

1. *If $\langle a, s' \rangle \in \delta(s)$, then there is some $\langle a, t' \rangle \in \delta(t)$ such that $s'Rt'$.*
2. *If $\langle a, t' \rangle \in \delta(t)$, then there is some $\langle a, s' \rangle \in \delta(s)$ such that $s'Rt'$.*

STRONG EXTENSIONALITY PRINCIPLE FOR $Can(Act)$: Let R be an lts-bisimulation on $Can(Act)$. If s and t are canonical states such that sRt, then $s = t$.

We also need the notion of a coalgebra morphism. As with bisimulations, there are two ways to get this. One could work with Δ_3-coalgebra morphisms from Chapter 16.[4] Alternatively, one could simply put down the classical notion. A morphism from $\mathcal{T}_1 = \langle S_1, \delta_1 \rangle$ to $\mathcal{T}_2 = \langle S_2, \delta_2 \rangle$ is a function $f :$ $S_1 \to S_2$ such that for all $s \in S_1$ and all $a \in Act$,

(3) $\qquad\qquad \langle a, t \rangle \in \delta_1(s) \quad \text{iff} \quad \langle a, f(t) \rangle \in \delta_2(f(s)).$

Theorem 18.6 *Let* $\mathcal{T} = \langle S, \delta \rangle$ *be an lts. There is a unique morphism* $\varphi : \mathcal{T} \to Can(Act)$. *Moreover, if* $\mathcal{T}$ *is strongly extensional, then* φ *is an isomorphism.*

Proof We use δ as a pump in the Corecursion Theorem. This gives a unique $\varphi : S \to Can(Act)$ such that

(4) $\qquad\qquad \varphi(s) \quad = \quad \{ \langle a, \varphi(t) \rangle \mid s \overset{a}{\longrightarrow} t \}.$

The image $\varphi[S]$ is a set of canonical lts states. Hence $\langle \varphi[S], i \rangle$ is a canonical lts. The definition of φ makes it a morphism of lts's. That is, (3) holds.

We next prove that φ is unique. Suppose that ψ were another morphism of lts's. Then (3) would hold and the image of ψ would be a set of canonical lts states. Therefore, (4) would also hold. By the uniqueness part of the Corecursion Theorem, $\psi = \varphi$.

We must also consider what happens when $\mathcal{T}$ is strongly extensional to begin with. We show that the map φ above is injective, since this implies that φ is an isomorphism. For this, we claim that the following relation R is a bisimulation on $\mathcal{T}$: sRt iff $\varphi(s) = \varphi(t)$. To see that this really is a bisimulation, suppose that sRt, so that $\varphi(s) = \varphi(t)$, and also that $\langle a, s' \rangle \in \delta(s)$. Then $\langle a, \varphi(s') \rangle \in \varphi(s)$, so also $\langle a, \varphi(s') \rangle \in \varphi(t)$. But then we must have some t' such that $\langle a, t' \rangle \in \delta(t)$, and such that $\varphi(t') = \varphi(s')$. This just tells us that $s'Rt'$. This concludes our proof that R is a bisimulation. $\dashv$

Exercise 18.12 Prove that there is a unique function $f : Act^\infty \to Can(Act)$ such that for every stream s from Act,

$$f(s) \quad = \quad \{ \langle 1^{st}(s), f(2^{nd}(s)) \rangle \}.$$

Thus every stream on Act can be interpreted as a canonical state. Which canonical states are of the form $f(s)$ for some stream s on Act?

18.4 Deterministic automata and languages

Automata theory is very important in computer science, both in its own right, and for the techniques it has generated which have been lifted to other parts of the subject. One of these techniques has to do with the optimization of a given automaton. The Myhill-Nerode Theorem tells us how to take any automaton

---

[4]There is a slight problem here in that our earlier definition used substitutions for the morphisms. So it is best suited to morphisms where the domain is *flat*. To get a notion that worked in all cases, it would be better to work with flat coalgebras first, and then use notation systems to get the notion for arbitrary coalgebras. This is what we did with generalized bisimulations in Section 17.4.

and find another automaton that accepts the same language and has the smallest number of possible states. We'll sketch a proof of this famous result by analogy with our work on lts's.

First, we want to move from labeled transitions to automata. The extra machinery involved is a specification of a set of *accepting* states. At the same time, we want to work with *deterministic* automata. These have the property that for all states s and acts a, there is exactly one t such that $\langle a, t \rangle \in \delta(s)$. These choices lead us to the operator $\Delta_4(b) = 2 \times (Act \to b)$. The 2 here is $\{0, 1\}$. The point of 0 and 1 will be explained shortly.

A *deterministic automaton* is a Δ_4-coalgebra $\mathcal{T} = \langle S, \delta \rangle$. That is, for each $s \in S$ and each $a \in Act$ there is at most one s' such that $\langle a, s' \rangle \in 2^{nd}(\delta(s))$. In this case we write $\delta(s, a) = s'$ instead of the more cumbersome $\langle a, s' \rangle \in 2^{nd}(\delta(s))$. Note that δ is now a two-place partial function.

The greatest fixed point of Δ_4 will be denoted by $Aut(Act)$, the set of *canonical deterministic automaton states over Act*. The idea is that in a deterministic automaton, the states are classified into accepting and non-accepting states. If $1^{st}(\delta(s)) = 1$, then s is an accepting state, and if $1^{st}(\delta(s)) = 0$, then s is not an accepting state.

At this point, we need definitions of bisimulation and of morphisms of automata. Let $\mathcal{A} = \langle S, \delta \rangle$ be an automaton. An *automata bisimulation on* $\mathcal{A}$ is a relation R on S with the following property: whenever sRt, then $1^{st}(\delta(s)) = 1^{st}(\delta(t))$, and also that for all $a \in Act$, $\delta(s, a) \, R \, \delta(t, a)$. The proof of this is an easy generalization of Proposition 18.5.

As you should expect by now, $Aut(Act)$ is strongly extensional: every automata bisimulation on $Can(Act)$ is a subrelation of the identity.

An *automata morphism* from $\mathcal{A}_1 = \langle S_1, \delta_1 \rangle$ to $\mathcal{A}_2 = \langle S_2, \delta_2 \rangle$ is a function $f : S_1 \to S_2$ such that for all $s \in S_1$, $1^{st}(\delta_2(f(s))) = 1^{st}(\delta_1(s))$, and all $a \in Act$,

$$f(\delta_1(s, a)) = \delta_2(f(s), a).$$

Theorem 18.7 *Let $\mathcal{A} = \langle S, \delta \rangle$ be an automaton. There is a morphism of automata $\varphi : S \to Aut(Act)$. Moreover, if $\mathcal{A}$ is strongly extensional, then φ is an isomorphism.*

Proof The map φ is given by

(5) $\varphi(s) = \langle 1^{st}(\delta(s)), \{\langle a, \varphi(\delta(s, a)) \rangle \mid \delta(s, a) \text{ is defined}\}\rangle.$

The details of the proof are as in Theorem 18.6.   ⊣

This result is essentially the construction in the Myhill-Nerode Theorem. The remainder of that result deals with languages and minimality. We will not go into minimality here, but we do want to make the connection of our work to languages. This will lead to our most important result in this section, the connection between languages and the greatest fixed point $\Delta_4^* = Aut(Act)$.

Definition Let Act be some set. Then Act^* is the set of finite strings from Act, including the empty string ϵ. We speak of the elements of Act^* as *words*, and use letters like v and w for them. A *language (over Act)* is a subset $L \subseteq Act^*$.

The idea of a deterministic automaton is that it reads the strings

$$w \quad = \quad a_0 a_1 \cdots a_k$$

from Act, beginning in some state s and changing states according to δ. If $\mathcal{A}$ is in an accepting state when it has read a string w, then we say $\mathcal{A}$ *accepts w at s.*[5] The empty string ϵ is accepted at s iff s is an accepting state. We define $lang_{\mathcal{A}}(s)$, the *language of A at s*, by

$$lang_{\mathcal{A}}(S) \quad = \quad \{w \in Act^* \mid \mathcal{A} \text{ accepts } w \text{ at } s\}.$$

Example 18.2 Suppose that $Act = \{a, b\}$, $S = \{s, t, u\}$, and δ is given as follows:

$$\begin{aligned}
\delta(s) &= \langle 0, \{\langle a, t\rangle, \langle b, u\rangle\}\rangle \\
\delta(t) &= \langle 0, \{\langle a, t\rangle, \langle b, u\rangle\}\rangle \\
\delta(u) &= \langle 1, \{\langle a, t\rangle, \langle b, t\rangle\}\rangle
\end{aligned}$$

This gives an automaton $\mathcal{A} = \langle S, \delta \rangle$. Note that u is the only accepting state here. It is not hard to see that $lang_{\mathcal{A}}(u)$ is the set of all words w which contain an even number of b's; this includes ϵ. Further, $lang_{\mathcal{A}}(t)$ is the set of all words w which contain an odd number of b's. $lang_{\mathcal{A}}(s)$ is the set of all words w such that w contains an odd number of b's iff w begins with a b.

Lemma 18.8 *Let $\mathcal{A} = \langle S, \delta \rangle$ be an automaton, and let R be the following relation on S:*

$$sRt \quad \text{iff} \quad lang_{\mathcal{A}}(s) = lang_{\mathcal{A}}(t).$$

Then R is an automaton bisimulation. Therefore, if $\mathcal{A} = Aut(Act)$, $s = t$.

Proof Suppose that sRt. We'll omit the subscript $\mathcal{A}$ when we write $lang(s)$ and $lang(t)$. Since $\epsilon \in lang(s)$ iff $\epsilon \in lang(t)$, we see that $1^{st}(\delta(s)) = 1^{st}(\delta(t))$. Let $a \in Act$. We need to show that $lang(\delta(s, a)) = lang(\delta(t, a))$. Let $w \in lang(\delta(s, a))$. Then $aw \in lang(s)$. So $aw \in lang(t)$ also, and we see that $w \in lang(\delta(t, a))$. The converse assertion is proved similarly. The last assertion follows from the first and Theorem 17.10. ⊣

Lemma 18.9 *Let $\mathcal{A} = \langle S, \delta \rangle$ be an automaton, and let $Aut(Act)$ be the canonical automaton. Let $\varphi : S \to Aut(Act)$ be given by (5). Then for all $s \in S$, the language of s in $\mathcal{A}$ is the same as the language of $\varphi(s)$ in $Aut(Act)$.*

---

[5]The reader may have noticed that we do not include a start state in our automata. To do so explicitly would involve "pointed coalgebras" in the manner of Chapter 9. There is no problem in doing this, but the statements of some of the results here become a bit more complicated. So we will speak of the language of an automaton $\mathcal{A}$ *at a state* rather than of *the* language of $\mathcal{A}$.

Proof By induction on n; a similar result appears in Lemma 18.10. ⊣

The next step in our work is to turn the set of languages into a deterministic automaton by the following definitions:

$$i_L = \begin{cases} 1 & \text{if } \epsilon \in L \\ 0 & \text{if } \epsilon \notin L \end{cases}$$
$$\delta_0(L, a) = \{w \in Act^* \mid wa \in L\}$$
$$\delta(L) = \langle i_L, \{\langle a, \delta_0(L, a) \mid a \in Act\rangle\}\rangle$$

We call this automaton $\mathcal{L}$.

Example 18.3 Refer back to Example 18.2. Let $L_s = lang_{\mathcal{A}}(s)$, and let L_t and L_u be defined similarly. Then in $\mathcal{L}$,

$$\delta(L_s) = \langle 0, \{\langle a, L_t\rangle, \langle b, L_u\rangle\}\rangle$$
$$\delta(L_t) = \langle 0, \{\langle a, L_t\rangle, \langle b, L_u\rangle\}\rangle$$
$$\delta(L_u) = \langle 1, \{\langle a, L_t\rangle, \langle b, L_t\rangle\}\rangle$$

It is important to check this just from the characterization of these languages found at the end of Example 18.2. At the same time, the fact that the function δ here behaves so much like the δ in Example 18.2 is suggestive of our results to come.

The main result of this section says that $\mathcal{L}$ is isomorphic to $Aut(Act)$. First, we need an important preliminary result.

Lemma 18.10 For all $L \subseteq Act^*$, $lang_{\mathcal{L}}(L) = L$.

Proof We show by induction on n that if w is a word of length n, then $w \in lang_{\mathcal{L}}(L)$ iff $w \in L$. For $n = 0$, note that $\epsilon \in lang_{\mathcal{L}}(L)$ iff $i_L(\epsilon) = 1$ iff $\epsilon \in L$. Assume this is true for n. Let w be a word of length n and let $a \in Act$. Then

$$\begin{aligned} aw \in L' \quad & \text{iff} \quad w \in lang_{\mathcal{L}}(\delta_{\mathcal{L}}(L, a)) \\ & \text{iff} \quad w \in \delta_{\mathcal{L}}(L, a) \quad \text{by induction hypothesis} \\ & \text{iff} \quad aw \in L \end{aligned}$$

This completes the induction, hence the proof. ⊣

We now establish the connection of $\mathcal{L}$ to $Aut(Act)$. By corecursion we get a map $\varphi : \mathcal{L} \to Aut(Act)$ such that

$$\varphi(L) = \langle i_L, \{\langle a, \varphi(\delta_0(L, a)) : a \in Act\rangle\}\rangle.$$

Lemma 18.11 For all $L \subseteq Act^*$, the language of $\varphi(L)$ in $Aut(Act)$ is L.

Proof This is immediate, by Lemmas 18.9 and 18.10. ⊣

Theorem 18.12 The map φ is an isomorphism of deterministic automata. That is, $\varphi : \mathcal{L} \to Aut(Act)$ is a bijection such that for all languages $L \subseteq Act^*$:

1. $i_L = 1^{st}(\varphi(L))$

2. $\delta(\varphi(L), a) = \varphi(\delta_0(L, a))$

Proof The main point of the proof is to show that φ is a bijection. Lemma 18.11 implies that it is one-to-one. We show that φ is surjective. Let s be a canonical state, and let L be the language of s in $Aut(Act)$. We claim that $\varphi(L) = s$. By Lemma 18.11 again, the language of $\varphi(L)$ is L. So s and $\varphi(L)$ are canonical states which accept the same language. So by Lemma 18.8, $\varphi(L) = s$.

The verification that φ is a morphism of automata, points (1) and (2) in the statement of the theorem, are easy. ⊣

One of the main applications of *AFA* has been providing semantics for various sorts of transitions systems. In this regard, automata are about the simplest example there is, just a bit more complicated than labeled transition systems since they have information about initial and accepting states. What we have shown in this section is that several of the key concepts of automata theory (bisimulation, morphism of automata, languages) are closely related to the work we did earlier in the book.

18.5 Labeled sets

We close this chapter by making a few remarks about the final operator Δ_5 and its fixed points.

In using set theory, it is frequently necessary to code up objects as sets, and this coding frequently requires us to label sets. For example, if we were coding up the formulas of the modal logic $\mathcal{L}_\infty A$ we would let $A' = A \cup \{\bigwedge, \neg, \diamond, \top\}$ where the latter are taken as urelements. Thus, for example, we would use $\langle \bigwedge, \Phi \rangle$ to code up the conjunction $\bigwedge \Phi$. If we take this approach, then the infinitary language $\mathcal{L}_\infty(A)$ would be a subclass of the least fixed point of the operator $\Delta_5(a) = A' \times \mathcal{P}(a)$.

This suggests variations on a theme. In the first place, we could of course look at greatest fixed points. Many of the results we have obtained above would have analogues for this greatest fixed point.

Another variation would be to consider the operator Δ whose least fixed point is exactly $\mathcal{L}(A)$ (or $\mathcal{L}_\infty(A)$). This would then suggest looking at its greatest fixed point. This would give us a language with circular sentences like $\varphi = \Box\varphi$ and $\psi = \diamond\psi$. These sentences have a natural semantics extending that given earlier for the wellfounded language. Under this semantics, the first of these sentences is true in all sets while the former is true in exactly the non-wellfounded sets. Alexandru Baltag has investigated this logic and obtained some very interesting results.

We conclude this section with a simple result which will be relevant to the next chapter. Suppose we want to code up the universe of sets within itself, so as to leave room for other things. Let $L = \{1\}$. The L-labeled sets form the largest collection M such that if a is an L-labeled set, then $a = \langle 1, b \rangle$ for some subset b of M. Define a binary relation E on M by cEa iff $a = \langle 1, b \rangle$ and $c \in b$. It is easy to see that $\langle M, E \rangle$ is isomorphic to $\langle V, \in \rangle$. The isomorphism d is given by

$$d(a) = \{ d(c) \mid cEa \},$$

an easy example of a corecursive definition.

Now if L were not just a singleton, but contained other labels as well, the L-labeled sets would of course be richer. The above remarks shows that we can think of them as containing an isomorphic copy of the universe of all sets.

Exercise 18.13 Suppose that A is some set of labels and $\mu : A \to A$ is some function from labels onto labels. Let $LSet(A)$ be the greatest fixed point of $\Delta_5(a) = A \times \mathcal{P}(a)$. Show how to extend μ uniquely to a function $\overline{\mu} : LSet(A) \to LSet(A)$ satisfying

$$\overline{\mu}(\langle p, b \rangle) \quad = \quad \langle \mu(p), \{ \overline{\mu}(c) \mid c \in b \} \rangle$$

The effect of $\overline{\mu}$ is to corecursively replace a label p by $\mu(p)$.

Historical Remarks

Perhaps the first source to contain the result of Exercise 18.5 that HF^1 is uncountable is Booth (1990). The connection of the Myhill-Nerode Theorem to the concept of finality in the sense of category theory is due to Goguen (1973). A result closer to our work appears in Fiore (1993). This connection is an important motivation for later work in areas such as abstract data types. In a broad sense, one can say that final objects embody a kind of intentionality, since two objects are equated in a final model just in case there is some *observable* difference between them. (The exact nature of this observation varies from setting to setting.) In contrast, initial models are closer to extensionality. In our setting, initial models correspond to least fixed points.

19

Modal logics from operators

The lesson of the past is that we cannot tell in advance how long we may have to wait to find a use for each new mathematical creation or where in the real world it might arise.

Robert Osserman
The Poetry of the Universe

As we hinted in Chapter 11, modal logic has had a rather checkered history. Despite the fact that such logical luminaries as Saul Kripke, Richard Montague, Jakko Hintikka, and Dana Scott contributed to it, the influence of the famous Harvard philosopher W. V. Quine led "possible worlds semantics" to be held in contempt in some logical circles. Indeed, in the late sixties, some of the editors of the prestigious *Journal of Symbolic Logic* began to reject papers in the field without further consideration of their content, a move that led to the formation of the *Journal of Philosophical Logic* and a split in the field of logic.

Imagine the surprise when, a few years later, the techniques and results of modal logic began to be important in computer science. The same or similar mathematical structures that had been used to talk about alternative possible worlds can also be used fruitfully to talk about the possible states of a computation, the possible paths through a network, or the possible forms of a linguistic expression. By now, modal logics have proliferated in computer science and continue to be a fruitful method for the analysis of the notion of possibility. The harmful split within the field of logic is gradually being overcome.

In this chapter we would like to show how to obtain all these modal logics in a uniform manner by looking at operators. That is, we would like to show how to obtain, from a more or less arbitrary operator Γ, an associated modal logic $\mathsf{L}(\Gamma)$. What should we hope for of a modal logic $\mathsf{L}(\Gamma)$ associated with an operator Γ? We studied the classical modal logic $\mathcal{L}_\infty(A)$ in the context of the operator $\Gamma(X) = \mathcal{P}(A \cup X)$, showing how one can look at modal logic as a logic for describing, and so giving information about, non-wellfounded sets in $V_{afa}[A]$. Bisimilarity of labeled graphs (Kripke structures) turned out to be the same as equivalence in the infinitary modal logic. Put another way, the modal logic is rich enough to distinguish between different sets; if $a \neq b$ then there is a sentence of the modal logic which is true of a and false of b. These various results show that the universe of non-wellfounded sets is a very natural setting for this theory. And so in a similar vein, the aim would be to have the sentences of $\mathsf{L}(\Gamma)$ describe elements of the largest fixed point Γ^*. The analogue of Kripke structures would be flat Γ-coalgebra. The logics should be strong enough to distinguish between distinct elements of Γ^*, or, equivalently, between coalgebras that are not bisimilar.

We said "would like to" at the start of the previous paragraph advisedly, since we are not in fact able to do all we would like. We think there should be such a general framework, and we begin the program of developing such a framework here. We obtain general results which apply in a case-by-case basis to the the important operators we have studied, and their associated logics, but we do not know how to generate a logic $\mathsf{L}(\Gamma)$ in a uniform fashion for an arbitrary uniform operator Γ. To follow the work of this chapter, you will need to be familiar with Chapters 11, 17, and 18.

19.1 Some example logics

To motivate and illustrate the definitions and results of this chapter, we use the operators of Chapter 18, seeing how each of them gives rise to a logic in a natural way. We recall the operators of interest in the table below:

Operator	Constants	Modal operators
$\Delta_1(a) = \mathcal{P}_{fin}(A \cup a)$	a for each $a \in A$	$\Diamond$
$\Delta_2(a) = A \times a \times a$	a for each $a \in A$	$left\!:,\ right\!:$
$\Delta_3(a) = \mathcal{P}(Act \times a)$	none	$\Diamond_a$ for each $a \in Act$
$\Delta_4(a) = 2 \times (Act \to a)$	$0, 1$	$a\!:$ for all $a \in Act$.
$\Delta_5(a) = A \times \mathcal{P}(a)$	a for each $a \in A$	$\Diamond$

With each operator we'll associate a logic L. Each logic will be determined by a specifying set of *constants* and a set of *(unary) modal operators*. The constants will be like the atomic propositions a of $\mathcal{L}_\infty(A)$; the modal operators will be generalizations of the $\Diamond$ operation. The constants and modal operators for the first few logics are shown in the table. You are invited to think about the other logics in Exercise 19.1 below.

All of our languages also have negation, full infinitary conjunction (and hence disjunction and the constants T and F).[1] Moreover, every 'existential' modal operator $\Diamond$ will give rise to a corresponding dual 'universal' modality $\Box$ defined by $\Box\varphi = \neg\Diamond\neg\varphi$. All of these are standard abbreviations, and they work exactly as in Chapter 11.

Example 19.1 Recall the operator $\Delta_1(a) = \mathcal{P}_{fin}(A \cup a)$. The largest fixed point of this operator is the set $HF^1[A]$ of hereditarily finite sets on A. We obtain a logic for describing elements of $HF^1[A]$. To define this logic, we must specify a set of constants and a set of modal operators. We take A as our set of constants. There is a single modal operator $\Diamond$. The semantics is given by

$$\begin{aligned} c &\models a & \text{iff} & \quad a \in c \\ c &\models \Diamond\varphi & \text{iff} & \quad \text{for some } b \in c, b \models \varphi \end{aligned}$$

Now there are two important remarks about this logic. First, once the two schemes above determine the semantics of the whole language, since negation and infinitary conjunction are automatic. This happens with all of our logics, and we won't repeat the point below.

Second, we have defined this semantics on elements of $HF^1[A]$, which are basically just canonical Δ_1-coalgebras, but we can also interpret this logic on flat Δ_1-coalgebras, as well as on the canonical ones. In this case, we'd have a Δ_1-coalgebra $\langle X, e \rangle$, where $e : X \to \Delta_1(X)$. The definition of the semantics would be

$$\begin{aligned} x &\models a & \text{iff} & \quad a \in e_x \\ x &\models \Diamond\varphi & \text{iff} & \quad \text{for some } y \in e_x \cap X, y \models \varphi \end{aligned}$$

As we have seen throughout this book, it will be useful to consider notions on flat coalgebras, even if we are really interested in the canonical ones. This is because the flat coalgebras are based on urelements, and so we have methods based on substitution.

Example 19.2 The next example has to do with the set A^{trees} of infinite binary trees. Recall that this set is the greatest fixed point of the operator $\Delta_2(a) = A \times a \times a$. For this operator, we take as atomic sentences the elements $a \in A$. There are two modalities which we will write as *left*: and *right*:. The semantics is

$$\begin{aligned} t &\models a & \text{iff} & \quad 1^{st}(t) = a \\ t &\models left : \varphi & \text{iff} & \quad 2^{nd}(t) \models \varphi \\ t &\models right : \varphi & \text{iff} & \quad 3^{rd}(t) \models \varphi \end{aligned}$$

---

[1] Actually, we will see later that the full infinitary conjunction and disjunction is not needed for many of our logics. See Exercises 19.8 and 19.9.

Thus, for example,

$$b \wedge \textit{left}: b \wedge \textit{right}: b$$

is true of a tree t iff t is headed by b, as are both of its immediate subtrees.

As before, we can just as well interpret this logic on an arbitrary Δ_2-coalgebra, say $\langle X, e \rangle$. In that case, the semantics would be

$$
\begin{array}{lll}
x \models a & \text{iff} & 1^{st}(e_x) = a \\
x \models \textit{left}: \varphi & \text{iff} & 2^{nd}(e_x) \models \varphi \\
x \models \textit{right}: \varphi & \text{iff} & 3^{rd}(e_x) \models \varphi
\end{array}
$$

Needless to say, there is nothing special about our choice of symbols for the operators. We chose it because it matches the idea of left and right subtrees.

Example 19.3 Next, we turn to a logic for the operator $\Delta_3(a) = \mathcal{P}(Act \times a)$ whose greatest fixed point is the set of canonical lts states. Here we take no constants whatsoever. But we have many modal operators. Indeed, each $a \in A$ gives us an operator $\Diamond_a$. The semantics on a flat Δ_3-coalgebra is given by

$$x \models \Diamond_a : \varphi \quad \text{iff} \quad \text{for some } y \text{ such that } x \xrightarrow{a} y, y \models \varphi.$$

(Recall that $x \xrightarrow{a} y$ means that $\langle a, y \rangle \in e_x$.) Thus, for example, the sentence

$$\Box_b \Diamond_a \mathsf{T} \wedge \Diamond_a \Box_b \mathsf{F}$$

will hold of a canonical state s iff every transition of type b can be followed by some transition of type a, but some transition of type a cannot be followed by any transition of type b.

Exercise 19.1 With the hint provided by the table, design logics for the operators Δ_4 and Δ_5.

Exercise 19.2 Consider Δ_4 and the logic from Exercise 19.1. Let $\mathcal{A}$ and $\mathcal{A}'$ be automata over Act, and let s and s' be states of these automata. Prove that $lang_\mathcal{A}(s) = lang_{\mathcal{A}'}(s')$ iff s and s' satisfy the same sentences of the language for Δ_4 with finite modal depth.

Exercise 19.3 Let B and C be any two fixed sets, and consider

$$\Delta_6(a) \quad = \quad B \times C \times \mathcal{P}(a) \times \mathcal{P}(a).$$

Design a logic for this operator.

19.2 Operator logics defined

For each of the operators Γ above, the associated operator logic $\mathsf{L}(\Gamma)$ is nicely behaved. It is strong enough to distinguish between distinct elements of the largest fixed point Γ^*, and it behaves well when we look at the relationship between flat Γ-coalgebras and their solutions. Rather than prove these one at time, though, such a fact calls out for some kind of uniform treatment. We present such a treatment here. We develop an axiomatic notion of a Γ-logic and present conditions under which such logics are well behaved in the ways just alluded to. Our definitions cover our examples, of course, as we show.

There are many choices that have to be made in setting up such an axiomatic framework, with different decisions that would still have guaranteed the results we are after and applied to the examples we have before us. We will not be surprised if the ultimate framework for such logics takes a different turn at one or more of these branch points.

Definition Let Γ be a smooth operator. A Γ-*logic* is a pair $\mathsf{L} = \langle \mathcal{L}, \models \rangle$ such that

1. $\mathcal{L}$ is a class of *formulas* (or *sentences*), built from *constants* and *modal operators*, and allowing all infinitary boolean operations.
2. $\models$ is a *semantics*, an assignment of truth values to all assertions of the form $b \models_{\mathcal{E}} \varphi$, where $\mathcal{E} = \langle c, e \rangle$ is a Γ-coalgebra, $b \in c$, and $\varphi \in \mathcal{L}$. We require that
   a. The semantics must treat the boolean operations in the classical way.
   b. The semantics must be invariant under isomorphisms: let $\langle c, e \rangle$ be a Γ-coalgebra, and let $den : X \to c$ be a Γ-notation scheme. Let $\bar{e}$ be the lift of e by den, so that $\mathcal{E}' = \langle X, \bar{e} \rangle$ is a flat Γ-coalgebra. Then for all $b \in X$ and all $\mathcal{L}$-sentences φ,

$$b \models_{\mathcal{E}} \varphi \quad \text{iff} \quad \ulcorner b \urcorner \models_{\mathcal{E}'} \varphi.$$

At this point in our work, we have a general definition of a logic, and we have a logic for each of the operators mentioned on page 284. We'll call these logics $\mathsf{L}_1, \ldots, \mathsf{L}_5$ for the rest of the chapter.

Expression maps

The main task in our axiomatization of operator logics is to find a way to relate the operator in question with sentences of the modal logic. We do this by what we call an "expression map." To motivate this idea, recall that sentences of the logics we consider give us information about canonical Γ-objects, that is, elements of Γ^*. If we start to write down natural sentences that such $a \in \Gamma^*$ satisfy, we can discover a connection with the operator Γ. Let's look at an example of this, with the operator Δ_1 with $A = \{a, b\}$. Suppose we have sets C and D satisfying

$$\begin{aligned} C &= \{a, D\} \\ D &= \{a, b, C, D\} \end{aligned}$$

At the first level of detail,

$$C \models \quad a \wedge \neg b$$
$$D \models \quad a \wedge b.$$

Having seen this, we see that

$$C \models \quad (a \wedge \neg b) \wedge \Diamond(a \wedge b) \wedge \Box(a \wedge b)$$
$$D \models \quad (a \wedge b) \wedge \Diamond(a \wedge \neg b) \wedge \Diamond(a \wedge b) \wedge \Box((a \wedge \neg b) \vee (a \wedge b)).$$

We would really like to think of the big sentences above as something like $\{a, a \wedge b\}$ and $\{a, b, a \wedge \neg b, a \wedge b\}$, respectively. Of course, these set expressions are not themselves sentences at all, but the sets seem to get at apt descriptions of the sentences, on the one hand, while being elements of $\Delta_1(\mathcal{L})$. Put the other way around, applying Δ_1 to sets of descriptions of our sets seems to get us something close one step better descriptions of our sets.

A bit more generally, let Z be an avoidance set for Δ_1. Suppose we had some $f \in \Delta_1(\mathcal{U} - Z)$, say

$$f = \{a, b, x_1, x_2, x_3\}.$$

We use the letter f since this set is a Δ_1-*form*. The idea is that f gives the form of what r "looks like" as an element of Δ_1^*. In general, we say that r is *of the form f* if there is a substitution

$$t : (\mathcal{U} - Z) \to \Delta_1^*$$

such that $f[t] = r$. Note that we take f to belong to $\Delta_1(\mathcal{U} - Z)$ instead of the larger class $\Delta_1(\mathcal{U})$, and we also take the domain of t to be $\mathcal{U} - Z$. This is to insure that $f[t]$ will belong to Δ_1^*. Indeed, by uniformity

$$f[t] \in (\Delta_1(\mathcal{U} - Z))[t] \quad \subseteq \quad \Delta_1(\Delta_1^*) \quad = \quad \Delta_1^*.$$

There are other, more technical reasons for these restrictions that we'll see later in this chapter. But for now, let's establish some terminology.

With each of our operators Γ, we'll fix at the outset some avoidance set Z_Γ. In nearly all the examples, Γ will have a smallest avoidance set, and we can take Z_Γ to be that one. (But even if Γ does not have a smallest one, we'll fix some avoidance set and pretend that it is given along with Γ.) A Γ-*form* will always be an element of $\Gamma(\mathcal{U} - Z_\Gamma)$, for this fixed avoidance set Z_Γ for Γ. A Γ-*substitution* will be a substitution defined on a subset of $\mathcal{U} - Z_\Gamma$.

To give us more information about r, suppose we had sentences ψ_1, ψ_2, ψ_3 about the values to be substituted for the indeterminates x_1, x_2, x_3. Now such an assignment is really nothing more than a substitution $s : \{x_1, x_2, x_3\} \to \mathcal{L}$ where we let $s(x_1) = \psi_1$, $s(x_2) = \psi_2$, and $s(x_3) = \psi_3$. (The action of s on other urelements is arbitrary in this discussion.) We can then think of the pair $\langle f, s \rangle$ as giving a description of a set r, namely that is is of the form $r[t]$ where $r(x_i) \models \psi_i$ for each i.

CHAPTER 19

Now notice that if we use s as a substitution in f we obtain

$$f[s] \quad = \quad \{a, b, \psi_1, \psi_2, \psi_3\}.$$

This is just the kind of set we saw before, closely associated with a real sentence satisfied by r itself, namely,

$$\theta(f, s) \quad = \quad a \wedge b \wedge \bigwedge_{1 \leq i \leq 3} \Diamond \psi_i \wedge \Box \bigvee_{1 \leq i \leq 3} \psi_i.$$

So far, we have only looked at one logic, L_1 and the canonical objects for the associated operator Δ_1. Examining the other example logics shows that the same kind of thing happens. In each case, there is some connection between the operator Γ and the sentences which happen to be satisfied by elements of Γ-coalgebras. Further, there is a map θ which takes each Γ-form f and each Γ-substitution s and returns a $\mathcal{L}$-sentence $\theta(f, s)$. This map will have special properties that we'll discuss in due course but let's put down explicit sentences for θ in our example logics.

First, we generalize our concrete work above for Δ_1. For all $f \in \Delta_1(X)$ and all $s : X \to \mathcal{L}_1$,

$$\theta(f, s) \quad = \quad \bigwedge\{p \mid p \in A \cap f\} \wedge \bigwedge\{\neg p \mid p \in A - f\}$$
$$\wedge \bigwedge\{\Diamond s_x \mid x \in f \cap X\}$$
$$\wedge \Box \bigvee\{s_x \mid x \in f \cap X\}$$

Note that we have in mind the case when X is new for Δ_1, so that $X \cap A = \emptyset$.

For Δ_2, we have

$$\theta(f, s) \quad = \quad 1^{st} f \wedge \mathit{left} \colon s(2^{nd} f) \wedge \mathit{right} \colon s(2^{nd} f).$$

Once again, the idea is that the Γ-form f should be something like $\langle a, z, w \rangle$ and that s_z and s_w are $\mathcal{L}_2$-sentences. Then $\theta(f, s)$ is a $\mathcal{L}_2$-sentence with the following property: for all Δ_2-coalgebras $\mathcal{E} = \langle X, e \rangle$, and all $x \in X$: $x \models_{\mathcal{E}} \theta(f, s)$ iff e_x and f have the same first element, the second element of the triple e_x satisfies the sentence assigned to the second element of f by s, and similarly for the third elements.

For Δ_3,

$$\theta(f, s) \quad = \quad \bigwedge_{a \in Act} \left[\bigwedge\{\Diamond_a s_x \mid \langle a, x \rangle \in f\} \wedge \Box_a \bigvee\{s_x \mid \langle a, x \rangle \in f\} \right].$$

Things are simpler for Δ_4, where we take

$$\theta(f, s) \quad = \quad 1^{st} f \wedge \bigwedge_{a \in Act} s((2^{nd} f)(a)).$$

Finally, for Δ_5 we set

$$\theta(f, s) \quad = \quad 1^{st} f \wedge \bigwedge\{\Diamond s_x \mid x \in 2^{nd} f\} \wedge \Box \bigvee\{s_x \mid x \in 2^{nd} f\}.$$

Up until now, we have only given definitions of expression maps for each of our example logics. The definitions were based more on intuition than on anything else. We have not really said what an expression map is, what properties they are supposed to have. We will have to rectify this in order to prove anything about them. This is our next goal. This will take a bit of work, and the actual requirements on expression maps are rather abstract. To get started, let's ask ourselves when $\theta(f,s) = \theta(g,t)$, for Γ-forms f, g and Γ-substitutions $s, t : (\mathcal{U} - Z_\Gamma) \to \mathcal{L}$.

For L_1, a direct calculation shows that $\theta(f,s) = \theta(g,t)$ iff the following conditions hold:

1. $f \cap A = g \cap A$.
2. For each $x \in f \cap X$ there is some $y \in g \cap X$ such that $s_x = t_y$.
3. For each $y \in g \cap X$ there is some $x \in f \cap X$ such that $s_x = t_y$.

This suggests that the question of whether $\theta(f,s) = \theta(g,t)$ has something to do with bisimulation. Let's try another case. For L_2, we see that $\theta(f,s) = \theta(g,t)$ iff

1. $1^{st} f = 1^{st} g$.
2. $s(2^{nd} f) = t(2^{nd} g)$.
3. $s(3^{rd} f) = t(3^{rd} g)$.

After looking at such examples, we finally come up with a definition of expression map.

Definition An *expression map* θ for the Γ-logic L is a function assigning sentences to pairs of Γ-forms $f \in \Gamma(X)$ and substitutions $s : X \to \mathcal{L}$ satisfying the following, known as the *equivalence condition*:

$$\theta(f,s) = \theta(g,t) \quad \text{iff} \quad f[s] = g[t].$$

Proposition 19.1 *All of our example logics* $\mathsf{L}_1, \ldots, \mathsf{L}_5$ *have expression maps.*

Proof These verifications are quite similar to what we did just above and our earlier work on generalized bisimulations. For variety, let's check the equivalence condition for Δ_5. Suppose that $\theta(f,s) = \theta(g,t)$. Then

1. $1^{st} f = 1^{st} g$.
2. For each $x \in 2^{nd} f$ there is some $y \in 2^{nd} g$ such that $s_x = t_y$.
3. For each $y \in 2^{nd} g$ there is some $x \in 2^{nd} f$ such that $s_x = t_y$.

This implies that $f[s] = g[t]$.

Going the other way, suppose that $f[s] = g[t]$. Then we check that conditions (1) – (3) just above hold. This implies that $\theta(f,s) = \theta(g,t)$. ⊣

Later on, the definition of an "expressive logic" will include the condition that there is an expression map θ. We'll eventually also need a consequence of the equivalence condition. In the proposition below, we use the notion of a cover preserving operator from Section 17.4. One of the reasons we spent time on that notion there was that we need it here. It is for such operators that our most general results hold.

It is also useful to introduce some notation at this point. For any substitution s, write $=_s$ for the class $\{\langle x,y \rangle \in (\mathcal{U} - Z_\Gamma) \times (\mathcal{U} - Z_\Gamma) \mid s_x = s_y\}$. The connection with covers is that $\langle =_s, \pi_1, \pi_2 \rangle$ is a cover for $\langle \mathcal{U} - Z_\Gamma, s \rangle$ (see Example 17.6 on page 262). So assuming that Γ preserves covers, we see that $\langle \Gamma(=_s), \pi_1, \pi_2 \rangle$ is a cover for $\langle \Gamma(\mathcal{U} - Z_\Gamma), s \rangle$.

Lemma 19.2 *Assume that Γ preserves covers, let L be a Γ-logic, and assume that θ is an expression map. Let f and g be Γ-forms, and let $s, t : \mathcal{U} - Z_\Gamma \to \mathcal{L}$. Suppose that s and t are substitutions such that $=_s$ is a subclass of $=_t$. If $\theta(f, s) = \theta(g, s)$, then $\theta(f, t) = \theta(g, t)$.*

Proof Assuming that $\theta(f, s) = \theta(g, s)$, so that $f[s] = g[s]$. Then by the cover preserving property, there is some $c \in \Gamma(=_s)$ such that $c[\pi_1] = f$ and $c[\pi_2] = g$. Our assumption on s and t is that $=_s$ is a subclass of $=_t$. By monotonicity of Γ, $c \in \Gamma(=_t)$. We also see that $\langle \Gamma(=_t), \pi_1, \pi_2 \rangle$ is a cover for $\langle \Gamma(\mathcal{U} - Z_\Gamma), t \rangle$. It follows that $c[\pi_1][t] = c[\pi_2][t]$. This implies $f[t] = g[t]$. Therefore, $\theta(f, t) = \theta(g, t)$. $\dashv$

Canonicalizing expression maps

The equivalence condition built into the definition of an expression map is a rather weak condition on θ. In particular, it does not give a sufficient condition to decide whether $x \models_\varepsilon \theta(f, s)$ or not.

In order to motivate the rather abstract property we need, we start with a simpler condition that turns out not to work. We say that an expression map θ satisfies the the *truthfulness condition* provided that for every flat Γ-coalgebra $\mathcal{E} = \langle X, e \rangle$, every $x \in X$ and every substitution $t : (\mathcal{U} - Z_\Gamma) \to X$ such that

1. for all $z \in \mathcal{U} - Z_\Gamma, t_z \models_\varepsilon s_z$,
2. $f[t] = e_x$

it is the case that $x \models_\varepsilon \theta(f, s)$.

All of our expression maps satisfy the truthfulness condition. For example, consider the case of L_1. Consider $f = \{b, z_1, z_2, z_3\}$ and $s(z_1) = \psi_1, s(z_2) = \psi_2$, and $s(z_3) = \psi_3$. So

$$\theta(f, s) \quad = \quad b \wedge \bigwedge_{c \in A - \{b\}} \neg c \wedge \bigwedge \Diamond \psi_i \wedge \Box \bigvee \psi_i.$$

Fix $\mathcal{E} = \langle X, e \rangle$ and $x \in X$, and suppose that there is some t with the two properties above. We want to show that indeed $x \models_{\mathcal{E}} \theta(f, s)$. Since $f[t] = e_x$, we can write e_x as $\{b\} \cup Y$, for some set $Y \subseteq X$. Furthermore, for each $y \in e_x \cap X$ there is some z_i such that $t(z_i) = y$. Finally, for each i, $t(z_i) \in e_x \cap X$. And since $t_z \models_{\mathcal{E}} s_z$, we see that indeed $x \models_{\mathcal{E}} \theta(f, s)$.

The problem with the truthfulness condition is that it gives us a sufficient condition for $x \models_{\mathcal{E}} \theta(f, s)$ but not a necessary one. It may well be the case that $x \models_{\mathcal{E}} \theta(f, s)$ and yet there is no substitution $t : (\mathcal{U} - Z_\Gamma) \to X$ satisfying the displayed conditions.

Example 19.4 Considering Δ_1, let $f = \{b, z\}$, $s_z = \mathsf{T}$, $X = \{x_1, x_2\}$, and let $\mathcal{E} = \langle X, e \rangle$ be the system with $e(x_1) = \{b, x_1, x_2\}$ and $e(x_2) = 0$. Then $x_1 \models_{\mathcal{E}} \theta(f, s)$, but there are no t such that $f[t] = e(x_1)$, simply because $f[t]$ would have at most two elements, and $e(x_1)$ has three elements.

Exercise 19.4 This exercise gives a condition which is both necessary and sufficient for for $x \models_{\mathcal{E}} \theta(f, s)$ for all of our example logics, though not in general. For a fixed $\mathcal{E} = \langle X, e \rangle$ and s, let

$$\models_{s, \mathcal{E}} = \{\langle y, z \rangle \in X \times (\mathcal{U} - Z_\Gamma) \mid y \models_{\mathcal{E}} s_z\}.$$

We say that a logic has the *satisfaction condition* provided: $x \models_{\mathcal{E}} \theta(f, s)$ iff there is some $c \in \Gamma(\models_{s, \mathcal{E}})$ such that $c[\pi_1] = e_x$ and $c[\pi_2] = f$. Prove that L_3 has the satisfaction condition.

As an introduction to our next condition on logics, we recall an important feature of the canonical sentences φ_α^a. For fixed α, the collection S_α be the collection of canonical sentences of $\mathcal{L}_\infty(A)$ of rank α has the following property: any two distinct sentences are inconsistent with one another. Thus, each set satisfies exactly one sentence in S_α. The fact that the triangle operator takes us from sets with this property to sets with this property turns out to be a generalization of our earlier result in that it is what we need to focus on in our attempt to define the right notion of expression map.

Definition Let $\mathsf{L} = \langle \mathcal{L}, \models \rangle$ be a Γ-logic. A class $S \subseteq \mathcal{L}$ is a *separator* if for every Γ-coalgebra $\langle \mathcal{E}, X \rangle$ and every $x \in X$, there is a unique $\varphi \in S$ such that $x \models_{\mathcal{E}} \varphi$.

Definition A substitution $s : X \to \mathcal{L}$ is *special for $\mathcal{E}$* if

1. for all $x \in X$, $x \models_{\mathcal{E}} s_x$.
2. for all $x, y \in X$, if $x \models_{\mathcal{E}} s_y$, then $s_x = s_y$.

For each separator S, we get a substitution $j_{\mathcal{E}, S}$ which is special for $\mathcal{E}$ simply by taking

$$j_{\mathcal{E}, S}(x) = \text{the unique } \varphi \in S \text{ such that } x \models_{\mathcal{E}} \varphi.$$

If θ satisfies the truthfulness condition, then for all flat coalgebras $\mathcal{E} = \langle X, e \rangle$, and all separators S, $x \models_\mathcal{E} \theta(e_x, j_{\mathcal{E},S})$. What we want of a good expression map is that it make canonical choices in the sense that it always picks out the sentence $\theta(e_x, j_{\mathcal{E},S})$ whenever possible.

Definition An expression map θ for a logic L is *canonicalizing* if for all Γ-forms f, all separators S, all substitutions $s : (\mathcal{U} - Z_\Gamma) \to S$, and all coalgebras $\mathcal{E}$, we have $x \models_\mathcal{E} \theta(f, s)$ iff $\theta(f, s) = \theta(e_x, j_{\mathcal{E},S})$.

Proposition 19.3 *The expression maps associated with all of our example logics are canonicalizing.*

Proof We'll only give the argument for L_1. Fix a Δ_1 form f, a separator S and a substitution $s : (\mathcal{U} - Z) \to S$. Suppose first that $x \models_\mathcal{E} \theta(f, s)$. Then:

1. $e_x \cap A = f \cap A$,
2. For all $y \in e_x \cap X$ there is some $z \in f \cap \mathcal{U}$ such that $y \models s_z$.
3. For all $z \in f \cap \mathcal{U}$ there is some $y \in e_x \cap X$ such that $y \models s_z$.

Since s takes values in a separator, we see that $s \upharpoonright e_x = j_{\mathcal{E},S} \upharpoonright e_x$. It follows that conditions (2) and (3) above hold when we replace s by $j_{\mathcal{E},S}$. But this implies that $\theta(f, s) = \theta(e_x, j_{\mathcal{E},S})$.

The other direction is easier and does not use the assumption that S is a separator. Assume that $\theta(f, s) = \theta(e_x, j_{\mathcal{E},S})$. We check that $x \models_\mathcal{E} \theta(f, s)$. Since $j_{\mathcal{E},S}$ "tells the truth," $x \models_\mathcal{E} \theta(e_x, j_{\mathcal{E},S})$. (This is by the truthfulness condition, a condition which all our examples satisfy.) Therefore, $x \models_\mathcal{E} \theta(f, s)$. $\dashv$

Here is an instructive observation which you might like to try to prove.

Exercise 19.5 Let S be any set of $\mathcal{L}$-sentences and let

$$S' = \{\theta(f, s) \mid f \text{ is a } \Gamma\text{-form and } s : (\mathcal{U} - Z_\Gamma) \to S\}.$$

Prove that if S is a separator, then so is S' is also a separator.

We are finally able to present our definition of an expressive operator logic.

Definition Let Γ be an operator that preserves covers. A Γ-logic L is *expressive* if it has a canonicalizing expression map.

By Propositions 19.1 and 19.3, all our example logics are expressive. Note also that neither the truthfulness condition nor the satisfaction condition of Exercise 19.4 is incorporated here. Although these are interesting and may well be needed for other theorems in this area, they are not needed for the results of the next section.

19.3   Characterization theorems

Throughout this section we let Γ be an operator and we suppose we have an expressive Γ-logic $L = \langle \mathcal{L}, \models \rangle$. Let θ be a fixed canonicalizing expression map for L. We will show that expressive logics live up to their name, namely, that they exhibit the same kind of connection between bisimulation and satisfaction by canonical sentences that we saw in our work on modal logic. First, we show how, using θ, we can define canonical sentences by analogy with what we did in the set case.

Definition Let $\mathcal{E} = \langle X, e \rangle$ be a flat Γ-coalgebra. Define $j_{\mathcal{E},\alpha} : X \to \mathcal{L}$ by recursion on α.

$$
\begin{aligned}
j_{\mathcal{E},0}(x) &= \top \\
j_{\mathcal{E},\alpha+1}(x) &= \theta(e_x, j_{\mathcal{E},\alpha}) \\
j_{\mathcal{E},\lambda}(x) &= \bigwedge\{j_{\mathcal{E},\beta}(x) \mid \beta < \lambda\}
\end{aligned}
$$

Further, we define the *canonical invariant* $\varphi_\alpha^{\mathcal{E},x}$ to be $j_{\mathcal{E},\alpha}(x)$. So φ is a *canonical sentence* if it is of the form $\varphi_\alpha^{\mathcal{E},x}$ for some coalgebra $\mathcal{E} = \langle X, e \rangle$, $x \in X$, and ordinal α. In this case, α is the *rank* of φ. We let S_α denote the collection of canonical sentences of rank α. (Whenever possible, we leave $\mathcal{E}$ out of the notation.)

We have seen the basic idea behind $j_{\alpha+1}(x)$ several times, but let's review it with the current notation. The idea is that we should take $e_x \in \Gamma(X)$ and substitute in according to j_α. For example, in the case of Δ_2, suppose that we have a coalgebra such that $e_x = \langle a, y, x \rangle$ and also φ_α^x and φ_α^y. We would like to set $\varphi_{\alpha+1}^x = \langle a, \varphi_\alpha^y, \varphi_\alpha^x \rangle$. However, this would in all cases give an element of $\Gamma(X)[j_\alpha] \subseteq \Gamma(\mathcal{L})$. And here is where the map θ comes in: $\theta(e_x, j_\alpha)$ is the desired element of $\mathcal{L}$, and we use it for $j_{\alpha+1}(x)$.

Notice that the definition is phrased in terms of flat Γ-coalgebras. If $\langle c, e \rangle$ is a coalgebra which is not flat, then we would define the canonical sentences by considering a Γ-notation scheme $den : X \to c$ and lifting e by den.

Theorem 19.4 *Let* $\mathcal{E} = \langle X, e \rangle$ *be a flat Γ-coalgebra and define a relation R by:*

$$x \, R \, y \quad \textit{iff} \quad \textit{for all canonical } \varphi \in \mathcal{L}, \, x \models_{\mathcal{E}} \varphi \textit{ iff } y \models_{\mathcal{E}} \varphi.$$

Then R is a Γ-bisimulation on $\mathcal{E}$. Conversely, if S is any Γ-bisimulation on $\mathcal{E}$ and $x \, S \, y$, then x and y satisfy the same canonical sentences.

Corollary 19.5 *Let a and b belong to Γ^*. Then $a = b$ iff a and b satisfy the same canonical sentences of $\mathcal{L}$.*

Proof Let $den : X \to \Gamma^*$ be a Γ-notation scheme. As we know from Chapter 17, the lift of the identity map on Γ^* is the canonical coalgebra $c : X \to \Gamma(X)$. Further, the solution to $\langle X, c \rangle$ is den. By condition 2b.2 in the definition of a logic, $\ulcorner a \urcorner$ satisfies the same sentences as a; the same goes for $\ulcorner b \urcorner$ and b. Now by Theorem 19.4, there is a Γ-bisimulation on the canonical coalgebra relating $\ulcorner a \urcorner$ and $\ulcorner b \urcorner$. By Proposition 17.9, $a = b$. ⊣

The corollary generalizes results which we have seen concerning Δ_4 and our logic for deterministic automata. Corollary 19.5 and Exercise 19.2 show that there is a bisimulation relating two states s and s' of an automaton iff the same languages of the two states are the same.

Note that Theorem 19.4, like the rest of this section, presupposes that Γ preserves covers. We discussed this condition in Section 17.4. The main result on such operators that we'll need is Theorem 17.11. This result gives a sufficient condition for two elements of Γ^* to be equal, based on substitutions defined on the canonical coalgebra.[2]

The next order of business is the proof of Theorem 19.4. We stress that Theorem 19.4 and Corollary 19.5 are stated in terms of *canonical* sentences. After we finish the proof of the theorem, we'll come back to this point.

For the remainder of this section we fix a flat Γ-coalgebra $\mathcal{E} = \langle X, e \rangle$. We follow the notation as it is used above.

Lemma 19.6 *For all α, the collection S_α of canonical sentences of rank α is a separating class. That is, if $\mathcal{E} = \langle X, e \rangle$ is a Γ-coalgebra then the substitution $j_{\mathcal{E},\alpha} : X \to \mathcal{L}$ is special for $\mathcal{E}$.*

Proof By induction on α. The cases of $\alpha = 0$ and α a limit are trivial. We therefore need only check the successor step. Fix $\mathcal{E} = \langle X, e \rangle$, and assume $j_{\mathcal{E},\alpha}$ is special for $\mathcal{E}$. In other words, $j_{\mathcal{E},S_\alpha} = j_{\mathcal{E},\alpha}$.

We check first that $x \models_\mathcal{E} \theta(e_x, j_{\mathcal{E},S_\alpha})$. This is an easy consequence of the fact that θ is canonicalizing, since $\theta(e_x, j_{\mathcal{E},S_\alpha}) = \theta(e_x, j_{\mathcal{E},S_\alpha})$.

For the uniqueness part of the specialness condition, assume that $x \models \theta(f, s)$ for some Γ-form f and some $s : (\mathcal{U} - Z_\Gamma) \to S_\alpha$. Then by the separator condition again, $\theta(f, s) = \theta(e_x, j_{\mathcal{E},S_\alpha})$. ⊣

Lemma 19.7 *If $\varphi^x_\alpha = \varphi^y_\alpha$, then for all $\beta < \alpha$, $\varphi^x_\beta = \varphi^y_\beta$.*

Proof By induction on α. We need only consider the case of a successor ordinal, so assume the result for α and also assume that $\varphi^x_{\alpha+1} = \varphi^y_{\alpha+1}$. That is, $\theta(e_x, j_\alpha) = \theta(e_y, j_\alpha)$. The point is to check that for $\beta < \alpha$, $\varphi^x_\beta = \varphi^y_\beta$. The proof is again by induction on β and we also only need to consider a successor ordinal $\beta + 1$. The induction hypothesis on α tells us that if $j_\alpha(z) = j_\alpha(w)$, then also $j_\beta(z) = j_\beta(w)$. We need only show that $\theta(e_x, j_\beta) = \theta(e_y, j_\beta)$. This follows from Lemma 19.2. ⊣

---

[2]On page 263 we discussed the operator $\Gamma(a) = \mathcal{P}(A \cup a)$. This operator is not smooth, and neither is the operator Δ_2 of this chapter. Nevertheles, the results of this section do hold for these operators.

Lemma 19.8 *Let $\mathcal{E} = \langle X, e \rangle$ and $\mathcal{E}' = \langle Y, e' \rangle$ be flat Γ-coalgebras. Suppose that $r : \mathcal{E} \to \mathcal{E}'$ is a Γ-morphism. Then for all $x \in X$ and all canonical invariants φ,*

$$x \models_\mathcal{E} \varphi \quad \textit{iff} \quad r_x \models_{\mathcal{E}'} \varphi.$$

Proof This result differs from the previous lemmas in that we have two coalgebras. In order to keep the distinction clear, consider the j_α function for $\mathcal{E}$. Let's write k_α for the corresponding function on Y. The lemmas above were stated for one coalgebra only, but the proofs go through for two, mutatis mutandis. We see that for all α, $x \models \varphi^y_{\alpha+1}$ iff $e_x[j_\alpha] = e'_{r(x)}[k_\alpha]$.

Another preliminary remark is that since S_α is a separator, we need only show that for all α, $r_x \models \varphi^x_\alpha$. We show this lemma by induction on α. Here again is the successor step. The induction hypothesis and the definition of a separator imply that for all $y \in X$, $j_\alpha(y) = k_\alpha(r_y)$. In other words, $j_\alpha = k_\alpha \circ r$.

We claim next that $a[k_\alpha \circ r] = (a[r])[k_\alpha]$ for all $a \in \Gamma(X)$. The reason is that X is new for Γ, so $support(a) \subseteq Z_\Gamma \cup X$ by Exercise 16.9. And the operations $[r] \circ [k_\alpha]$ and $[k_\alpha \circ r]$ are substitution-like and agree on the urelements in $support(a)$.

Applying this observation to $e_x \in \Gamma(X)$,

$$
\begin{aligned}
e_x[j_\alpha] &= e_x[k_\alpha \circ r] \\
&= e_x[r][k_\alpha] \\
&= e'_{r(x)}[k_\alpha]
\end{aligned}
$$

By the equivalence condition, $\theta(e_x, j_\alpha) = \theta(e'_{r(x)}, k_\alpha)$. Therefore, $\varphi^{\mathcal{E},x}_{\alpha+1} = \varphi^{\mathcal{E}',r(x)}_{\alpha+1}$. As we saw in Lemma 19.6,

$$r_x \models \varphi^x_{\alpha+1} \quad = \quad \varphi^{\mathcal{E}',r(x)}_{\alpha+1}.$$

Therefore, $r_x \models \varphi^{\mathcal{E},x}_{\alpha+1}$. ⊣

Lemma 19.9 *Suppose that R is a Γ-bisimulation on $\mathcal{E}_1$, and $x \, R \, y$. Then for all canonical φ, $x \models \varphi$ iff $y \models \varphi$.*

Proof Owing to the connection of Γ-bisimulation with reorganization, this follows immediately from Lemma 19.8. ⊣

Lemma 19.10 *Let $\mathcal{E} = \langle X, e \rangle$ be a flat Γ-coalgebra with solution s. Suppose that x and y satisfy the same sentences of $\mathcal{L}$. Then $s_x = s_y$.*

Proof Although X can be a proper class, we first give the proof where X is a set. Let α be so large so that for all $x, y \in X$, if $\varphi^x_\alpha = \varphi^y_\alpha$, then $\varphi^x_{\alpha+1} = \varphi^y_{\alpha+1}$. We only need to show that the following relation R is a Γ-bisimulation on $\mathcal{E}$:

$$x \, R \, y \quad \textit{iff} \quad \varphi^x_\alpha = \varphi^y_\alpha.$$

At this point once again we use the assumption that Γ preserves covers. By Theorem 17.11, we need only show that there is some substitution, say t, so that $e_x[t] = e_y[t]$ whenever $x \, R \, y$. We take $j_{\alpha+1}$ for t. Since j_α is special for $\mathcal{E}$, $\theta(e_x, j_\alpha) = \theta(e_y, j_\alpha)$. By the equivalence condition, $e_x[j_\alpha] = e_y[j_\alpha]$. This concludes the proof in the case X is a set.

If X is a proper class, then for a given x and y there still is a set $X_0 \subseteq X$ so that $e_0 = e \restriction X_0$ maps into $\Gamma(X_0)$. In this case, we would have a flat coalgebra $\mathcal{E}_0 = \langle X_0, e_0 \rangle$ whose solution s_0 is the restriction of s to X_0. By what we already know, $s_0(x) = s_0(y)$. Therefore, $s_x = s_y$ as well. $\dashv$

This also concludes the proof of Theorem 19.4.

Invariance under reorganization

Recall that the main hypothesis in Theorem 19.4 was expressivity. If we add another hypothesis, we can get a stronger conclusion.

Definition A Γ-logic L is *invariant under reorganizations* if for every reorganization $r : \mathcal{E}_1 \to \mathcal{E}_2$ of a flat Γ-coalgebra $\mathcal{E}_1$ onto an arbitrary Γ-coalgebra $\mathcal{E}_2$, the following holds: for all $\varphi \in \mathcal{L}$ and all $x \in X = dom(\mathcal{E}_1)$,

$$x \models_{\mathcal{E}_1} \varphi \quad \text{iff} \quad r_x \models_{\mathcal{E}_2} \varphi$$

Here is the point of this condition: Up until now, our definitions are so general that a logic L could contain a constant C with the property that for all coalgebras $\mathcal{E} = \langle a, e \rangle$ and all $b \in a$,

(1) $\qquad\qquad b \models C \quad$ iff $\quad e_b$ has exactly two elements.

That is, the semantics could look at the specific coalgebra and not at the canonical one the coalgebra describes. The reorganization condition above rectifies this problem. To evaluate whether $b \models C$, or whether $b \models \varphi$ for an arbitrary φ, we can take a Γ-notation scheme $den : X \to a$ to get a flat coalgebra $\langle X, e' \rangle$; we then see whether $\ulcorner b \urcorner \models \varphi$. So (1) could not possibly hold for a logic invariant under reorganizations.

Exercise 19.6 Show that the logics for the operators $\Delta_1, \ldots, \Delta_5$ introduced in Section 19.1 are invariant under reorganization.

As in Lemma 19.9, we have the following easy result.

Lemma 19.11 *Let $\mathsf{L} = \langle \mathcal{L}, \models \rangle$ be a Γ-logic for an operator which is also invariant under reorganization. Suppose that R is a Γ-bisimulation on $\mathcal{E}$, and let $a \, R \, b$. Then for all $\varphi \in \mathcal{L}$, $a \models \varphi$ iff $b \models \varphi$.*

Theorem 19.12 *Let Γ be an operator which preserves covers, and let $\mathsf{L} = \langle \mathcal{L}, \models \rangle$ be a logic for Γ which is expressive and also invariant under reorgani-*

zation. Let $\mathcal{E} = \langle c, e \rangle$ be a Γ-coalgebra. If $a, b \in c$ satisfy the same canonical $\mathcal{L}$-sentences, then they satisfy all of the same $\mathcal{L}$-sentences.

Proof Let $\mathcal{E}' = \langle X, e \rangle$ be a flat coalgebra isomorphic to $\mathcal{E}$ via some map $den : X \to c$. Then $\ulcorner a \urcorner$ and $\ulcorner b \urcorner$ satisfy the same canonical sentences. By Theorem 19.12, there is a Γ-bisimulation on $\mathcal{E}'$ relating $\ulcorner a \urcorner$ and $\ulcorner b \urcorner$. So by Lemma 19.11, $\ulcorner a \urcorner$ and $\ulcorner b \urcorner$ satisfy all of the same $\mathcal{L}$-sentences. By the reorganization condition again, so do a and b. ⊣

Characterization by finitary formulas

Theorems 19.4 and 19.12 are not optimal results. These say that every $a \in \Gamma^*$ is characterized by the class of (canonical) formulas which it satisfies.[3] However, most of our operators have the property that if $b \in \Gamma^*$ satisfies φ_ω^a, then $b = a$. That is, elements of Γ^* are characterized by *finitary* canonical formulas. We say "finitary" instead of just finite or hereditarily finite because this weaker condition can hold, as the following important example shows.

Exercise 19.7 Consider the operator for deterministic automata, $\Delta_4(a) = 2 \times (Act \to a)$. Suppose that Act is an infinite set.

1. Show that each $c \in \Gamma^*$ is characterized by φ_ω^c.
2. Show that each φ_n^c is not hereditarily finite.

The goal of the next exercises is to present an abstract form of this result. Similar results hold for all infinite regular cardinals κ, not just ω, but we will not put these down. Also, for the rest of this discussion, fix a cover preserving Γ and an expressive Γ-logic.

Definition Let X be any set. A relation Q on X is *finite-to-finite* if the following conditions hold:

1. For every $y \in X$, there are only finitely many z such that $y \, Q \, z$.
2. For every $y \in X$, there are only finitely many z such that $z \, Q \, y$.

Let $\mathcal{E} = \langle X, e \rangle$ be a Γ-coalgebra. We say that a Γ-coalgebra $\mathcal{E}$ has the *strong separator condition* if for all Γ-forms $f \in \Gamma(X)$, and all $x \in X$, there is a finite-to-finite relation $Q \subseteq X \times X$ such that for any separator S, and any $s : X \to S$, the following are equivalent,

1. $x \models_\mathcal{E} \theta(e_x, s)$
2. For each $y \in X$ there is some $z \in X$ such that $y \, Q \, z$ and $y \models_\mathcal{E} s_z$, and for every $z \in X$ there is some $y \in X$ such that $y \, Q \, z$ and $y \models_\mathcal{E} s_z$.

---

[3] Indeed, our logics have an even stronger property: For a given $a \in \Gamma^*$, there is an ordinal α so that φ_α^c characterizes c. This property is weaker than the characterization by finitary formulas that we present in Exercises 19.8 and 19.9.

Exercise 19.8 Show that every coalgebra for each of the following logics has the strong separator condition.

1. L_1, for the operator $\Delta_1(a) = \mathcal{P}_{fin}(A \cup a)$.
2. L_4, for the operator $\Delta_4(a) = 2 \times (Act \rightarrow a)$.

Exercise 19.9 Let $\mathcal{E} = \langle X, e \rangle$ be a Γ-coalgebra with the strong separator condition.

1. Prove that if $j_\omega(x) = j_\omega(y)$, then $\theta(e_x, j_\omega) = \theta(e_y, j_\omega)$.
2. Let R be the relation on X given by

$$x \, R \, y \quad \text{iff} \quad j_\omega(x) = j_\omega(y).$$

Prove that R is a Γ-bisimulation on $\mathcal{E}$. It follows that if x and y satisfy the same finitary canonical sentences, then $s_x = s_y$, where s is the solution to $\mathcal{E}$.

A logic for Δ_2

We close this chapter with some exercises on the completeness of the logic for Δ_2. Although we only work with this example, the basic pattern in Exercise 19.11 gives completeness results for other logics, too.

Exercise 19.10 Consider the operator $\Delta_2(a) = A \times a \times a$ for trees, and also the logic L_2 for it.

1. Find some axioms concerning the modal operators *left*: and *right*: which are sound for L_2.
2. Assuming that the set A is finite, find a sound axiom for the constants of L_2.
3. In analogy with the modal logic K from Chapter 11, find some valid rules of inference for these modal operators.

Suppose we take the axioms and rules of inference of propositional logic, together with the axioms and rules from Exercise 19.10. We would get a logical system for L_2, including a notion of provability $T \vdash \varphi$. In particular, we would have definitions pertaining to the finitary part of L_2. Since our axioms of propositional logic are finitary, it makes sense to restrict attention to the finitary part of the logic, too. Now we ask about the completeness of this logical system. Let Th be the set of maximal consistent subsets of $\mathcal{L}_2$. An argument using Zorn's Lemma shows that every consistent T has a maximal consistent extension T'. To prove completeness, we need only show that every maximal consistent $T \in Th$ is satisfiable.

Exercise 19.11 Prove the completeness of our logical system by filling in the details of the following outline:

1. Prove that for every finitary sentence φ there is a finitary canonical sentence ψ such that either $\vdash \psi \to \varphi$ or $\vdash \psi \to \neg\varphi$.

2. Find a natural map $\alpha : Th \to \Delta_2(Th)$. This gives us a Δ_2-coalgebra, which we also denote by Th.

3. Prove a "truth lemma" for the coalgebra Th and thereby show that the logic is complete.

CHAPTER 19

Wanted: A strongly extensional theory of classes

The theory *ZFA* boot-straps on *ZFC* in the sense that the universe of hypersets can be looked at as an enrichment of the universe of sets axiomatized by *ZFC*. It is a richer universe, in that every system of equations has a solution. But it shares much with the wellfounded universe. In particular, its conceptual apparatus contains the same basic dichotomy of classes into the small and the large.

Actually, of course, neither theory is really a theory of classes at all. In these theories, talk of proper classes has to be understood as a *façon de parler*, since such classes are not actual objects about which the theory has anything to say. But to the extent that the theories can be viewed as having an implicit theory of classes, both have the same basic dichotomy.

20.1   Paradise lost

The large classes are the ones that are not sets. They can be characterized in the following equivalent ways in either *ZFC* or *ZFA*: A class C is "large" if and only if

vN1: C is not a member of any set;

vN2: C is not a member of any class;

vN3: C is not the same size as any ordinal; and

vN4: C *is* the same size as the universe.

The equivalence of these four statements is built into *ZFC* in response to the paradoxes.[1] It responds to them in that it seeks to pin the blame for the paradoxes on a failure to distinguish between small and large classes. What

---

[1] Actually, the equivalence of (vN4) with the remainder is equivalent to the Global Axiom of Choice (see p. 28), a version of *AC* somewhat stronger than what is usually included in *ZFC*.

started out as a paradox turns into a proof that the Russell class is large, that is, not a set, and so not a candidate for membership in itself.

Looking at these statements, one cannot fail to be struck by the quite different character of (vN1) and (vN2) from (vN3) and (vN4). While the latter are clearly about the size of C, there is nothing on the face of it that makes either of the former about size.

The linkage of these two different ideas arose in the work of John von Neumann. In his development of set theory, von Neumann explicitly postulated the equivalence of (vN1) and (vN4). His idea seems to have been that objects that can be treated as a totality should be able to be members of sets, and that the universe and other problematic multiplicities were simply too large to be grasped as a completed totality. But if the universe itself is too large, then any multiplicity of the same size should also be too large. Elegance suggests that anything not the same size as the universe should be "small enough" to be a considered as a single, unitary object. In von Neumann's development of set theory, there was a universal class, that is a class of all sets. However, it could not be a member of itself, or of any other set, since it was clearly the same size as the universe.[2]

In the more recent history of set theory, it has become standard to justify the axioms of ZFC, including Foundation, not by recourse to the size distinction, but by reference to the so-called *iterative (or cumulative) conception* of set, which says that the only sets there are are those that can be "collected together" out of "previously constructed" sets. This conception of set is based on a temporal metaphor, that of "building up" the universe of sets in "stages." "Before" one can form a wellfounded set, one has to have all of the elements of the set one is going to "collect together."

Notice that there is a quasi-constructivist flavor to this idea, the idea of being able to take a multiplicity and collect it together into a single, unitary object. This seems, at least on the face of it, quite a different idea than von Neumann's, which admitted the universe as a completed totality, but took it to be too large to be an element of any other totality.

An advantage of the iterative conception is that it gives an *explanation* as to why certain classes are too large to be elements of any set. A class X is "too large" if and only if its elements are "formed" at arbitrarily large stages in the set formation process. Granting this, there will never come a stage at which X itself gets "formed," and hence X will never be a candidate for membership in a set.

This explanation has an intuitive appeal, and it has resulted in the ascendancy of the cumulative conception as the dominant one in modern set theory.

---

[2] See section 8.3 of Hallett (1984) for a discussion of von Neumann's thinking behind his introduction of this principle.

Set theorists have not adopted the details of von Neumann's theory, but its spirit remains in *ZFC*, built into the equivalence of two very different ideas. This spirit has been inherited by the richer *ZFA*.

However, now that we are used to the idea that non-wellfounded sets are perfectly sensible objects, we have lost the cumulative picture and so the explanation of the paradoxes it made available. We have also lost the equivalence of (vN1) with the rest of (vN2) through (vN4). Consequently, we have lost the explanation as to why the universe is too large to be a completed totality. Thus, we need to reconsider the ban on large classes being elements of classes. That is what we propose to do in this chapter.

20.2   What are *ZFC* and *ZFA* axiomatizations of?

Many of the axioms of *ZFC* can be justified, informally, on the iterative conception. In particular, the Axiom of Foundation can be so justified, since on this conception, non-wellfounded sets cannot arise. For suppose you have a descending sequence of sets. You can never "construct" any member of the sequence, because first you have to construct all of its members, which leads to an infinite regress. (And once again, the argument basically assumes that the "before" relation on the construction process has no infinite descending sequences to argue that the iteratively constructed sets have the same property.)

Other axioms that can be justified on this conception are Power Set, Pairing, Union, and Infinity. Replacement (and likewise Collection), one of the most characteristic axioms of the theory, cannot be so justified, however. Recall that Replacement says that any operation F that defines a unique object $F(x)$ for each x in some set b, the predicate

being an $F(x)$ for some $x \in b$

has a set as its extension. There is nothing in the iterative conception by itself that supports this claim. What prevents these various $F(x)$ from occurring at arbitrarily large stages?

The standard response is that since we can imagine a new stage beyond all the old stages, this stage can be added to the old stages. At this new stage we can collect these sets into a new set c, and add it to the universe and keep on going.

It is well known that this response is flawed.[3] Recall that the term "operation" really is short-hand for "definable operation," where the definition takes place in first-order logic with quantifiers ranging over the universe of sets. Adding these new stages and c to the universe can drastically change the way the function F works. That is, it can happen that for some $x \in b$, $F(x)$ takes on a different value in the new universe. This is not just a theoretical possibility.

---

[3]This point has been made by several authors. See, for example, Hallett (1984) and van Aken (1986). Barwise first read it in the early 70's in an unpublished manuscript by Harvey Friedman.

The definition of the Hanf number for second-order logic is a very concrete example of this. This cardinal can be shown to exist, but its size changes as one extends the iteration process; see Barwise (1972).

If we return to the metaphor of size, rather than time, though, we get a solution to this problem. That is, the size conception *does* provide a justification for Replacement. Suppose we are given our set b and an operation F that assigns an object $F(x)$ to each $x \in b$. Since b is small, its image c under the operation F must also be small, since it cannot be larger than b itself. Hence it must be a set. The same reasoning justifies Separation as well.

The size metaphor provides an alternative justification for several of the axioms of ZFC that have justifications via the iterative conception: Pair and Infinity, for example. As van Aken (1986) notes, though, it does not justify all of them. Most obviously, it does not justify Foundation. The set $\Omega = \{\Omega\}$ is clearly not "too big" to be a set. It only has one element in its whole transitive closure.

The size metaphor fails to justify the Union Axiom. There is no reason to think that because an object b is small that its union $\bigcup b$ is small. For example, if X is large but $b = \{X\}$ exists, the latter is certainly small, having only one element. But its union $\bigcup b$ is large, since $\bigcup b = X$. Of course this cannot happen if we adopt von Neumann's strategy of denying elementhood to sets that are large, for then $\{X\}$ will not exist. But it is exactly this move which we are reconsidering.

Notice that if it is assumed that a large class is never a member of any other class, then every small class is hereditarily small. As a result, the distinction between small class and hereditarily small class vanishes. But if we allow large classes to be members of other classes, then the distinction becomes significant. If we consider not just small classes in the Union Axiom, but hereditarily small classes, the this axiom is more reasonable. If b is a small class of small classes, then analogy with cardinal arithmetic suggests that $\bigcup b$ is small.

One axiom is borderline: the Power Set Axiom. Is it true that if b is small then $\mathcal{P}(b)$ is small? Many authors have thought not, especially since Cohen's independence results concerning the size of the $2^{\aleph_0}$. However, there is another way to look at it. If "b is small" means having fewer members than the universe, then it is perhaps not unreasonable to suppose that if b is small, so is $\mathcal{P}(b)$.[4] This dichotomy about the meaning of small will return later.

Suppose we start out with some conception of pre-existing classes, thought of as multiplicities, candidates for those completed totalities we want to calls sets. We have two possible schemes for classifying them: those multiplicities that appear in the cumulative process, that is, those in the least fixed point of $V = \mathcal{P}(V)$, and those multiplicities that are hereditarily smaller than the

---

[4]The power set axiom is, admittedly, the axiom whose truth is least clear under this conception.

universe. The family of all multiplicities of the first sort satisfies some but not all of the axioms of *ZFC*. The family of all multiplicities of the second sort satisfy a different group of axioms. Each axiom of *ZFC* is satisfied by one or the other family, but neither satisfies all the axioms. What are we to make of a theory that seems to require two distinct conceptions of set to justify its axioms? Is it simply a confused hodge-podge?

The natural answer to this question, it seems to us, is to construe the axioms of *ZFC* as axiomatizing those multiplicities that are simultaneously sets under *both* conceptions. That is, think of *ZFC* as axiomatizing the notion of set thought of as the hereditarily small, cumulative classes. To justify all the axioms of *ZFC*, we need to add additional assumptions. One is that the iterative process itself is large. The other is that the powerset of a small class is small. With these two assumptions, it is possible to give a reasonably convincing informal justification of all the axioms of *ZFC*, on the basis of this conception of set.

	Arbitrary	Cumulative
Arbitrary Hereditarily small		*ZFC*

In this way *ZFC* can be argued to be a sensible, coherent theory, the theory of hereditarily small, cumulative classes. Experience shows that it is rich enough to model most of the structures of classical mathematics, structures like the reals, complexes, Hilbert spaces, function spaces, and so on. But thinking of things in this way raises unanswered questions. For example, if *ZFC* is the theory of hereditarily small, cumulative classes, shouldn't there be a theory of hereditarily small collections more generally, one that includes the iteratively collected small classes but also hereditarily small classes like Ω that are outside in the cumulative hierarchy?

While this is not the way that *ZFA* arose, we propose that *ZFA* is a natural axiomatization of the notion of hereditarily small class.

	Arbitrary	Cumulative
Arbitrary Hereditarily small	*ZFA*	*ZFC*

If this is so, then it suggests that we are missing two other theories of collections:

	Arbitrary	Cumulative
Arbitrary	*SEC*	*CC*
Hereditarily small	*ZFA*	*ZFC*

To explain this table we need to review the notion of relativizing a set theory to a (definable) class. Recall that the relativization ψ^C of a sentence ψ in the language of set theory to a definable class C results by replacing all quantifiers $\exists y \varphi(y)$ by $\exists y(y \in C \wedge \varphi(y))$ and all quantifiers $\forall y \varphi(y)$ by $\forall y(y \in C \to \varphi(y))$. (The requirement that C be definable results from the need to write down the theory.) The relativization T^C of a theory T consists in the set of all relativizations ψ^C of the axioms $\psi \in T$. If we also add the sentence $\forall x(x \in C)$, then we obtain a stronger theory $T^{(C)}$; it essentially restricts attention to the members of C. This is what we will mean by relativizing a theory to C in what follows.

As an example, consider the class WF of wellfounded sets. If we relativize ZFC^- (for example) to WF we obtain a theory $ZFC^{-(WF)}$ that says the wellfounded sets satisfy ZFC^-, plus the added axiom that $\forall x(x \in WF)$. This is equivalent to ZFC, of course.

The table is intended to suggest that there should be a natural theory SEC of *arbitrary* classes such that ZFA is equivalent to the relativization of SEC to the hereditarily small classes, and ZFC is equivalent to the relativization of SEC to the hereditarily small, wellfounded classes.

In order for this to make sense, SEC will have to provide us with a notion of small class. The equivalences (vN3) and (vN4) suggest two possibilities. We might define the small classes as those whose size is at most that of some ordinal, or we might define them to be those that are smaller than the universe. There also could be other ways of defining them. Once we have defined the small classes, we define the hereditarily small classes to be those which are small, have small elements, and so on.

The table also suggests that if such a theory were available, we could relativize it to the wellfounded classes obtaining a theory CC. This relativization should be a straightforward matter, if we could find the theory SEC of arbitrary classes. So we concentrate on this.

What evidence is there that such a theory of arbitrary classes might be possible? Or, to put it another way, doesn't Russell's Paradox show us that such a theory is impossible?

There have certainly been many logicians over the years who have felt dissatisfied with the official ZFC line on classes, people who have attempted to develop theories that allow for a universal class, for example. Indeed, there are far too many such theories; it is hard to choose between them. None have gained a wide following. The present authors do not have anything like a thorough knowledge of the various theories that have been developed. Indeed, we confess that until prompted by the above line of thought, we were not very interested in such theories.

It may well be that one of the existing theories exactly fills the bill. If so, perhaps the independent considerations we raise here will help that theory gain some credibility over its competitors.

In the remainder of this chapter we want to do two things. First, present some criteria that would govern the theory SEC of arbitrary classes that we would like to see developed. Second, suggest a way of exploring models in search of such a theory.

20.3 Four criteria

We start out with four criteria on the theory SEC of classes we are after. (We will add a fifth later.)

1. **Usefulness:** The theory should be useful.
2. **Neatness of fit:** The theory of classes should mesh neatly with the theories *ZFC* and *ZFA*.
3. **Paradise regained:** There should be a principled way of explaining the paradoxes.
4. **The universe regained:** There should be a class U of all classes.

We amplify these in turn.

Usefulness. The theory SEC should allow us enough classes to do things we want to be able to do, things we expect to be able to do from our experience in *ZFA*. For example, all definable collections of (hereditarily small) sets should be classes. From this it would follow that greatest fixed points of definable monotone operators would be classes of this theory. It would also follow that the class *On* of all (small) ordinals would exist as a class. (It would not be an ordinal, of course, since if it were it would be a member of itself and so not wellordered. Hence *On* will have to be a large class, as one would expect.) Since it is not uncommon to want to form ordered pairs of classes, this requires that we be able to form unordered pairs of classes.

One of the places where a theory of classes would be most useful would be in category theory. Thus, for example, intuitively the class of all categories, with functors as morphisms, forms a category. It would be nice to have the existence of this category be a consequence of the theory SEC.

Neatness of fit. By this, we mean several things. First of all, we want to be able to show that *ZFA* is the relativization to the hereditarily small classes. (Notice this rules out theories like Quine's NF which have more than one set a such that $a = \{a\}$.) This requires, first of all, that we have a way to define the large classes. Recalling (vN1)–(vN4), we see that there might be several options. Clearly (vN1) will not do. Neither will (vN2), since we want a class $\{C\}$ for every class C. This leaves (vN3) and (vN4) as possible definitions. Which of these one takes depends on other considerations which we will take up later. Whichever one takes, it should then follow that *ZFC* is the relativization to the hereditarily small, cumulative classes.

By neatness of fit, we also mean that the theory *SEC* should be strongly extensional, just as *ZFC* and *ZFA* are (hence the abbreviation "*SEC*" for the sought-for theory of Strongly Extensional Classes). That is, in the theory of classes we are after, bisimilar classes are to be considered identical.

This condition takes us outside of many of the theories of classes that have been proposed as alternatives to *ZFC*, so we say a bit more about it. It could be that there was a theory of classes so that you could prove the hereditarily small classes were strongly extensional without the other classes being strongly extensional, though it is hard to see how this would come about by size considerations. But there are two points to be made. The first is that if one had a universe of classes which was not strongly extensional, one could always identify bisimilar classes and so obtain a universe of the sort we are after. The second and more important point is that our experience with *ZFA* and *ZFC* suggests that set and class theories which are not strongly extensional do not really have a useful criterion of class identity.

Paradise regained. Giving up Foundation gives up a principled explanation for why there is no Russell class, so we need a different analysis of this and the other paradoxes of naïve set (and class) theory. In *ZFC* or *ZFA* we have a principled way of knowing which collections are first-class objects: the small ones, that is, the one's which can be put into a one-one correspondence with some cardinal (equivalently ordinal) number. In the theory we are after, we lose this neat characterization. We do not want to replace this principled account with an *ad hoc* restriction on which collections are first-class objects.

Universe regained. Since we have seen many examples of reflexive sets, it seems reasonable to require that the universe U itself be reflexive, that is, a member of itself. Hence we would also have classes like $\{U\}$, $\langle U, On \rangle$, etc.

In view of the Difference Lemma, this puts a real constraint on our theory. We know that if the universe U is reflexive and strongly extensional, then the collection of all sets distinct from U will not be a class. Thus U cannot be closed under complements. The requirement of strong extensionality rules out most of the theories of classes admitting a universal class, most of which have taken it as given that the classes should be closed under complements.

What intuition can there be for a notion of class that is not closed under complements? We return to this later, but for now we would simply point to a couple of analogies. In recursion theory, we are used to dealing with the class of recursively enumerable sets, which are not closed under complements. In Proposition 5.2 we saw that no self-sufficient set of formulas in the language of set theory can be closed under negation. Similarly, in topology we are used to dealing with open sets, which are likewise not in general closed under complements.

We now turn to the question of how to define the notion of large class. Should we use (vN3) or (vN4)?

Suppose we adopt (vN3) and define small to mean the same size as some ordinal. But what do we mean by an ordinal? Presumably, ordinals are classes wellordered by membership. It would not do to require ordinals to be small, since we are trying to use (vN3) to define small. But then the class *On* of all ordinals (if it exists) would be a small class. This would contradict the Burali-Forti Paradox. So we see that *On* would not be a class at all. This would be contrary to one of the hopes discussed under the criterion of usefulness.[5]

We want to take seriously the desire to have a (necessarily large) class *On* of ordinals and so adopt (vN4) as our definition of large class. Thus, we propose the following additional criterion on the theory we are after.

5. **One large size:** Within *SEC* one should be able to prove that every class have size at most that of the (large) class *On* of (small) ordinal numbers.[6]

Since we don't know just what principles *SEC* will have, we have to be a bit careful in how we define the notion of size here. Let us say that a class C has size at most that of class D if there is a class F which is an operation mapping D onto C. Then to say that a class C has size at most that of *On* is to say that there is a class operation whose domain consists of (small) ordinals and whose range is C. And we say that C is small if there is some ordinal β such that C has size at most β.

We hasten to add that we do not have a fully developed theory *SEC* to propose here. What we have, instead, is a viewpoint that suggests some axioms that should be present in (or provable from) the theory. We call this theory SEC_0. We also have a method for building models which suggests that the viewpoint is coherent and shows that the theory SEC_0 is consistent.

20.4   Classes as a *façon de parler*

In ordinary set theory it is commonplace to justify the use of classes by saying that they are just ways of talking about families of sets, when these families are identified up to extensional equivalence. Let's take this seriously but, in line with our abandonment of the cumulative picture, allow that we talk not just of "previously constructed sets," but rather of any objects at all. And in line with strong extensionality, we want to identify families not just up to extensional equivalence but up to bisimulation.

---

[5] See the historical remarks at the end of this chapter for discussion of some work which takes this general line.

[6] At first glance, this looks different from (vN4), but really they are equivalent, given that both *On* and *U* must be large. This formulation is more convenient for our purposes.

It would seem reasonable to suppose that these ways of talking can be coded within the class V of all sets. So we want to see if we can build a universe $U \supseteq V$ of classes using objects in V to talk about, or as we will say, to *name*, classes. Notice that if the universe U can be coded by names in V then U will be no larger in size than V. Given the axiom of Global Choice, U will have the same size as On.

Now if we are a bit free-wheeling, we can rephrase all this in more familiar terms. First, there should be a largest cardinal number κ, and the universe U will be of this size. So will any large class. The small classes will be those of size less than κ. Each small class will have a power class, but (by Cantor's Theorem) a large class cannot have a power class. It follows that κ will have to be a strong limit cardinal. (This means that for all cardinals λ, if $\lambda < \kappa$ then $2^\lambda < \kappa$.) But since we want Replacement to hold of the hereditarily small classes, κ will also have to be regular. Hence the largest cardinal κ will be a regular, strong limit cardinal. Such cardinals are called *strongly inaccessible*.

Since within SEC we will be able to prove that κ is a strongly inaccessible cardinal, to carry out our construction we must assume that there is a strongly inaccessible κ. Let $V = \langle H(\kappa), \in, < \rangle$ where $<$ is some wellordering of $H(\kappa)$ of order type κ. We think of V as a surrogate for the universe of all pure sets, and κ as a surrogate for On in our model building. It is easy to see that V is a model of ZFA plus global choice. (This is sketched in the solution to Exercise 18.8.)

We first discuss a general method for building universes $U \supseteq V$ which are candidates for models of SEC. (We will later apply this method to show that a particular theory SEC_0 is consistent.) By a V-*set* we mean an element of V and by a V-*class* we mean a subset of V.

Definition A *class notation system* is a partial operation $\mathcal{N}$ from V-sets to V-classes. For a fixed notation system we usually write B_b for $\mathcal{N}(b)$ and say that b codes B_b. The domain of this operation is called the class of *codes*. Given a notation system $\mathcal{N}$, we use *AFA* to define, for each code b,

$$Ext(b) \quad = \quad \{Ext(a) \mid a \in B_b, \ a \text{ a code}\}.$$

We call b a *name* of $Ext(b)$ and $Ext(b)$ the *canonical class* named by b, relative to the notation system $\mathcal{N}$. We take $U_{\mathcal{N}}$ to be the collection of all such canonical classes for the notation system under consideration, dropping the subscript when a notation system is fixed by context.

In applications, the codes are going to be formulas in some kind of a language coded up in set theory. They might be first-order formulas, infinitary formulas, second-order formulas, whatever ways of talking we might want to allow. We call the classes named by these codes *canonical* classes for the same reason we used the term "canonical" earlier in the book. Each formula (code) gets assigned a class which is independent of the form of the formula: bisimilar formulas are assigned the very same canonical class. Notice that U is transitive, since every element of a canonical class is itself a canonical class. We call an element of U *small* if it has size less than κ; otherwise we call it *large*.

We can make our idea that the classes should be just ways of talking up to bisimulation a bit more precise. Namely, we suggest that SEC should be the theory of canonical classes, relative to sufficiently rich class notation systems. We are going to apply this basic picture to show the consistency of a particular theory of classes, our current best approximation of the theory we are after.

There is, of course, a big question as to just what one should mean by a "sufficiently rich" class notation system. One's first temptation is to require that the language be quite expressive. However, there is a tension between expressive power and other considerations that makes the whole matter more subtle than one might first imagine. We will see that in order for this proposal to be plausible, we will need to require that the modes of expression be self-sufficient (in the sense of Section 5.5). This requirement does not prevent the language from being quite expressive, but it does put other constraints on it, as we saw there. In particular, it cannot possibly be closed under negation!

20.5   The theory SEC_0

Based on the intuitive ideas presented above, we propose the following collection of axioms and dub it SEC_0. For ease of exposition, we will treat this theory as a theory of pure sets. We formulate it in the language of set theory, whose only relation symbol is $\in$ (and the logical relation of identity) augmented by a constant symbol κ. We state the axioms informally, except when there might be some ambiguity about what we mean. We will show that this theory is consistent.

Strong Extensionality: For each formula $\varphi(x, y)$, the universal closure of:

$$Bisim(\varphi) \wedge \varphi(A, B). \rightarrow A = B.$$

Here $Bisim(\varphi)$ is the natural formula expressing that $\varphi(x, y)$ defines a bisimulation:

$$(\forall x, y)[\varphi(x, y) \rightarrow .(\forall z \in x)(\exists w \in y)\varphi(z, w) \wedge (\forall w \in y)(\exists z \in x)\varphi(z, w)].$$

Pair: For any classes A, B there is a class $C = \{A, B\}$.

Union: For any class A, $\bigcup A$ exists.

Product: For any classes A, B there is a class $C = A \times B$ of all pairs $\langle x, y \rangle$ with $x \in A$ and $y \in B$.

The Universe Exists: There is a class U such that for all a, $a \in U$.

Size of the Universe κ is a transitive class which is wellordered by $\in$, and U has size at most κ. (That is, there is a class $F : \kappa \to U$ which is functional and has range U.)

We call a class C *small* if there is a $\beta < \kappa$ and a class $F : \beta \to C$ which is a surjective function with range C; otherwise A is large. (It follows of course that every $\beta < \kappa$ is small. It will follow from Power that U is large, and from Collection that κ is large.) The remaining axioms all make use of the notion of small class. We call A a *set* if $TC(A)$ is small. (Note that $TC(A)$ will exist by Union and Infinity.)

Anti-Foundation: Every small graph has a unique decoration.

Infinity: There is a set $\omega < \kappa$ containing $\emptyset$ and closed under successor.

Power: Every small class A has a power class $\mathcal{P}(A)$, and $\mathcal{P}(A)$ is small.

V Exists: There is a class V of all sets.

Comprehension for Classes: For each formula $\varphi(x)$, the universal closure of:

$$\exists A \forall x [x \in A \leftrightarrow .x \in V \wedge \varphi^V(x)]$$

Comprehension for Small Classes: For every small class A and class B, $A \cap B$ is a (small) class.

Collection: For each formula $\varphi(x,y)$ in the language of set theory the universal closure of: if A is a small class and $(\forall x \in A)(\exists y)\varphi(x,y)$ then there is a small class B such that $(\forall x \in A)(\exists y \in B)\varphi(x,y)$.

We do not propose Collection for the theory SEC. It is easy to see that SEC_0 does satisfy the other criteria we have set forth, but we do not have enough experience with it to be at all confident that it meets our first criterion: usefulness. This theory is simply the set of those principles we see how to justify on the basis of our informal proposal made earlier, sufficient to meet the remaining criteria. It could well be that there are other principles that can be justified and need to be added to get a truly useful theory. In particular, we are not convinced that we expressed all the comprehension principles on classes that are justified on the conception. (We return to this issue later.)

The consistency of SEC_0

It is easy to see that within SEC_0 we can prove that κ is a strongly inaccessible cardinal, so to prove the consistency of this theory, we will need a strongly inaccessible κ. In these days, this assumption is considered relatively unproblematic, but it does go beyond ZFC.

Let $V = \langle H(\kappa), \in, < \rangle$ where $<$ is some wellordering of $H(\kappa)$ of order type κ as before. We now specialize the general notion of a class notation system to some particular instances, those needed for proving SEC_0 consistent. First a preliminary notion. We will assume some standard method of formalizing formulas in the language of set theory, allowing parameters for elements of V, as sets in V. We write $\ulcorner\varphi\urcorner$ for the set-theoretic formalization of φ.

We recall the definition of a *self-sufficient* set of formulas from page 62. In our context, we want to allow the atomic formulas to mention the ordering symbol $<$ but only allow constant symbols for sets from $H(\kappa)$. This change does not affect any of the results from before, and the extra expressive power will be put to good use later. We also assume that each such formula φ is formalized as an element $\ulcorner\varphi\urcorner$ of $H(\kappa)$.

There is an important hierarchy of self-sufficient sets of first-order formulas:

$$\Sigma_1 \subseteq \Sigma_2 \subseteq \cdots \Sigma_k \subseteq \cdots,$$

A formula is in Σ_k if it is equivalent to one written in the form $\exists y_1 \forall y_2 \exists y_3 \ldots y_n \psi$, where all the quantifiers in ψ are bounded quantifiers. This hierarchy is known as the *arithmetical hierarchy*. Its union is the set of all first order formulas. An analogous hierarchy was well known in arithmetic. It was developed for set theory, and applied to get a number of very interesting results by Azriel Lévy in Lévy (1965). The ideas were important in the general development of admissible set theory, as detailed in Barwise (1974). Lévy works in *ZFC* but everything works equally well for *ZFA*, since we included the Collection scheme in this theory. (Collection is needed in order to know that one can move a bounded quantifier inside an unbounded quantifier without increasing the place of the formula in the hierarchy.) There is an analogous hierarchy of self-sufficient sets of second-order formulas, each of which contains all the first-order formulas.

Definition Let $\mathcal{F}$ be any self-sufficient set of formulas. The $\mathcal{F}$-*class notation system* uses as codes sets of the form $\ulcorner\varphi\urcorner$ where $\varphi \in \mathcal{F}$ has at most one free variables. The set $\ulcorner\varphi\urcorner$ codes the V-class

$$\{a \in V \mid V \models \varphi[a]\}.[7]$$

An $\mathcal{F}$-*canonical class* is a canonical class relative to the $\mathcal{F}$-class notation system. We let $U = U_{\mathcal{F}}$ be the collection of these canonical classes.

In other words, $U_{\mathcal{F}}$ is given as follows:

$$U_{\mathcal{F}} = \{Ext(\ulcorner\varphi\urcorner) \mid \varphi \in \mathcal{F}\},$$

where

$$Ext(\ulcorner\varphi\urcorner) = \{Ext(a) \mid V \models \varphi[a]\}.$$

The Recursion Theorem for self-sufficient sets has the following useful application. Given a formula $\varphi(x_1, \ldots, x_{m+n})$ of $\mathcal{F}$ and elements $a_1, \ldots, a_n \in$

---

[7] The notation $V \models \varphi[a]$ means that the context which assigns a to the one free variable satisfies the formula φ in the structure V.

V, let

$$\ulcorner\varphi[a_1,\ldots,a_n]\urcorner \quad = \quad S_n^m(\ulcorner\varphi\urcorner, a_1,\ldots,a_n).$$

By the S_n^m property, this is the set-theoretic formalization of the formula which results by substituting constants denoting $a_1,\ldots,a_n$ for $x_1,\ldots,x_n$ in φ. Given a formula $\varphi(x_1,x_2)$ of $\mathcal{F}$, we say that a formula $\varphi^*(x_1,x_2)$ of $\mathcal{F}$ is a *hereditary canonicalization of ϕ* if for all a,

$$Ext(\ulcorner\varphi^*[a]\urcorner) \quad = \quad \{Ext(\ulcorner\varphi^*[b]\urcorner) \mid V \models \varphi[a,b]\}.$$

Proposition 20.1 *Let $\mathcal{F}$ be a self-sufficient notation system. Then every formula $\varphi(x_1,x_2)$ in $\mathcal{F}$ has a hereditary canonicalization in $\mathcal{F}$.*

Proof Apply the Recursion Theorem to the formula $\theta(x_0,x_1,x_2)$ given by

$$\exists z[\varphi(x_1,z) \wedge x_2 = S_1^1(x_0,z)]$$

to obtain a formula $\varphi^*(x_1,x_2)$ satisfying

$$\varphi^*(x_1,x_2) \quad \leftrightarrow \quad \exists z[\varphi(x_1,z) \wedge x_2 = S_1^1(\ulcorner\varphi^*\urcorner,z)].$$

Thus, for each set $b \in V$, $\varphi^*[a](= S_1^1(\varphi^*,a))$ codes $\{\ulcorner\varphi^*[b]\urcorner \mid \varphi[a,b]\}$. Hence it names the class

$$\{Ext(\ulcorner\varphi^*[b]\urcorner) \mid V \models \varphi[a,b]\},$$

as desired. $\dashv$

As an application, consider the formula $In = \text{``}(x_2 \in x_1)^*\text{,''}$ the hereditarily canonicalization of $\text{``}(x_2 \in x_1).\text{''}$ Then for any $a \in V$, $Ext(\ulcorner In[a]\urcorner) = \{Ext(\ulcorner In[b]\urcorner) \mid b \in a\}$. By the uniqueness part of AFA, there is only one function that satisfies this condition, the identity function. That is, for each a, $Ext(\ulcorner In[a]\urcorner) = a$. Hence each $a \in V$ has a name $\ulcorner In[a]\urcorner$. From now on we use $\bar{a}$ for the name $\ulcorner In[a]\urcorner$ of a.

For any set $\mathcal{F}$ of first-order formulas, we let $SEC_0(\mathcal{F})$ be SEC_0 except that we only require Class Comprehension for formulas $\varphi \in \mathcal{F}$. We will prove the following theorem:

Theorem 20.2 *Let $\mathcal{F}$ be any self-sufficient set of first-order formulas. The $\mathcal{F}$ canonical classes form a model of $SEC_0(\mathcal{F})$.*

The consistency of SEC_0 follows immediately from this, the compactness theorem of first-order logic, and the fact that the Lévy hierarchy exhausts the first-order formulas. (We could avoid the use of the Compactness Theorem and get a single standard model by working with a self-sufficient set of second-order formulas containing all first-order formulas, for example the Σ_1^1 formulas.)

A function $f : U^n \to U$ is said to be *representable* by the formula $\varphi(x_1,\ldots,x_n,x_{n+1})$ in $\mathcal{F}$ if for all $\mathcal{F}$ codes $b_1,\ldots,b_n$,

$$f(Ext(b_1),\ldots,Ext(b_n)) \quad = \quad Ext(\ulcorner\varphi[b_1,\ldots,b_n]\urcorner).$$

The following result shows that most of SEC_0 follows from self-sufficiency.

Theorem 20.3 *Let $\mathcal{F}$ be any self-sufficient set of formulas, and let U be the transitive class of $\mathcal{F}$-canonical classes.*

1. *$H(\kappa) \subseteq U$.*
2. *Every $\mathcal{F}$-definable class of V is a canonical class. In particular, $H(\kappa) \in U$ and $\kappa \in U$.*
3. *$U \in U$, that is, U is a canonical class.*
4. *If C is a collection of less than κ canonical classes then C is a canonical class.*
5. *If C is a canonical class, so is $\bigcup C$.*
6. *U is closed under ordered pairs. Moreover, the ordered pair operation on U is representable in $\mathcal{F}$.*
7. *U is closed under cartesian products.*
8. *Every element of U has size $\leq \kappa$. Indeed, for every canonical class C there is a function F whose domain is some ordinal $\leq \kappa$ and whose range is C such that F is itself a canonical class.*

Proof Item (1) follows from the fact that every $a \in V$ has a name $\bar{a}$. Item (2) follows from the fact plus the following observation:

If $\varphi(x_1, x_2)$ is any $\mathcal{F}$ formula and $b \in V$ then there is a code for the V-class

$$\{\bar{a} \mid V \models \varphi[a, b]\}$$

This observation follows from the fact that this class is also $\mathcal{F}$ definable since $\mathcal{F}$ is self-sufficient.

Item (3) follows by considering any $\mathcal{F}$ formula $\varphi(x_1)$ (like $x_1 = x_1$) that defines a class containing all codes. For then

$$Ext(\ulcorner \varphi \urcorner) \quad = \quad \{Ext(b) \mid b \text{ a code}\} \quad = \quad U.$$

To prove (4), assume that C is a set of less than κ canonical classes. Let c be a set that has at least one name for each of these and no other elements. Then $c \in H(\kappa)$ because every subset of $H(\kappa)$ of size smaller than κ belongs to $H(\kappa)$. Then $\ulcorner (x_2 \in x_1)[c] \urcorner$ is a code for c and hence is a name for C.

To prove (5), we must use the formula $Sat_1(x_0, x_1)$ in $\mathcal{F}$ formula that captures the satisfaction relation for $\mathcal{F}$ formulas in one free variable. Now given any $\mathcal{F}$ formula $\varphi(x_1)$, let $\theta(x_1)$ be the following $\mathcal{F}$ formula:

$$\exists w[\varphi(x_1/w) \wedge Sat_1(w, x_1)].$$

Then $Ext(\theta) = \bigcup C$.

The first part of (6) follows by two applications of (4) since $\langle a, b \rangle = \{\{a\}, \{a, b\}\}$. This means that U is closed under pairs. To prove the fact that the ordered pair operation is representable takes a bit more work. First, for any n, let Set_n be the formula

$$x_{n+1} = x_n \lor \ldots \lor x_{n+1} = x_n$$

We claim that Set_n represents the function f defined by $f(a_1, \ldots, a_n) = \{a_1, \ldots, a_n\}$. After all,

$$
\begin{aligned}
Ext(\ulcorner Set_n[a_1, \ldots, a_n] \urcorner) &= \{Ext(a) \mid V \models Set_n[a_1, \ldots, a_n, a]\} \\
&= \{Ext(a_1), \ldots, Ext(a_n)\}
\end{aligned}
$$

which is what we needed to show. In particular, Set_1 represents the operation of taking singletons and Set_2 represents the unordered pair operation. Now, let $OP(x_1, x_2, x_3)$ be

$$x_3 = S_1^2(\ulcorner Set_2 \urcorner, x_1, x_2) \lor x_3 = S_1^1(\ulcorner Set_1 \urcorner, x_1)$$

Then

$$
\begin{aligned}
Ext(\ulcorner OP[a_1, a_2] \urcorner) &= \{Ext(b) \mid V \models OP[a_1, a_2, b]\} \\
&= \{Ext(S_1^2(\ulcorner Set_2 \urcorner, a_1, a_2)), Ext(S_1^1(\ulcorner Set_1 \urcorner, a_1))\} \\
&= \{\{Ext(a_1), Ext(a_2)\}, \{Ext(a_1)\}\} \\
&= \langle Ext(a_1), Ext(a_2) \rangle
\end{aligned}
$$

This shows that OP represents the order pairing operation.

To prove the first part of (7), we apply the fact that the ordered pair operation is representable. Let C_1, C_2 be canonical classes, say $C_1 = Ext(\ulcorner \varphi_1(x_1) \urcorner)$ and $C_2 = Ext(\ulcorner \varphi_2(x_2) \urcorner)$. Let $\psi(x)$ be

$$\exists x_1, x_2[\varphi_1(x_1) \land \varphi_1(x_2) \land OP(x_1, x_2, x)].$$

Then it is easy to see that $Ext(\psi) = C_1 \times C_2$.

To prove the first part of (8), we note that since κ is strongly inaccessible, $H(\kappa)$ has size κ, so there are at most κ codes. Hence, every canonical class C has size at most κ. However, we want more than that. We want to show that there is a canonical class F that witnesses the fact that C has size at most κ. We will do this for the class U, the general case being similar. First, since we have included the wellordering $<$ in our structure V, there is a $\mathcal{F}$-definable bijection f from κ onto $H(\kappa)$. Let $\varphi(x_1)$ define the following:

$$\{b \mid \exists \beta < \kappa[OP(\overline{\beta}, f(\beta), b)]\}$$

where OP is the formula representing the ordered pair operation. It is easy to see that $\varphi \in \mathcal{F}$ since $\mathcal{F}$ is self-sufficient. Then $\ulcorner \varphi \urcorner$ names the class

$$\{\langle \beta, Ext(b) \rangle \mid b = f(\beta), \beta < \kappa\},$$

so this is a canonical class which is a function with domain κ and range U. $\dashv$

This result allows us to establish the following proposition, which finishes our proof of the consistency of SEC_0.

Corollary 20.4 *Let $\mathcal{F}$ be any self-sufficient set of formulas, and let U be the class of $\mathcal{F}$-canonical classes. Then the structure $\langle U, \in_U, \kappa \rangle$ is a model of $SEC_0(\mathcal{F})$.*

Proof We use the results Theorem 20.3. U is a transitive set closed under pairing and containing ω, so we get the axioms of Strong Extensionality and Pair for free. The Axiom of Union follows from (20.3.6), while the Axiom of Products exists by (20.3.7). The Universe Exists axiom follows from (20.3.3). The Size of the Universe axiom follows from part (8).

We now turn to the axioms involving the notion of a small class. That the Anti-Foundation Axiom holds in U follows from the observation that every small graph G is isomorphic to some graph $H \in H(\kappa)$. H has a decoration d by *AFA*. It is not hard to transfer H by isomorphism to get a decoration of the original G. For the uniqueness, assume that

$$U \models \text{``} d' \text{ is a decoration of } H \text{''}.$$

Then as a set, d' belongs to $H(\kappa)$. By absoluteness, d' really is a decoration of H. So $d = d'$ by *AFA*.

We turn to the verification of the Power axiom. Let C be a small class. Then C has size $< \kappa$. Let $B = \mathcal{P}(C) \cap U$. By (20.3.4) and the fact that κ is a strong limit cardinal, we see that $B \in U$. Since U is transitive,

$$U \models B \text{ is the power set of } C.$$

(See also the proof of the Comprehension Axiom for Small Classes below.)

Next, we show the axiom V Exists. We need to know that there is a class V containing every set, that is, every canonical class whose transitive closure is small in U. Let a be a set in U. Then

$$U \models \text{``} TC(a) \text{ has size } < \kappa \text{''}.$$

By absoluteness, $TC(a)$ really has size $< \kappa$. So $a \in H(\kappa)$. Thus $a \in U$ by (20.3.1). It follows that to prove that V exists we need only show that $H(\kappa) \in U$. But this holds by (20.3.2).

Comprehension for Classes using formulas in $\mathcal{F}$ follows from (20.3.2). Comprehension for Small Classes follows from (20.3.4). That is, if B is small then *every* subclass of B is the union of $< \kappa$ singletons and thus belongs to U. Finally, Collection also follows from (20.3.4). $\dashv$

We conclude this section by discussing an open question. Is there a self-sufficient set $\mathcal{F}$ of formulas so that the $\mathcal{F}$-canonical classes are closed under intersection: if $C_1, C_2 \in U$ then $C_1 \cap C_2 \in U$? More generally, is there a self-sufficient set $\mathcal{F}$ of formulas so that the $\mathcal{F}$-canonical classes satisfy the Generalized Positive Comprehension Principle. (This principle, called GPK in the work of Forti and Honsell, is described in page 322.) If so, we could add GPK to SEC and a number of our other axioms would become redundant. The key step in establishing either the special case or the general result would seem to be finding a self-contained class $\mathcal{F}$ and a formula $(x_1 \equiv x_2)$ in $\mathcal{F}$ such that $V \models (a \equiv b)$ iff $Ext(a) = Ext(b)$. The natural thing is have $(x_1 \equiv x_2)$ express the condition of being bisimilar. However, when one writes out this condition, one has Sat_1 occurring negatively as well as positively. Since $\neg Sat_1$ cannot be in $\mathcal{F}$ (for reasons similar to the reason $\neg Sat_0$ cannot be in $\mathcal{F}$), it seems unlikely that GPK can be justified for any self-sufficient set $\mathcal{F}$. But we do not have a proof that this is impossible.

Exercise 20.1 Let $\mathcal{F}$ be self-sufficient and let f be any $\mathcal{F}$ definable function on $H(\kappa)$. Prove that there is a formula $\psi(x_1, x_2)$ in $\mathcal{F}$ so that for all b. $Ext(\ulcorner\psi[b]\urcorner) = \langle b, f(b)\rangle$.

Overspill results While we have seen that the class of non-reflexive *sets* can be construed as a canonical class, we have not considered the class of non-reflexive classes, and other possible problematic classes. What do these constructions tell us about our theory?

We know that self-sufficient classes of formulas cannot be closed under negation. This gives us some hint as to why there cannot be a canonical class of all non-reflexive canonical classes. But we can do better than this by proving a simple result that holds for class notation systems much more generally.

Theorem 20.5 *Let $\mathcal{N}$ be any class notation system and let B be a V-class which has a code relative to $\mathcal{N}$.*

1. *If every canonical class other than U has a name in B then U also has a name in B.*

2. *If every non-reflexive canonical class has a name in B then some reflexive canonical class also has a name in B.*

Proof Let b be a code for B. The first claim is a direct consequence of the Difference Lemma: for otherwise b would name $U - \{U\}$ and so the latter would be a member of U. The second is an immediate consequence of Russell's Paradox, since otherwise b would name the Russell class. ⊣

Corollary 20.6 *Let $\mathcal{F}$ be a self-sufficient class of formulas and consider the $\mathcal{F}$-notation scheme.*

1. *The V-class of all names for classes other than U is not definable by any formula in $\mathcal{F}$.*

2. *The V-class of all names for non-reflexive classes is not definable by any formula in $\mathcal{F}$.*

Proof Immediate. ⊣

Other paradoxes also give rise to analogous results.

We take the moral of these results to be that certain definitions of classes do not work the way you might think if you restrict attention to strongly extensional classes. You can try to define something like the Russell class, but the (strong) extension of the predicate is not going to be what you expect. We illustrate this point further in some of the exercises below.

Exercise 20.2 Show that if every wellfounded canonical class has a name in B then some non-wellfounded canonical class also has a name in B. Let $\mathcal{F}$ be a self-sufficient set of formulas. Show that the V-class of names of wellfounded canonical classes is not V-definable.

More generally speaking, what these results suggest is that the paradoxes of class theory can be seen as semantical paradoxes of naming. Classes which we feel we can name cannot in fact be named in any systematic way. Why, we ask ourselves? We return to discuss the answer to the question in the final section.

Exercise 20.3 Show that for any class notation system, if $U \in U$ then U does not satisfy the Axiom of Separation.

We trust the remarks make plausible our belief that there could be a useful strongly extensional theory of classes waiting to be fleshed out, based on the conception of class as ways of speaking about sets and classes, up to bisimulation. However, the details also make pretty clear that to have this be usable, the ways of talking are going to need to be self-sufficient, which means that they are not going to be closed under negation.

20.6 Parting thoughts on the paradoxes

In the conclusion to his thought-provoking book, Michael Hallett writes of ZFC:

> this system works both in that it seems to avoid contradiction and in that it gives mathematics enormous freedom. Yet we have no satisfactory simple heuristic explanation as to *why* it works, why the axioms should hang together as a system. Our only plausible candidates, the limitation of size theory and the derivative iterative theory of set, do not work. (1984)

In this chapter we have suggested a response to this lament. We have suggested that ZFC is the theory of hereditarily small, cumulative classes, and that ZFA is the theory of hereditarily small classes.

We feel that this is an improvement in the situation, but the proposal would be much more convincing if one had an independently motivated theory of classes such that meets at least our four criteria. In the preceding section we presented a conception of an arbitrary class: start with a sufficiently rich, self-sufficient class of formulas and identify them up to bisimulation. We showed that this conception gives rise to models satisfying the theory SEC_0. But this theory does not have the feel of being particularly principled in that we have no sense that we have captured everything that follows from the conception, or that the theory is particularly useful. Still, it is a start.

This problem is, of course, closely related to the one that beset Frege, Russell, von Neumann, and so on: what should we make of the paradoxes? In the preceding section we proposed that within a theory of strongly extensional classes, the intuitive paradoxes can be looked at as semantical paradoxes of naming. This should bring to mind our earlier treatment of the semantical paradoxes. To see why this might be an attractive way to go, let's try going the other way, first, by applying the standard solutions to a semantical version of the Russell Paradox.

There are many versions of Russell's Paradox. One of the easiest to explain is known as Grelling's Paradox. Let R be the following predicate of English:

is a predicate of English which does not hold of itself.

For example, the predicate "contains more than four words" contains more than four words, so it holds of itself, and hence R does not hold of "contains more than four words." On the other hand, the predicate "contains more than ten words" does not contain more than ten words so R does hold of it. Similarly, "is a predicate of English" is a predicate of English, whereas "is a predicate of German" is not a predicate of German. Hence R holds of the latter but not the former. This all seems straightforward until one asks the embarrassing question: does R hold of itself or not? If it does, it doesn't; if it doesn't, it does.

The usual version of the Russell Paradox is just the extensional version of this paradox, so one would hope that a solution to the set-theoretic version of the Russell Paradox would shed some light on it. But neither of the standard solutions seems to. The solution embodied in the distinction between small and large classes doesn't seem to throw any light on the matter since there is no obvious sense to be made of the claim that this predicate is somehow "too large." Nor does the solution contained in the cumulative hierarchy seem appropriate, since there does not seem to be anything problematic in saying that "is a predicate of English" is a predicate of English.

It seems to us that current received wisdoms about the set theoretical paradoxes are far from satisfactory. The distinction between "logical" and "semantical" paradoxes is poorly motivated at best. And neither the large-small story nor the cumulative story makes much sense when applied to Grelling Paradox, the paradox of denoting, or the grandparent of them all, the Liar Paradox. Thus it is hard not to feel that the set-theoretic paradoxes have only been avoided, not really understood.

Earlier in this book we proposed analyses of two semantical paradoxes, The Liar and the Paradox of Denoting. We could also treat the Grelling version of the Russell Paradox as kin to the Liar Paradox and the paradoxes of denotation. The basic line would be that we can talk about predicates holding of expressions, and we could ask whether the predicate R held of itself in the full, total world. The answer would be "No." But if we tried to assert this, it would result in some kind of subtle context shift.

Now what if we turn the tables and apply this idea to set theory? What results? Suppose we have some universe U of classes. It might even be a member of itself. And so might $\{U\}$. The Difference Lemma tells us that $U - \{U\}$ is not a member of U. Russell's Paradox tells us that

$$\{x \in U \mid x \text{ is non-reflexive}\}$$

is also not in U. But then *where are these classes*? We have just made reference to them and asserted that they are not in the universe U. Why can't we coherently add suitable names to our way of talking about classes, that is to our class notation system? The moral of the semantical paradoxes was that we need to be aware of extremely subtle shifts in context as we speak. Theorem 20.5 and its corollary points to a similar moral under the view of classes as ways of talking up to observational equivalence.

Only time will tell what history has to say about classes and the final treatment of the paradoxes, but we suggest that in the case of set theory the story is far from told. Paradoxes are a rich source of puzzle and inspiration for thinking about difficult issues. We think that one of many morals to be drawn from them is the subtle and not-so-subtle context-sensitivity in language and thought, even mathematical thought about the universe of sets and classes.

Historical Remarks

The gist of this chapter is original, though as we have indicated, there has been a lot of work done on theories of classes containing a universal class. Most of the models of these theories are not strongly extensional, however. An exception may be found in some work of Forti and Honsell.

Beginning in 1989, Forti and Honsell began a study of collections which they call *hyperuniverses*. Forti and Honsell had been first formulators of axioms like *AFA*, and their work on hyperuniverses was motivated by interest in the set-class problem that we are discussing in this chapter. A hyperuniverse is a collection of a certain type: the definition is in topological terms, and to provide all the details would take us too far afield. Instead, we'll summarize some of the properties and invite the reader to seek out some articles. A summary of the basic construction may be found in Forti and Honsell (1992), and their work is also summarized in Forti and Honsell (to appear) . In addition, Forti, Honsell, and Lenisa (1994); is an application to the study of transition systems.

The construction begins with a *weakly-compact* cardinal κ; this is a cardinal bigger than a strongly inaccessible cardinal (but from the point of view of modern set theory, the difference is a trifle). Using this and an equivalence relation E, they construct a structure $N = N_{\kappa,E}$ with some very strong properties. $N \in N$, and N is closed under operations like singleton, binary union and intersection (but not complement). Further N is "big" in the sense that every subset of N of size $< \kappa$ is an element of N, and the ordinals of N are just the elements of κ. This shows that we can define large and small for hyperuniverses, using κ as a measure of size. Another important fact is that it is possible for hyperuniverses to be strongly extensional and to satisfy *AFA*.

Perhaps the most interesting property is the following Generalized Positive Comprehension principle. To state it, we need a definition: A *generalized positive formula* is one built from atomic formulas $x \in y$ and $\mathcal{U}(x)$ using $\wedge$, $\vee$ (but not $\neg$), the quantifiers $\forall$ and $\exists$, and also closed in the following way: if φ is generalized positive and θ is any formula with just x free, then both

$$(\forall x \in y)\varphi \qquad \text{and} \qquad (\forall x)(\theta \to \varphi)$$

are generalized positive.

The Generalized Positive Comprehension Principle (GPK) is the schema

$$(\exists y)(\forall x)(x \in a \leftrightarrow \varphi(x)),$$

where φ is a generalized positive formula not containing y. It turns out that hyperuniverses satisfy this principle. To get a feel for what this means, note that $x = x$ is a generalized positive formula. This immediately gives the existence of a universal set.

Hyperuniverses seem like interesting mathematical structures, but we do not see a direct connection to theories of the kind we are after. That is, as far as we know, there is no background theory which offers a principled explanation for the paradoxes, or one which motivates the GPK on independent intuitive grounds. In particular, as we have noted earlier, we have not been able to see that any conception of class as a *façon de parler*, up to bisimulation, supports this principle. On the other hand, we have not been able to show that no conception of class could support it.

21

Past, present, and future

Southward blowing, turning northward,
Ever turning blows the wind;
On its round the wind returns. ...
All such things are wearisome:
No one can ever state them;
The eye never has enough of seeing,
Nor the ear of hearing.
Only that shall happen which has happened,
Only that occur which has occurred;
There is nothing completely new under the sun.

Ecclesiastes 1:6, 8–9

We began this book with the aim of writing a short tutorial that would summarize in an elegant way what we knew about hypersets and their applications outside of mathematics. The finished product is both larger and less "finished" than we expected. It is larger because, as often happens, we came across many new problems and applications. It is less finished because we fully expect others to find more elegant ways of handling some of the "new" material presented here, to point out that some of what we think is new isn't, and to solve some of the problems we have left open. In this final chapter we want to make miscellaneous remarks that seem pertinent but did not fit in elsewhere, and to point out a few directions for investigation which seem to us likely to be of interest.

21.1 The past

Though this book was written at the end of the twentieth century, from a
mathematical point of view, much of it could have been written at the beginning
of the century, when Cantor was working on transfinite numbers and discovering
the subject we now call set theory. A few years later, in 1908, Zermelo's
landmark paper appeared with axioms of set theory. Essentially everything in
this book can be done with these axioms, and all of our definitions and methods
have been common in set theory for at least sixty years. So why are hypersets
only coming into their own now, after a hundred years of work in set theory?

One could answer by saying that the theory of hypersets involves a more
inclusive concept of set than that from within mainstream twentieth century
set`theory. But the question is: why is the older cumulative conception,
considered sufficient for so long by so many of us, only now giving way to this
richer conception?

While various set-theorists over the century have explored various axioms
which guarantee the existence of non-wellfounded sets, the axioms were poorly
received by working logicians until Aczel's work. The main reason for the
change in climate was that Aczel's theory, unlike the earlier work, was inspired
by the need to model real phenomena. Aczel's own work grew out of his work
trying to provide a set-theoretic model of Milner's calculus of communicating
systems (CCS). This required him to come up with a richer conception of set.

While we did not go into CCS in this book, we did treat simpler related
notions, including labeled transition systems and deterministic automata. Once
Aczel pointed out the basic circularity implicit in CCS, and how to model it
in set theory with *AFA*, it became obvious that similar circularity had been
around for a long time in computer science (and in other areas) and could also
be modeled with *AFA*.

There is one thing that seems to tie all the examples of circularity together.
It is that some sort of intentional phenomena is involved. This is most obvious
in the case of the semantical paradoxes, and the phenomena discussed in
Chapter 4. However, even in the examples from computer science, one could
argue that intentionality is present.

To make this point, let's consider one of the cases where this claim is, on
the face of it, least plausible. Recall our labeled transition system describing
clocks:

$$clock \quad = \quad tick.clock \ + \ break.stuckclock$$

A clock is a purely physical device. It sits there ticking away, until it breaks.
There is nothing particularly intentional about it. Nor is its physical mechanism
mysteriously circular. Circularity arises from our desire to characterize the
clock's states up to observational equivalence. And that is also where the
intentionality comes in. Observation is an intentional notion, and it is we

designers, users, and observers of clocks who find it convenient to analyze them in these terms.

Labeled transition systems and similar state-based formalisms come to the fore in the process of design, both of hardware and software. We know we want a device or program that behaves in a certain way. We describe it in terms of states and their relationships. Later, once we have the design in place, we worry about the process of implementing the states in physical devices. The point here is that *design*, like observation, is a highly intentional phenomena. We know what we seek before we have it in hand, and building in circularity can be a particularly efficient method of design.

It seems to us that a new family of sciences is emerging to take their place alongside the physical, life, and social sciences. These sciences deal with cognition, perception, computation, information, representation, and reasoning. Indeed, it would make sense to call them the "intentional sciences." We suggest that the reason hypersets were somewhat slow in coming was that it was not until the rise of the intentional sciences that people felt the need to provide rigorous mathematical tools for the analysis of intentional phenomena.

21.2 The present

Ten years or so ago, when Barwise started giving lectures about hypersets and their uses, the reception among set-theorists was chilly. "Granted that the theory is coherent, still, are these things really 'sets'?" they would ask. But as more and more applications have been found, and more people have explored the theory, it has gained an increasingly wide acceptance. Two recent mainstream textbooks on set theory, Devlin (1993) and Moschovakis (1994), each devote a chapter to hypersets.

There is one way in which the richer theory was disappointing to set theorists. Historically, most of the work in twentieth century set theory has involved two things: properties of sets of real numbers, and cardinality issues. These two areas intersect in questions like the continuum hypothesis. Since the pioneering work of Paul Cohen in the early 1960's, many of these questions are now known to be independent of set theory as it is axiomatized in ZFC. Set theorists have wanted a richer conception of set, richer in that it helps us settle some of these basic questions left open by ZFC.

It has to be admitted that ZFA does not do this. The consistency proof we have given shows this to be the case. We showed how to start with any M model of ZFC^- and enrich it to a model M_{afa} where all equations have solutions. In doing so, we do not effect the sets that exist in M at all; none of the old sets get any new subsets. There are no new sets of integers, reals, or what have you. In particular, all computations of cardinality in M will have the same results when carried out in M_{afa}. (Exercise 9.7 discusses this at length.)

So this conception does not, on its own, give us answers to the questions left open by *ZFC* that have dominated the work of set theorists.

It seems possible, though, that in freeing ourselves of the straight-jacket imposed on us by the Axiom of Foundation, progress could be made. Perhaps thinking of things in terms of systems of equations or decorations of graphs might suggest additional principles that would be plausible on this conception of set but would settle some of the issues left open by *ZFC*. Time will tell. But even if it doesn't, the theory is exciting both for its own intrinsic beauty and for its applications.

21.3 The future

In working on this book over the past few years, we have been struck by the many interesting problems that emerge from the perspective of hypersets. There were far too many for us to pursue them all, and some we did pursue turned out to elude us. So we conclude this book by pointing to some of these problems and research topics that we think could be rich veins to be mined.

Operators on sets

Operators on sets have been used quite extensively in this book. We are not aware of any systematic study of the kinds of operators which we have used. This is perhaps due to our use of urelements, substitution, and the like. However, the time is ripe to begin such a study. One thought-provoking observation to get things going is that all our conditions on operators such as uniformity and smoothness, hold for all our examples and also all operators built from them using composition.

Open Problem 21.1 *Is there a natural condition C on operators with the property that there is some finite set $\Gamma_1, \ldots, \Gamma_n$ of operators, including all our examples, such that an arbitrary operator Δ satisfies C iff Δ is a composition of operators from $\Gamma_1, \ldots, \Gamma_n$?*

There is nothing special about composition here: any result which shows how to generate all of the operators satisfying some condition C in terms of a finite set of initial operators would be interesting.

A second problem relates to the generalized notion of bisimulation due to Aczel and Mendler (1989) and studied in Chapter 17. We saw in that chapter that the notion of a Γ-bisimulation generalized all the known examples. However, in specific cases it is not always so easy to specialize the definition of a Γ-bisimulation to a particular operator.

Open Problem 21.2 *Is there an effective procedure which, when given a definition of a smooth operator Γ will give an explicit statement of the definition of a Γ-bisimulation?*

By an explicit statement, we mean a definition of bisimulation that refers to one coalgebra, not to Γ-morphisms from some other coalgebra.

Finally, we want to mention a shortcoming in our theory that we noticed just before the book went to press. We had intended the theory of smooth operators to cover all the operators of interest in this book, and for the results on smooth operators to generalize most if not all of the results which we know for specific operators. As it happens, our theory does not cover the case of $\Gamma(a) = \mathcal{P}(A \cup a)$: even when $A \subseteq \mathcal{U}$, this operator is not smooth. On the other hand, as we point out on page 263, our results for smooth operators have slightly weaker versions for this operator. Now, it is not hard to see that we could re-work our theory of smooth operators to cover this operator: we would need to assume an avoidance set on the ranges of substitutions in addition to one on the domains. But this seems ad-hoc, and so we pose a problem for the further development of our theory:

Open Problem 21.3 *What is the most easily defined and naturally studied class of operators which includes the smooth operators and* $\Gamma(a) = \mathcal{P}(A \cup a)$*?*

Corecursion theory

Chapter 17 is clearly just the beginning of a study of corecursion. We can think of more questions than we should devote space to. We list some of the more pressing ones.

One of the most vexing problems is the very statement of the Corecursion Theorem. To state it, we had to resort to notation systems. Ideally, we would like to be able to state the Corecursion Theorem in the following form:

Let Γ be a smooth operator, and let $\pi : C \to \Gamma(C)$. There is a unique map

$$\varphi : C \to \Gamma^*$$

such that for every $c \in C$:

$$\varphi(c) \quad = \quad \pi(c)[\varphi]^*$$

Here $[\ ]^*$ would be some kind of substitution operation, one that would substitute φ into $\pi(c)$ in the right way; it would have to depend on both C and Γ. With the particular operators Γ we have looked at, it is clear how to do this, but we have not been able to find a general way to do it.

Open Problem 21.4 *Find a way to define* $[\ ]^*$ *in general so as to make the above a theorem.*

Another open question on corecursion concerns the relation of corecursion as we have it to other ways of defining functions. In particular, we wonder about the relation to least fixed point definitions on directed complete partial orders (dcpo's).[1] Take streams, for example. The set A^∞ might be viewed as the set of maximal elements of the dcpo $P = \bigcup_{n \le \omega}(n \to A)$ of finite and infinite sequences of elements of A. That is, there is a natural map $i : A^\infty \to P$. For each $\pi : b \to A \times b$ we get a map of a natural map $H^\pi : (b \to P) \to (b \to P)$; we leave it to you to figure out how this works in this case. Now as it happens, the set of functions from $b \to P$ to itself is a dcpo, and H^π is a monotone map. So classical results tell us that H^π will have a least fixed point, say H^π_*. And it will be the case that $i \circ \varphi = H^\pi_*$. The overall point is that corecursion by π has been reduced to finding least fixed points of monotone maps on dcpo's. Our question is whether this reduction can *always* be made.

Open Problem 21.5 *Let Γ be a smooth operator. Is it the case that there exists a dcpo P which allows us to interpret Γ-corecursion in terms of least fixed points in P? That is, such that there is a map $i : \Gamma^* \to P$, and a way of taking each Γ-coalgebra $\pi : C \to \Gamma(C)$ to a monotone map*

$$H^\pi : (C \to P) \to (C \to P)$$

with the following property: $i \circ \varphi = H^\pi_*$, where $\varphi : C \to \Gamma^*$ is given by the Corecursion Theorem for Γ and π, and H^π_* is the least fixed point of H^π.

We have been able to prove this for a class of smooth operators which contains all the operators of interest in this book and is closed under composition. However, the general case is open. This problem also may also be closely related to Open Problem 21.2. For more on this problem concerning approximations, see Moss and Danner (to appear).

The finitary case: Ordinary computability theory can be thought of as the theory of recursion over the set HF^0. It seems that to include things like computable streams and computable binary trees, there should be an analogous corecursion theory. The natural setting for this theory would be the either $HF^{1/2}$ or HF^1. The existence of computable streams like $\langle 0, \langle 1, \langle 2, \ldots \rangle \rangle \rangle$ that are not in $HF^{1/2}$ suggests this set is too small.

Open Problem 21.6 *Develop the corecursion theory of HF^1, or perhaps some more effective version of it.*

This should relate to definability questions over this structure as well as to real programming issues, since we can certainly write circular programs. We have not investigated this topic at all but consider it extremely natural. We also did not take up issues like the implementation of corecursion, or other ways in which the method would be used in practice.

---

[1]You will need to know a bit about dcpo's and fixed point theorems to understand this discussion. One source is Moschovakis (1994). The only departure from the standard treatment is that we allow dcpo's to be proper classes.

Anti-founded model theory

Built into the very bedrock of our thinking about classical logic is the hierarchy of first-order languages, second-order languages, third-order languages, and so on. Non-wellfounded sets raise the possibility that this framework is too constrictive. Anti-founded model theory takes non-wellfoundedness seriously by allowing models in which this hierarchy breaks down, models containing other models in a possibly non-wellfounded manner. Both the chapter on modal logic and that on the semantical paradoxes are examples of anti-founded model theory. There are open problems associated with each.

Modal logic: The work on modal logic in Chapter 11 suggests other results connecting modal theories with classes of sets. See Baltag (to appear) and also Barwise and Moss (1996) for some work in this direction. We also remind the reader of two questions left open there.

Open Problem 21.7 *Characterize those first-order classes of Kripke structures closed under decorations.*

Another is more vague but more important:

Open Problem 21.8 *Find general ways to relate classes W which are transitive on sets with their associated modal logics.*

The reader who has read both the chapter on modal logic and that on the semantical paradoxes will realize that the definition of accessible model in the latter means that we could do for quantificational modal logic what we did earlier for propositional modal logic in the former.

Open Problem 21.9 *Develop quantificational modal logic within anti-founded model theory.*

The basic idea would be to add sentence operators $\Box$ and $\Diamond$ to the language and define truth for sentences involving these operators by adding to the earlier clauses the following:

$$M \models \Box\varphi \quad \text{iff} \quad N \models \varphi \text{ for all models } N \in D_M$$
$$M \models^- \Box\varphi \quad \text{iff} \quad N \models^- \varphi \text{ for some model } N \in D_M$$
$$M \models \Diamond\varphi \quad \text{iff} \quad N \models \varphi \text{ for some model } N \in D_M$$
$$M \models^- \Diamond\varphi \quad \text{iff} \quad N \models^- \varphi \text{ for all models } N \in D_M$$

It seems like an interesting exercise to work through the development of modal logic in this context, in a manner similar to that of Chapter 11. We have not explored this so have no idea what surprises lie in wait.

Anti-founded approaches to semantics: In retrospect, we can see that our work on the semantical paradoxes in Chapter 13 is not terribly systematic. We simply investigated two problematic topics, truth and reference. Focusing on these in turn, we developed a model theoretic framework that allowed us to draw conclusions from the reasoning involved with the simplest paradoxes associated with each. The aim of that chapter was to suggest that the paradoxes do not force one into non-standard logics, partiality, or Tarski-like hierarchies, but rather can be accounted for by subtle shifts in context.

With this experience to guide us, it seems it would be interesting to develop a metatheory for doing syntax and semantics, including truth and reference, but also topics like definability. This metatheory should be developed within *ZFA* and admit non-wellfounded models. In particular, it should allow certain reflexive natural models. By "natural," we mean that all syntactic and semantic predicates should behave as expected. These models would, for example, be total, truth-complete and denotation complete, and every object would have a name. The hope would be that such a model could be taken as a model of the intuitive world, and that no paradoxes would infect this world.

Open Problem 21.10 *Investigate the logic of such reflexive, natural models.*

For example, such models would satisfy statements like the following:

$$\varphi \rightarrow \exists x \mathsf{True}_x(\ulcorner \varphi \urcorner)$$

$$\forall y \, [\mathsf{True}_y(\ulcorner \exists x \varphi \urcorner) \leftrightarrow \exists z(z =_x y \wedge \mathsf{True}_z(\ulcorner \varphi \urcorner))]$$

Is there a natural, complete set of axioms and rules of inference for this logic?[2]

An interesting question is what kinds of partial models can be expanded to reflexive, natural models. A concrete problem suggested by Chapter 13 is the following.

Open Problem 21.11 *Does every truth-correct model have an expansion to a total, truth-correct (hence truth-complete), reflexive model?*

This problem is motivated by the examples on the Liar Paradox in Chapter 13. We have not been able to find a counter-example to it but a proof seems difficult. It could be that a positive answer will require a bit more by way of assumptions, so as to make the starting model better behaved with respect to elementary syntactic and semantic constructions. If so, that would not bother us. Or, to put it the other way around, a negative answer to Open Problem 21.11 that resulted from syntactic or semantic shortcomings of an elementary sort would not interest us nearly as much as one that showed some intrinsic semantic obstacle to total, reflexive, truth-correct models.

Generalized modal logic

In Chapter 19, we developed a general operator-theoretic framework for studying modal logics on a wide variety of structures. We know that this chapter barely scratches the surface of the work to be done. For example, although there are known completeness results for the individual logics, we have no uniform way of generating the axioms and rules of inference directly from those

---

[2]Matthias Scheutz has initiated a promising study in this direction.

logics. Even more critically, we would like to know if there is a way to generate our logics on the basis of operators alone, in such a way as to generalize the examples. We have not pursued this far enough to list any other open problems, so much as to point to the subject as one that is ripe for further development.

Notations for sets and classes

The preceding chapter started out as a subsection of this chapter. While it grew too large for this role, it is still very programmatic. So let us state for the record the problem posed there:

Open Problem 21.12 *Find a strongly extensional theory SEC of classes that satisfies the criteria set out in the preceding chapter.*

We suspect that Open Problems 21.6 and 21.12 may be related. Rather than to try to explain this remark in any detail, let us just raise a couple of specific questions and leave it to the reader to see how they bridge the two topics.

Anti-admissible sets: The theory of admissible sets (cf. Barwise (1974) for example) provides a setting for a general study of definability, one that includes computability theory as one limiting case, namely the case HF^0.

Open Problem 21.13 *Develop a theory of anti-admissible sets and a corecursion theory for them.*

It seems, intuitively, that HF^1 should be an example of a corresponding anti-admissible set. Other examples would be H^1_κ for other cardinals κ. One would expect that the theory of anti-admissible sets would be obtained by taking the theory KP of admissible sets and replacing the foundation scheme by some form of Flat Solution Lemma. The question is: what form? If we take the Flat Solution Lemma as formulated here, it is not strong enough since $HF^{1/2}$ is a model of this theory. So it seems that what we want would require that any "A-corecursive" set of equations has a solution, not just any finite set. One would want to be able to show that the solution itself was A-corecursive.

The hereditarily r.e. sets: Consider the following relation E on natural numbers:

$$n \, E \, m \quad \text{iff} \quad n \in W_m,$$

where W_m is the recursively enumerable (r.e.) set with code m, that is, the domain of the m-th computable function. To illustrate the power of the Second Recursion Theorem, Rogers (see Section 11.4 of Rogers 1967) investigates this as a kind of membership relation and shows that it satisfies the axioms of Pairing, Union, Choice, and Infinity. He also shows that it does not satisfy the Axiom of Foundation. For example, he shows that there is an e such that nEe iff $n = e$. Such an e would be analogous to Ω. It is clear that there is a u such that nEu for every n since the set of all natural numbers is r.e. Since every recursive function has infinitely many distinct codes, the relation E does not satisfy the Axiom of Extensionality.

Using *AFA* there is a natural remedy for this: define

$$d(n) \quad = \quad \{d(m) \mid mEn\}.$$

We call $d(n)$ the n-th *hereditarily r.e. set*, and call n a name for $d(n)$. Let *HRE* be the set of all hereditarily r.e. sets. *HRE* is a transitive set of pure sets and is independent of the particulars of coding the r.e. sets. Being transitive, it is strongly extensional. Since every finite set is r.e., a simple inductive proof shows that $HF^0 \subseteq HRE$. $\Omega \in HRE$, because of the code e mentioned above. *HRE* is reflexive, because of the code u mentioned above.

Most of the results about canonical classes established in Proposition 20.3 have analogues for the *HRE* sets. Similarly, the overspill results proved in Theorem 20.5 have analogues here. For example, if W is an r.e. set of integers such that every set in *HRE* other than *HRE* has a name in W, then *HRE* also has a name in W, by the Difference Lemma. Similarly, if every non-reflexive set in *HRE* has a name in W, then so does some reflexive set, by Russell's Paradox.

Open Problem 21.14 *Characterize the set of hereditarily r.e. sets in some independent, insightful way.*

This problem is based on ordinary recursion theory. An equivalent way to look at it would be in terms of Σ_1 subsets of HF^0. But this suggests a question with respect to any admissible set A. Consider the Σ_1 formulas (with parameters from A) as ways of specifying Σ_1 subsets of A and form a class notation system using just the Σ_1 formulas. We can then collapse this notation system and obtain a natural reflexive, transitive set A^* extending A. Then A^* is related to A as U was related to V in the preceding chapter, and as *HRE* is related to HF^0. Using the A-recursion theorem, one can show that A^* is closed under the union operation. We would like to understand A^* in general. To make this vague desire into a problem, we state the following:

Open Problem 21.15 *Find a theory T such that the various A^*, for admissible sets A, are the transitive models of T.*

We trust the reader sees that this problem is related to the problem of finding the theory *SEC* proposed in the preceding chapter. In this connection, we remind the reader of the problem left open in that chapter. We can generalize it as follows.

Open Problem 21.16 *Given a regular (even a weakly compact) cardinal κ is there a self-sufficient set $\mathcal{F}$ of formulas so that the $\mathcal{F}$-canonical classes over $H(\kappa)$ satisfy the Generalized Positive Comprehension Principle?*

Other applications

In Chapter 5, we expressed the conviction that there are many potential applications of hypersets to philosophical problems besides the ones we actually worked out in the book. All too often one hears an argument in philosophy end with the punch line, "But that would be circular." We are not speaking of self-justifying arguments, but rather situations where someone fails to pursue a line of thought because the mathematical models involved are circular. This is unfortunate, because it means that a decision taken in regard to the axioms of set theory dictates philosophical debate on very different issues. Just as we think the axiomatization of set theory is not a fixed matter, all the more so do we feel that people who understand different mathematical models should feel free to use them.

The tongue-in-cheek title of our book is intended to suggest that circularity has an undeservedly bad reputation in philosophical circles. On the other hand, we certainly do not think that every proposal or argument using circularity bears close scrutiny. For example, one of the morals of our resolution of the Hypergame Paradox is that certain kinds of circular definitions really are incoherent.

On the positive side, though, our experience with hypersets shows that some phenomena are better understood in terms of circular rather than as hierarchical. We have tried to suggest some other potential application areas in Chapter 5. We also feel there are interesting topics to be explored in applying *AFA* in game theory and knowledge representation. These convictions, first expressed long before we had many of the results in the second half of this book, have only increased over the past couple of enjoyable years spent exploring the antifounded universe.

Appendix: definitions and results on operators

An *operator* is an operation $\Gamma : V_{afa}[\mathcal{U}] \to V_{afa}[\mathcal{U}]$. That is, Γ is an operation taking sets to sets. An operator extends to classes by taking

$$\Gamma(C) = \bigcup \{\Gamma(a) \mid a \subseteq C\}.$$

Our example operators Here are the example operators that we see throughout the text:

$\Gamma_1(a) = \mathcal{P}(a)$
$\Gamma_2(a) = \mathcal{P}_{fin}(a)$
$\Gamma_3(a) = A \times a$ for some fixed set A.
$\Gamma_4(a) = \{f \mid f$ is a partial function from a to $a\}$.
$\Gamma_5(a) = \mathcal{P}(Act \times a)$ for some fixed set Act.

Conditions on Operators A *monotone operator* is an operator Γ which satisfies the condition that for all sets a and b: if $a \subseteq b$ then $\Gamma(a) \subseteq \Gamma(b)$.

Γ is *proper* if for all a, $\Gamma(a) \subseteq V_{afa}[\mathcal{U}]$.

A set $X \subseteq \mathcal{U}$ is *new for* Γ if for all substitutions t with domain X, and all sets a, $\Gamma(a[t]) = \Gamma(a)[t]$. X is *very new for* Γ if in addition, whenever $[s]_a : a \to a[s]$ is a bijection, then also $[s]_{\Gamma(a)} : \Gamma(a) \to \Gamma(a[s])$ is a bijection.

We say that *almost all urelements are new for* Γ if there is a *set* $X_{\Gamma} \subseteq \mathcal{U}$ such that for all $Y \subseteq \mathcal{U}$ with $Y \cap X = \emptyset$, the elements of Y are new for $\mathcal{U}$.

335

Similarly, *almost all urelements are very new for* Γ if there is a set $X \subseteq \mathcal{U}$ such that every subset of $\mathcal{U} - X$ is very new for $\mathcal{U}$.

Γ is *uniform* (*smooth*) if Γ is monotone, proper, and if almost all urelements are (very) new for Γ.

Monotone operators are introduced in Chapter 15, proper and uniform operators appear first in Chapter 16, and smooth operators in Chapter 17.

The condition of monotonicity insures that the least fixed point Γ_* and the greatest fixed point Γ^* exist. These fixed points will in general be proper classes.

The motivation for proper operators is that for proper Γ, every flat Γ-coalgebra $\mathcal{E} = \langle X, e \rangle$ has a unique solution s. If Γ is monotone and proper, and if X is new for Γ, then s maps X into Γ^*. (Without the newness condition, s might not map into Γ^*.) Further, we have results for uniform operators such as the Representation Theorem for Γ^* (page 235) and the Solution Lemma Lemma (page 236).

The notion of a smooth operator is needed in order to get the Corecursion Theorem (page 250).

Comments on the conditions

1. All of the operators studied in this book are monotone. The operator $\Gamma(a) = \{\{a\}\}$ is not monotone, but it is proper and all sets of urelements are very new for it.

2. If Γ and Δ are monotone, then $\Gamma \circ \Delta$ is monotone.

3. The identity operator $\Gamma(a) = a$ is not proper. The unary union operation $\Gamma(a) = \bigcup a$ is also not proper. However, all of the operators of central concern in the book are proper.

4. More generally, Γ is not proper iff for some set set a, $\Gamma(a) \cap \mathcal{U} \neq \emptyset$. If A is any set, then $\Gamma(a) = A \cup a$ will not be proper, but it is monotone, and almost all sets of urelements are very new for it.

5. If Γ is a proper operator and Δ is any operator, then $\Gamma \circ \Delta$ is proper.

6. Every set which is very new for an operator Γ is also new for Γ. Hence every smooth operator is uniform.

7. Concerning our example operators, all but Γ_4 are smooth. Γ_4 is monotone and proper, but not uniform and hence not smooth.

8. The operator $\Gamma(a) = \mathcal{P}(a) \cup \mathcal{P}(\mathcal{P}(a))$ is uniform but not smooth. The operator $\Gamma(a) = \mathcal{P}(A \cup a)$ is also uniform but not smooth (but see the discussion on page 263).

9. If Γ and Δ are uniform (smooth), then so is $\Gamma \circ \Delta$. In fact, Δ can satisfy all of the properties except properness, and $\Gamma \circ \Delta$ will still be uniform (smooth).

Answers to the Exercises

2.1. Recall that the empty set is a subset of every set. Since a is a singleton, its power set contains just the empty set $\emptyset$ and a itself. So $\mathcal{P}(a) = b$. Thus $a = \{b\} = \{\mathcal{P}(a)\}$.

2.2. Notice that a is a singleton, so $\mathcal{P}(a) = \{\emptyset, a\} = c$. Further,

$$\mathcal{P}(\mathcal{P}(a)) = \mathcal{P}(\{\emptyset, a\}) = \{\emptyset, \{\emptyset\}, c, \{a\}\} = \{\emptyset, \{\emptyset\}, c, d\} = b.$$

So $a = \{b\} = \{\mathcal{P}(\mathcal{P}(a))\}$, as desired.

2.3. $a = \{b, \{b, \emptyset\}\}$, $\bigcup a = \{b, \emptyset, \{\{a\}\}\}$, $\bigcup\bigcup a = \{\{\{a\}\}, \{a\}\}$, etc. After doing this a while one stops adding new sets, and the resulting union is therefore transitive. We get

$$TC(a) = \{b, \{b, \emptyset\}, \emptyset, \{\{a\}\}, \{a\}, a\}$$

Further, $TC(b) = TC(a)$. In general, if c, d are any two pure sets such that $c \in TC(c)$ and $d \in TC(c)$ then $TC(c) = TC(d)$.

2.4. No, we cannot. Extensionality would just tell us that if we knew that a and b have the same elements, they would be equal. But in order to know that they had the same elements, we would have to know in particular that every element of a (namely a) equals some element of b (which would have to be b). So before we could apply Extensionality we would already have to know that $a = b$.

In fact, a number of non-wellfounded set theorists have shown that if standard set theory (in which sets like a and b simply do not exist) is consistent, then so is set theory with the assumption that there are many different self-singletons. For more on this, see Boffa (1968) or Aczel (1988).

2.5. If C were small, its union would be small. But $\bigcup C$ contains the proper class of all sets. This is because for all sets a, $\{a\} \in C$. Since $\bigcup C$ is both large and small, we have a contradiction.

2.6. For each set a, let $a^* = \{\cdots\{\{a\}\}\cdots\}$; we use $n + 1$ pairs of set braces. Each a^* belongs to C_n, since $f_n(a^*)$ has at most $n + 1$ elements. Also, $a \in f_n(a^*)$. So if C_n were a set, $\bigcup f_n[C_n]$ would contain all sets.

2.7. Assume toward a contradiction that On is small. We use Proposition 2.1 to show that On would be an ordinal. All parts of the definition are easy to check. For example, we check that $\in_{On}$ is antisymmetric. Suppose that α and β are ordinals and $\alpha \in \beta \in \alpha$. Since ordinals are transitive, α and β would be subsets of each other. Hence they would be equal.

So now that we know that On is an ordinal, we see that it must be an element of On. But Proposition 2.1.2 tells us that no ordinal is an element of itself.

2.8. Suppose that a is a class for which there is a bijection f between a and the cardinal $|a|$. Then $f^{-1} : |a| \to a$ is a bijection. Its image is a. Since $|a|$ is a set, and since the sets are closed under images of functions, a is a set.

Going the other way, let a be a set. Let $<$ be a wellorder of a. Let α be such that $\langle \alpha, \in_\alpha \rangle \cong \langle a, < \rangle$. Let β be the least such that for some wellorder $\prec$ of a, $\langle \beta, \in_\beta \rangle \cong \langle a, \prec \rangle$. We claim that β is a cardinal. For if $\gamma < \beta$ were equinumerous to β, then we could get a wellorder $\prec'$ of a such that $\langle \gamma, \in_\gamma \rangle \cong \langle a, \prec' \rangle$. This would contradict the minimality of β.

2.9. The idea is to build up a universe of two sorts of objects by tagging. First, let U be the class of all pairs $\langle 0, a \rangle$, as a ranges over all sets. Let

$$
\begin{aligned}
W_0 &= \emptyset \\
W_{\alpha+1} &= W_\alpha \cup (\{1\} \times \mathcal{P}(W_\alpha \cup U)) \\
W_\lambda &= \bigcup_{\beta < \lambda} W_\beta
\end{aligned}
$$

Let $W = \bigcup_\alpha W_\alpha$. We interpret the language of set theory with urelements on W by taking $x \in y$ to mean $\langle 1, x \rangle \in y$. Further, we use U as the class of urelements.

To define the operation new(a, b) in M, we set

$$
\text{new}(a, b) = \langle 0, \langle a, \{p \in TC(b) : p \notin p\} \rangle \rangle.
$$

The set $\{p \in TC(b) : p \notin p\}$ is guaranteed not to belong to $TC(b)$. Hence new(a, b) will not belong to b. Since a can be read off from new(a, b), if

$a \neq a'$, then $\mathsf{new}(a,b) \neq \mathsf{new}(a',b)$. This operation shows that the Strong Axiom of Plenitude will hold in W.

It is fairly standard to show that W satisfies all of the axioms of *ZFC*, and we omit this.

2.10. We take $\varphi(a)$ to be the statement that a is wellfounded. To prove that $\varphi(a)$ holds for all a, we take some particular a and verify $\varphi(a)$ on the assumption that for all $b \in a$, $\varphi(b)$ holds. Thus we assume that for all $b \in a$, b is wellfounded. If a were non-wellfounded, then there would be a sequence $a = a_0 \ni a_1 \ni \cdots$. But this gives an infinite sequence starting with the element $a_1 \in a$. This contradiction shows that a really is wellfounded.

2.11. Let X be $N \to A$, the set of functions from the natural numbers to A. For each $f \in X$, let $f^+ : N \to A$ be given by $f^+(n) = f(n+1)$. Then we get a map $i : X \to A \times X$ by $i(f) = \langle f(0), f^+ \rangle$. It is easy to check that i is a bijection.

2.12. For part (1), assume first that $\mathcal{U}$ is a proper class. Let S be any set, and let κ be the cardinality of S. Then $\mathcal{U} - S$ is a proper class, so it has a subset T of size κ. Since S and T have the same size, there is a bijection f between them. This verifies the Axiom of Plenitude.

Going the other way, we assume the Axiom of Plenitude. If $\mathcal{U}$ were a set, then we would have a function $f : \mathcal{U} \to \mathcal{U}$ whose image $f[\mathcal{U}]$ were disjoint from $\mathcal{U}$. But this is impossible.

For part (2), assume the Strong Axiom of Plenitude. To get a wellordered proper class of urelements, we use transfinite recursion (see Proposition 2.4) to get an operation F on the ordinals with the property that for all α,

$$F(\alpha) \quad = \quad \mathsf{new}(\emptyset, \{F(\beta) \mid \beta < \alpha\}).$$

Then F is one-to-one, and so its image of F is a proper class. The class C consists of the pairs $\langle F(\alpha), F(\beta) \rangle$ where $\alpha < \beta$.

Finally, for (3), let $C \subseteq \mathcal{U}$. The Global AC gives a wellorder of the class of all sets. But since $\mathcal{U}$ is in bijective correspondence with the class of all singletons of urelements, we also get a wellorder of $\mathcal{U}$. Let $<$ be a wellorder of C. We define $H : On \times V_{afa}[\mathcal{U}] \to \mathcal{U}$ by transfinite recursion so that

$$H(\alpha, b) \quad = \quad \text{the } <\text{-least } x \in C \text{ such that } x \notin b \cup \bigcup_{\beta < \alpha} H(\beta, b).$$

The assumption that C is a proper class is needed to prove that H defines a total operation. By Global AC, let $G : On \to V_{afa}[\mathcal{U}]$ be a definable bijection. To verify the Strong Axiom of Plenitude, we let $\mathsf{new}(a,b) = H(G(\alpha), b)$. This shows that if we assume the Global AC and the existence of a proper class of urelements, we get the Strong Axiom of Plenitude. As we showed in part (1), the Axiom of Plenitude implies that there is a proper class of urelements.

3.1. Recall the formal definition of a function from page 12. Each state is a set of pairs $\langle a, s \rangle$ from $Act \times \mathcal{T}$. This is exactly the definition of a relation. If $\mathcal{T}$ is deterministic, then this relation must be a function. The fact that the states are relations like this is not a general feature of labeled transition systems. In fact, it is equivalent to the labeled transition system being canonical.

3.2. (1) A bisimulation between automata $\mathcal{A}$ and $\mathcal{B}$ should be a relation R on the state sets which preserves the initial states, the set of final states, and which is a bisimulation on the associated labeled transition systems. In more detail, this means that $q_0^A R q_0^B$; that if sRs', then $s \in F^A$ iff $s' \in F^B$; and finally that R is a bisimulation between the labeled transition systems $\langle S^A, \delta^A \rangle$ and $\langle S^B, \delta^B \rangle$.

The main fact in the result on recognition of finite sequences (words) is the following: Suppose that R is a bisimulation between $\mathcal{A}$ and $\mathcal{B}$. Suppose that sRt. If $\mathcal{A}$ is in state s and can arrive at the state s' by reading the word w and making transitions non-deterministically according to δ^A, then in $\mathcal{B}$ we can start at t, read w following δ^B non-deterministically, and arrive at some state t' such that $s'Rt'$. The proof is by induction on the length of w.

Turning to (2), we won't write out the proof of the theorem here. In short, though, one considers an equivalence relation defined on the states. The desired automaton is the quotient of the original. And the natural map $s \mapsto [s]$ of a state to its equivalence classes is easily verified to be a bisimulation relation.

For (3), consider an automaton $\mathcal{A}$ with one state s which is both the initial and final state. Suppose also that $s \xrightarrow{a} s$ but that no other transitions are possible from s. We claim that $\mathcal{A}$ is not bisimilar to any canonical automaton. For suppose it were, and let u be the state corresponding to s. Then $u = \{\langle 1, a, u \rangle\}$. But now u is an element of its own transitive closure, and this contradicts the Foundation Axiom.

The equation needed in (4) is $Can = 2 \times \mathcal{P}(Act \times Can)$, where $2 = \{0, 1\}$.

3.3. For (1), let K be a program of two inputs so that $[\![K]\!](Q, R)$ is defined iff $Q = R$. By the Recursion Theorem, there is a fixed Q^* so that for all R, $[\![Q^*]\!](R) \simeq [\![K]\!](Q^*, R)$. So for all R, $[\![Q^*]\!](R)$ is defined iff $R = Q^*$.

Part (2) is a re-play of a result we saw in Chapter 2, namely Proposition 2.7.2. Suppose that Q exists. We claim that Q is in the wellfounded part of H. To see this, suppose that $H(R, Q)$. Then by definition of H, $[\![Q]\!](R)$ is defined. This means that R is in the wellfounded part of H, too. But if H is wellfounded below all such R's, it is wellfounded below Q, too. We therefore know that Q is in the wellfounded part of H. Thus $[\![Q]\!](Q)$ is defined. In other words $H(Q, Q)$. And so Q is not in the wellfounded part of H, a contradiction.

3.4. To get an interpreter, note that the function $f(P, Q) = [\![P]\!]_M(Q)$ is effectively computable. So by completeness of L we have some int such that for all P, $[\![int]\!]_L(P, Q) \simeq [\![P]\!]_M(Q)$. For a compiler, let S_1^1 be as in the definition for the language L. By completeness again, let $comp$ be such that $[\![comp]\!]_L(P) \simeq [\![S_1^1]\!]_L(int, P)$. We see that

$$[\![[\![comp]\!]_L(P)]\!]_L(Q) \simeq [\![[\![S_1^1]\!]_L(int, P)]\!]_L(Q) \simeq [\![int]\!]_L(P, Q).$$

For mix, we first get int' such that for all P, Q, and R, $[\![int']\!]_L(P, Q, R) \simeq [\![P]\!]_M(Q, R)$. Then we get mix so that $[\![mix]\!]_L(P, Q) \simeq [\![S_1^2]\!]_L(int', P, Q)$.

4.1. The point is that first order structures must "contain" the things they talk about. More precisely, if $M \models (a \ bel \ p)$, then p belongs to $TC(M)$. If a has a true belief of p in M, then $M \in p$. But then we have $M \in p \in TC(M)$, $p \in \cdots \in M \in p$, and this means that both M and p are non-wellfounded.

5.1. Sentence 1 on page 59 is known as the Knower Paradox. Could it be known? If it were, then it would be true, since everything known is true. But if it is true, then what it claims is true, and it claims it could not be known, so, indeed, it could not be known. But we have just now established that what it claims is the case. Hence we know it, and so it can be known after all.

Sentence 2 on page 59 is known as the Prover Paradox. It is basically the same, except that it can be made mathematically rigorous. It leads to a proof of the famous Gödel Incompleteness Theorem.

The next example is called the Richard Paradox. First, notice that there are infinitely many natural numbers, but only a finite number of English expressions involving fewer than twenty-three words. Hence there are numbers which cannot be defined by any such expression, and so at least one n. But this seems paradoxical in that we seem to have given a definition of a particular number, but the definition uses fewer than twenty-three words.

Sentence 4 is a version of what is called Löb's Paradox. Using this sentence, we seem able to prove that this book deserves a Pulitzer Prize. Here is the argument.

Let us first try to prove that sentence (4) is true. Since it is a conditional, we assume the premise and try to prove the conclusion. The premise is that (4) is true. But then apply modus ponens, we get that the conclusion is true, as desired. Then we have established (4). But then we can apply modus ponens again, to obtain the conclusion of (4), namely, that this book deserves a Pulitzer Prize.

While the conclusion may true (who are we to deny it?), the argument seems dubious, since the same argument would let us prove anything at all.

5.2. $R_a = a$, $R_b = 2 = \{0, 1\}$, and $R_c = 1 = \{0\}$. For the verifications that $R_d \notin d$ in these concrete cases, we'll use a standard observation based on the rank operation for well-founded sets. Let $d \in \{a, b, c\}$. Note that R_d is wellfounded, and that the rank of R_d is greater than the rank of any wellfounded element of d. This proves that $R_d \notin d$.

6.1. Here $b_x = \{x, y\}$, $b_y = \{y, z\}$, $b_z = \{x, y\}$, $c_x = \emptyset$, $c_y = \{p, q\}$, and $c_z = \{p\}$. A solution would give sets s_x, s_y and s_z, and they would have the properties that

$$
\begin{aligned}
s_x &= \{s_x, s_y\} \\
s_y &= \{p, q, s_y, s_z\} \\
s_z &= \{p, s_x, s_y\}
\end{aligned}
$$

6.2. Take $X = \{x\}$, $A = \emptyset$, and let $e_x = \{x\}$. Then a solution is a function s on X so that $s_x = \{s_x\}$. This s_x is the set we want, and we call it Ω. Now if a is a set so that $a = \{a\}$, then we would have a solution t to our system: $t_x = a$ would do it. So by uniqueness of solutions $t_x = s_x$. Therefore, $a = \Omega$.

6.3. Take indeterminates $x_0, x_1, \ldots$ and then write

$$
\begin{aligned}
x_0 &= \{0, x_1\} \\
x_1 &= \{1, x_2\} \\
&\vdots
\end{aligned}
$$

This system has a solution s, and $s(x_0)$ is the set we want.

6.4. We take two infinite lists of indeterminates, and use the system:

$$
\begin{array}{llll}
x_0 &= \{y_0, x_1\} & y_0 &= \emptyset \\
x_1 &= \{y_1, x_2\} & y_1 &= \{y_0\} \\
x_2 &= \{y_2, x_3\} & y_2 &= \{y_0, y_1\} \\
&\vdots & &\vdots \\
x_n &= \{y_n, x_{n+1}\} & y_n &= \{y_0, y_1, \ldots, y_{n-1}\} \\
&\vdots & &\vdots
\end{array}
$$

The solution s assigns to y_n the von Neumann n.

7.1. Note that $s_v = \Omega$ *is* a solution, since for each v, b_v is a non-empty set containing only indeterminates. But then by the uniqueness of solutions, this s is the solution to $\mathcal{E}$.

7.2. Given any system $\mathcal{E} = \langle X, A, e \rangle$, the relation
$$R = \{\langle x, x \rangle \mid x \in X\}$$
is an A-bisimulation between $\mathcal{E}$ and itself. This shows that bisimulation is a reflexive relation. To see that it is symmetric, suppose that R is an A-bisimulation between $\mathcal{E}$ and $\mathcal{E}'$. Let R' be the converse of R:
$$R' = \{\langle x, y \rangle \mid \langle y, x \rangle \in R\}.$$
Then R' is an A-bisimulation between R' and R. Finally suppose that R is a A-bisimulation between $\mathcal{E}_1 = \langle X, A, e_1 \rangle$ and $\mathcal{E}_2 = \langle Y, B, e_2 \rangle$, and also that S is an A-bisimulation between $\mathcal{E}_2$ and $\mathcal{E}_3 = \langle Z, C, e_3 \rangle$. Then we take T to be
$$\{\langle x, z \rangle \in X \times Z \ : \ \text{for some } y \in Y, \langle x, y \rangle \in R \text{ and } \langle y, z \rangle \in S\}.$$
We omit the verification that T is an A-bisimulation.

7.3. Yes, the wellfounded sets also have the strong extensionality property. To see this, we use the principle of Proof by $\in$-Induction (see section 2.5). Let $\phi(x)$ be a formula which says

Every set bisimilar to x is equal to x.

We want to show that all x have this property. Fix some x, and assume that for all $y \in x$, $\phi(y)$ holds. Now suppose that this x is bisimilar to some set, say x'. Check that for all $y \in x$, there is some $y' \in x'$ such that y and y' are bisimilar; and vice-versa. (That is, describe these bisimulations in terms of the bisimulation between x and x'.) Then by the assumption that $\phi(y)$ holds for all $y \in x$, we see that for all $y \in x$ there is some $y' \in x'$ so that $y = y'$; and vice-versa. But now $x = x'$, by the Axiom of Extensionality.

7.4. Replace R by
$$S = \{\langle c, d \rangle \in TC(\{a\}) \times TC(\{b\}) \mid cRd\}.$$
The definition of TC implies that S is a bisimulation.

7.5. (1) Ω is reflexive and closed under singletons.

(2) The set $b = \{\emptyset, b\}$ is reflexive and closed under differences.

(3) Suppose b were reflexive, transitive, and closed under singletons. As $b \in b$, we have $\{b\} \in b$ by closure under singletons. So $b - \{b\} \in b$ by closure under differences. But then the set b contradicts the Difference Lemma (Proposition 7.7).

7.6. Let $c = \emptyset$, let $V = \{V, c, b\}$ and let $b = \{V, b\}$.

7.7. Let $b = a \cap WF$. Then every element of b is an ordinal, and b is transitive. So b is an ordinal. We must show that $a - b = \emptyset$; suppose toward a contradiction that $a - b \neq \emptyset$. Let $c \in a - b$. Then c is not wellfounded, so c cannot be equal to, or a member of, any member of b. Since $\in$ is a linear order of a, $b \subseteq c$. Also, c must have some member c' which is non-wellfounded and hence itself belongs to $a - b$. We use these observations to show that all elements of $a - b$ are bisimilar and hence equal. The bisimulation is

$$I_b \cup ((a - b) \times (a - b)),$$

where I_b is the identity relation on b. At this point we know $a - b$ is a singleton $\{c\}$. But now c must belong to itself. This contradicts our assumption that a has no reflexive members.

8.1. For each set a, let $\mathcal{E}_a = \langle X, \emptyset, e \rangle$ be the generalized flat system defined as follows: $X = TC(a) \cap V_{afa}[\mathcal{U}]$, and for all $b \in X$, let $e_b = TC(b) \cap V_{afa}[\mathcal{U}]$. Let s_a be the solution to $\mathcal{E}_a$ Then we set $Sk(a) = solution\text{-}set(\mathcal{E}_a)$.

To see that this works, fix a set a. For all sets $b \in TC(a)$, $s_a \restriction TC(b)$ is a solution to $\mathcal{E}_b$. Therefore, $s_a \restriction TC(b) = s_b$. It follows that $Sk(b) = s_a(b)$. This implies that

$$Sk(a) = \{Sk(b) \mid b \in a \text{ is a set}\}$$

as desired. The uniqueness is proved using bisimulations, exactly as in Theorem 8.1.

8.2. Let a be a set. So $G(a) = \{G(p) \mid p \in a\}$. Then

$$
\begin{aligned}
F(G(a)) &= F(\{(G(p)) \mid p \in a\}) \\
&= \{F(q) \mid q \in \{G(p) \mid p \in a\}\} \\
&= \{F(G(p)) \mid p \in a\}
\end{aligned}
$$

8.3. Consider the relation $F(a) \, RG(a)$, as a runs through $V_{afa}[A]$. This is an A-bisimulation.

8.4. To say that s is a solution means that $s_x = \{p, s_x, s_y\}$, and similarly for y and z. The first part is immediate from the Solution Lemma. We check that $s_x = e'_x[s]$, and the same calculations show that $s_y = e'_x[y]$ and $s_z = e'_x[z]$. We calculate

$$
\begin{aligned}
e'_x[s] &= \{p, \{p, x, y\}, \{q, x, z\}\}[s] \\
&= \{p, \{p, x, y\}[s], \{q, x, z\}[s]\} \\
&= \{p, \{p, s_x, s_y\}, \{q, s_x, s_z\}\} \\
&= \{p, s_x, s_y\} \\
&= s_x
\end{aligned}
$$

Similar calculations work for y and z.

8.5. In the notation from the proof of Theorem 8.2,

$$X^+ \;=\; \{x,y\} \,\cup\, \{\{\{y\},\emptyset\},\{y\},\emptyset\}.$$

So Y in this case would be x, y, and fresh indeterminates $x_{\{\{y\},\emptyset\}}$, $x_{\{y\}}$, and $x_\emptyset$. For convenience, we call these u, v, and w. $\mathcal{E}^\flat$ is the following flat system over Y:

$$
\begin{aligned}
x &= \{u,x\} \\
u &= \{v,w\} \\
v &= \{y\} \\
w &= \emptyset \\
y &= \{v,p\}
\end{aligned}
$$

8.6. Let $X = \{x,y\}$. Suppose that we want to solve the system $x = e_x$, $y = e_y$. First, we solve $x = e_x$. This gives a substitution r defined only on x with the property that $r_x = e_x[r]$. Second, solve the equation $y = e_y[r]$. This gives a substitution s defined only on y such that $s_y = e_y[r][s]$.

Let t be the substitution defined on X by $t_x = r_x[s]$ and $t_y = s_y$. We claim that for all sets a, $a[t] = a[r][s]$. To see this, note that for all urelements z, $z[t] = z[r][s]$. (For this, we need three cases: $z = x$, $z = y$, and $z \neq x,y$.) Our claim now follows from Exercise 8.3. Now it follows that

$$
\begin{aligned}
e_x[t] &= e_x[r][s] \\
&= r_x[s] \\
&= t_x
\end{aligned}
$$

And $e_y[t] = e_y[r][s] = s_y = t_y$.

We now prove the uniqueness of solutions for $\mathcal{E}$. Suppose u is a substitution with domain X such that $u_x = e_x[u]$ and $u_y = e_y[u]$. We'll consider also the following two substitutions:

$$v \;=\; \{\langle x, u_x\rangle\} \qquad\qquad w \;=\; \{\langle y, u_y\rangle\}$$

Note that $[u] = [w][v]$, since $x \notin support(w_y)$. We also have the substitution r defined above, and we'll need the substitution r' with domain y which solves the system $y = e_y$.

Since $r_x = e_x[r]$, $r_x[w] = e_x[r][w] = e_x[w][r^w]$, where r^w is the substitution $\{\langle x, r_x[w]\rangle\}$. It follows that r^w is the solution to the equation $x = e_x[w]$. But this equation also has v as a solution, since $v_x = e_x[u] = e_x[w][v]$. So by uniqueness of solutions of equations in one variable, $r_x[w] = (r^w)_x = v_x$.

By the same reasoning as in the last paragraph, $r'_y[v] = w_y$.

As we have just seen, $r_x[w] = v_x$. This implies by Exercise 8.3 that for all sets $a \in V_{afa}[\mathcal{U} - \{y\}]$, $a[r][w] = a[v]$. Therefore $r'_y[r][w] = r'_y[v] = w_y$.

This tells us that w is the solution of $y = r'_y[r]$. This equation is the key step for uniqueness, since it involves only r and r' and the fact that w comes from u, a solution to the original $\mathcal{E}$. That is, if we started with t instead of u, we would see that $\{\langle y, t_y \rangle\}$ is also a solution to $y = r'_y[r]$. Therefore $w = \{\langle y, t_y \rangle\} = s$. So $u_x = v_x = r_x[w] = r_x[s] = t_x$, and $u_y = w_y = s_y = t_y$. This proves that $u = t$.

8.7. Consider the operations $F(p) = p[t \star s]$ and $G(p) = (p[s])[t]$. These are both substitution-like: F is by the definition of substitution, and G is by Exercise 8.3. We claim that F and G agree on all urelements; this and Exercise 8.2 would imply the desired result. If $x \in dom(s)$, then $F(x) = s_x[t] = (x[s])[t] = G(x)$. If $x \notin dom(s)$, then $x \notin dom(t)$; hence $F(x) = x$ and also $G(x) = (x[s])[t] = x$.

8.8. First, we check easily that the domains of both $u \star (t \star s)$ and $(u \star t) \star s$ are $dom(s)$. And for $x \in dom(s)$, we use Exercise 8.7 to calculate:

$$(u \star (t \star s))_x = (t \star s)_x[u] = (s_x[t])[u] = s_x[u \star t] = ((u \star t) \star s)_x.$$

Second, the left-identity element is the identity substitution: $dom(i) = \emptyset$. This i has the property that for all a, $[a]i = a$. So $i \star s = s$ for all s. To see that i is the unique such substitution, let $s \neq i$. Let $x \in \mathcal{U}$ be such that $s_x \neq x$. Let t be the substitution with domain $\{x\}$ such that $t_x = x$. Then $(s \star t)_x = s_x \neq x$. Hence $s \star t \neq t$.

Incidentally, the left identity element i is not a two-sided identity, since for all s, $s \star i = i$.

8.9. For 1, $e \star e$ is exactly the second system described in Exercise 8.4. That is, $(e \star e)_x = \{p, \{p, x, y\}, \{q, x, z\}\}$, etc. In that exercise, we saw that the solution s to e is the solution to $e \star e$. Now we are in a position to prove this on the basis of the abstract formulation of the Solution Lemma (Theorem 8.5) and the earlier exercises on properties of the $\star$ operation. Note that

$$s \star (e \star e) = (s \star e) \star e = s \star e = s.$$

(We are using Exercise 8.8; note that s and e have the same domain.) It follows that s is a solution to $e \star e$. So by uniqueness, it is *the* solution.

9.1. Let t be defined on X_x by $t_y = s_y$. In other words, t is the restriction to X_x of the solution s of $\mathcal{E}$. We show that t is a solution to the system $\mathcal{E}_x$. Recall that if $y \in X_x$, then $b_y \subseteq X_x$ and $e_y \cap A = e_y \cap A_x$. Thus

$$
\begin{aligned}
s_y &= \{s_z : z \in e_y \cap X\} \cup (e_x \cap A) \\
&= \{s_z : z \in e_y \cap X_x\} \cup (e_x \cap A_x)
\end{aligned}
$$

This means that t really is a solution to $\mathcal{E}_x$. Applying the uniqueness part of the Solution Lemma to $\mathcal{E}_x$, we see that $t = s'$.

9.2. Write $\{a, e_y\}^+$ as the pointed system

$$\langle\langle\{x, y, z\}, \{a, b\}, e'\rangle, z\rangle,$$

where $e'_x = \{a, y\}$, $e'_y = \{b, x\}$, and $e'_z = \{a, y\}$. The bisimulation that we want is $\{\langle x, x\rangle, \langle y, y\rangle, \langle x, z\rangle\}$.

Second, write $\{b, e_x\}^+$ as $\langle\langle\{x, y, z\}, \{a, b\}, e'\rangle, z\rangle$, where $e'_x = \{a, y\}$, $e'_y = \{b, x\}$, and $e'_z = \{b, x\}$. This time, the bisimulation is

$$\{\langle x, x\rangle, \langle y, y\rangle, \langle y, z\rangle\}.$$

9.3. First, a general definition: Call a pointed system e_1 an *E-subset* of e_2 if every E-element of e_1 is $\equiv$ to some E-element of e_2. In other words $(e_1 \subseteq e_2)^{tr}$.

Suppose that $e = \langle \mathcal{E}, x\rangle$ is a pointed system, where $\mathcal{E} = \langle X, A, e\rangle$. We need e whose E-members are the E-subsets of e. For each $Y \subseteq X$ and $B \subseteq A$, let

$$e_{Y,B} \quad = \quad (\{e_y \mid y \in Y\} \cup B)^+.$$

Each $e_{Y,B}$ is an E-subset of e. More importantly, every E-subset of e is of the form $e_{Y,B}$ for some $Y \subseteq X$ and $B \subseteq A$. This is by our definitions, and by Proposition 9.2. By Collection, we can also consider

$$f \quad = \quad \{e_{Y,B} \mid Y \subseteq X, B \subseteq A\}^+.$$

This f is a pointed system, and

$$[(\forall g)(g \subseteq e \to g \in f]^{tr}.$$

Since e was an arbitrary pointed system, we have checked that translation of the Powerset Axiom.

9.4. We take $n(e, f) = \mathsf{new}([e], \mathit{support}([f]))$. (Recall here that $[e]$ is the set of standard pointed systems of minimal rank bisimilar to e.) This map n has the property that if $e \equiv e'$ and $f \equiv f'$, then $n(e, f) \equiv n(e', f')$. This makes it appropriate for translating the function new. If $e \not\equiv e'$, then $n(e, f) \neq n(e', f)$ by the analogous property of new. Finally, if $(f \subseteq \mathcal{U})^{tr}$, then the support of f is the set of all E-elements of f. So $(n(e, f) \notin f)^{tr}$.

9.5. Suppose that ("f is a relation")^{tr}. As in the first part of the proof of Lemma 9.5.1, let S be a set containing a pair $\langle p_y, q_y\rangle$ for each $y \in e_x$. By the Axiom of Choice, we can find a subset $f \subseteq S$ with properties (1) and (2) of the lemma. Then using the right-to-left direction of Lemma 9.5.1,

("e is a function with the same domain as f")^{tr}.

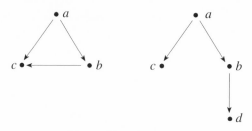

FIGURE 16 Two Pictures of 3

9.6. Recall that we have an operation $a \mapsto \overline{a}$ taking M to M_{afa}. This operation is one-to-one, and it preserves membership both ways. The Solution Lemma tells us that every flat system is bisimilar to a set. Hence every pointed flat system is bisimilar to the canonical flat system set of some set. Since the elements of M_{afa} are equivalence classes, each of them of the form $\overline{a}$, where $a \in M$. Therefore the operation $a \mapsto \overline{a}$ is an isomorphism.

9.7. In part (1), we show by induction that every pointed system in the wellfounded part of E is bisimilar to a system of the form $\overline{a}$ for some wellfounded a.

For part (2) assume that $ZFA \vdash \phi^{wf}$ but that $ZFC^- \not\vdash \phi^{wf}$. Let M be a model of ZFC^- satisfying $\neg\phi$. As we know, the extended model M_{afa} satisfies ZFA. But also, M_{afa} satisfies ϕ^{wf}, by part (1). That is, every set which is wellfounded in M_{afa} belongs to M. And now we have contradicted the assumption that $ZFA \vdash \phi^{wf}$.

In part (3), we claim that $ZFC^- \vdash CH \leftrightarrow CH^{wf}$. The crux of the matter is that a function between two wellfounded sets is itself wellfounded. Omitting some standard details having to do with absoluteness, we see that if $ZFA \vdash CH$, then $ZFA \vdash CH^{wf}$. So by part (1), $ZFC^- \vdash CH^{wf}$. And then by our observation again, $ZFC^- \vdash CH$.

10.1. Two pictures of 3 are shown in Figure 16. The nodes in them may be any sets.

10.2. Let d be the inverse of the function f from (3). Although we have six equations to check, we'll only give the reasoning for the element 76. So we show that $d(76) = \{d(6), d(0)\}$. Since d is the inverse of f, $d(76) = x$,

$d(6) = y$, and $d(0) = z$. The equation to check is now $x = \{y, z\}$. This holds because it is part of (5).

10.3. Let G have the identity function as its decoration. For each $a \in G$, $d(a) = a$. So each $a \in G$ is a set, never an urelement. The definition of a decoration implies that each $d(a)$ is a set of other $d(b)$, so this tells us that each $a = d(a)$ is a set of sets, each also with this property. It follows from this that each $a \in G$ is pure.

Next, we show that G is transitive. Suppose $a \in G$ and $b \in a$. Then since $a = d(a)$, $b \in d(a)$. So for some $c \in G$ such that $a \to c$, $b = d(c)$. But since d is the identity, we see that $b = c$. Also, $b = d(c) \in d(a) = a$. So $b \in a$ as desired.

10.4. This is just a replay of an argument from Proposition 10.4. Let X be a set of urelements in correspondence with A, say via the bijection $x : A \to X$. Let Y be another set, in correspondence by $y : A \to Y$. map $e : X \to \mathcal{P}(X)$ be given by

$$e(x_a) \;=\; \{x_b \mid aRb\} \cup \{y_a\}.$$

By the Flat Solution Lemma, let s be the solution to $\langle X, e \rangle$. Then s is injective, since for $a \neq a'$, $s(x_a)$ contains y_a and $s(x_{a'})$ does not. Also, the solution condition implies that s is an isomorphism.

10.5. The wellfounded relations are exactly the relations representable as restriction of $\in$ if we are working in *ZFC*.

10.6. By the proof of Proposition 10.5, we may assume that $\langle A, R \rangle$ is a binary relation so that each $a \in A$ is a subset of $A \cup \mathcal{U}$. Indeed, we may assume that a contains exactly one urelement x_a, and so that the map $a \mapsto x_a$ is one-to-one. Let $X = \{x_a \mid a \in A\}$. Let κ be an infinite cardinal greater than the cardinality of A. Let $s : X \to \mathcal{P}(\kappa)$ be a substitution with the property that each $s(x_a)$ is a subset of κ of size κ. Also, take s to be one-to-one. (The standard theory of cardinals implies that such s exist.)

We claim first that for all $a \in A$, $a[s] \notin X[s]$. The reason is that each $a[s]$ will have size $< \kappa$, since $|a| < \kappa$ has this property. This implies that $A[s] \cap X[s] = \emptyset$. Next, note that $a[s] \cap X[s] = \{s(x_a)\}$. This is because x_a is the only urelement in a, and the other elements of a are elements of A, and finally because $A[s] \cap X[s] = \emptyset$. This observation tells us that $[s]$ is one-to-one on A, since if $a[s] = b[s]$, then $s(x_a) = s(x_b)$. Since s is one-to-one on X, we would have $x_a = x_b$ and hence $a = b$. Now that we know $[s]$ is one-to-one on A, we can let $B = A[s]$. B is a pure set because each $s(x_a)$ is pure. The same argument as in Proposition 10.5 shows that $[s]$ is an isomorphism.

10.7. We work in *ZFA* and show how to get a class model of the theory in question. Let V_{afa} be the class of pure sets. For each set $A \subseteq \mathcal{U}$, define $W_\alpha[A]$ by recursion as follows:

$$
\begin{array}{rcl}
W_0[A] & = & V_{afa} \\
W_{\alpha+1}[A] & = & \mathcal{P}(A \cup W_\alpha[A]) \\
W_\lambda[A] & = & \bigcup_{\beta<\lambda} W_\beta[A]
\end{array}
$$

Let W be the class of all sets a such that for some set $A \subseteq \mathcal{U}$ and ordinal α, $a \in W_\alpha[A]$. Then W is a transitive class of sets closed under power set, pairing, and union. Thus it satisfies the corresponding axioms of set theory, and also the Axioms of Infinity, Choice, and Separation. It is not hard to check the Axiom of Collection, and we omit this. To get the Axiom of Plenitude, we define new in W to be the same as in the universe.

If $p \in \mathcal{U}$, then W satisfies the sentence saying that there no solution to the equation $x = \{x, p\}$. For a solution of this in W would really give the set $a = \{a, p\}$, and for each A it is easy to check by induction on α that for all α, $a \notin W_\alpha[A]$.

Finally, we check that $W \models AFA_0$. Let G be an unlabeled graph. Then the decoration d of G belongs to $V_{afa} \subseteq W$. And since W is a model of the rest of set theory, W satisfies the sentence "d is the decoration of W."

11.1. In one direction, assume $a \models \neg(\neg\varphi)$. This means that a does not satisfy $\neg\varphi$; this is what we write as $a \not\models \varphi$ Since a does not satisfy $\neg\varphi$, it *does* satisfy φ. (Note that we are using the logical law of double negation.)

Going the other way, assume that $a \models \varphi$. Then it certainly is not true that $a \models \neg\varphi$. And since "$a \models \varphi$" is false, $a \models \neg\neg\varphi$ does indeed hold.

11.2. Suppose that $\models^W \varphi$. To prove that $\models^W \Box\varphi$, take any set $a \in W$. Let $b \in a$. Then $b \models \varphi$, since $\models^W \varphi$ and $b \in W$ by transitivity. As b is an arbitrary set in a, we see that $a \models \Box\varphi$. And since $a \in W$ is arbitrary, $\models^W \Box\varphi$.

11.3. Not every such W is modally normal. Let W be the class of all sets, for example. Then $\Diamond\mathsf{T} \models \Diamond\mathsf{T}$, but $\Diamond\mathsf{T} \not\models \Box\Diamond\mathsf{T}$. The set $\{\emptyset\}$ satisfies $\Diamond t$ but not $\Box\Diamond\mathsf{T}$.

For the second part, let $W \subseteq V_{afa}[A]$ be transitive on sets. We claim that W is modally normal if and only if the following condition holds: for all sets $b \in W$, either $b \subseteq \mathcal{U}$, or there is a set $S \subseteq A$ such that $b = S \cup \{b\}$. (This result was discovered by the MOC Workshop.)

To see this, first suppose that W meets this condition. Assume that $T \models^W \varphi$. Let $b \in W$ be such that $b \models T$. So $b \models \varphi$ as well. If b has no sets, then trivially $b \models \Box\varphi$. In the other case, b would be the only set belonging to b, and so $b \models \Box\varphi$.

Conversely, let's assume that W is modally normal. Let b be any set in W. Let

$$\varphi = \bigwedge_{p \in b} p \wedge \bigwedge_{p \in A - b} \neg p.$$

Let T be $\{\varphi\}$. Note that $T \models^W \varphi$. Also $b \models T$. So $b \models \Box\varphi$. This means that every set $a \in b$ has exactly the same urelements as b. Let $c = TC(\{b\})$. Therefore every set $a \in c$ has the same urelements as b. It follows that all sets in c are bisimilar. Thus either $c = \emptyset$ or c is the singleton $\{b\}$. This implies that $b = S \cup \{b\}$, where $S = support(b)$. (This proof makes it appear that the infinitary language is needed for the proof, but this is not so. We leave it to the reader to work out a proof that only uses the finitary language.)

11.4. Assume we have ψ_b for all sets $b \in a$. We set ψ_a to be the following sentence:

$$\bigwedge_{p \in a} p \bigwedge_{p \notin a} \neg p \wedge \triangle\{\psi_b \mid b \in a\}.$$

(As usual, p ranges over urelements and b over sets.) We show by induction on $WF[A]$ that for all $b \in V_{afa}[A]$, $b \models \psi_a$ iff $b = a$. Assume that this holds for all sets $a' \in a$. Let $b \in V_{afa}[A]$. Then b has the same urelements as a. By induction hypothesis, $b \models \psi_a$ iff every set $b' \in b$ equals some set $a' \in a$, and for every set $a' \in a$ there is some set $b' \in b$ equal to a'. This just means that for $b \in V_{afa}[A]$, $b \models \psi_a$ iff $b = a$.

When A is finite, we show by induction on $HF^0[A]$ that each ψ_a belongs to $\mathcal{L}$.

11.5. First, we check that each sentence θ^a is maximal. Since $a \models \theta^a$ but $a \not\models \mathsf{F}$, $\theta^a < \mathsf{F}$. Suppose that $\theta^a < \psi$. Then the class C of sets satisfying ψ must be a proper subclass of the sets satisfying θ^a. So we must have $C = \emptyset$. Thus $\psi \equiv \mathsf{F}$.

Second, we need to see that every maximal sentence is of this form. Let φ be maximal. We claim that the class C of sets satisfying φ must be a singleton. For if a and b were distinct elements of C, then $\varphi \wedge \neg\theta^a$ would be satisfied by b. We would have

$$\varphi < \varphi \wedge \neg\theta^a < \mathsf{F},$$

and this contradicts the maximality of φ. Hence C is of the form $\{a\}$ for some a. This means that $\varphi \equiv \theta^a$.

11.6. For (1), note that the following are axioms of $\boldsymbol{K}$:

$$\Box(\psi \to (\varphi \to \psi)) \to (\Box\psi \to \Box(\varphi \to \psi))$$
$$\Box(\varphi \to (\psi \to (\varphi \wedge \psi))) \to (\Box\varphi \to \Box(\psi \to (\varphi \wedge \psi)))$$

Also, $\varphi \to (\psi \to (\varphi \wedge \psi))$ is a tautology. Using necessitation, the antecedent of the second sentence above is provable. So the consequent is provable. Using it and the first sentence, we see that

$$\vdash \Box\varphi \to (\Box\psi \to \Box(\varphi \wedge \psi)).$$

From this we use propositional logic to get what we want.

For (2), note that $\vdash \Box(\varphi \wedge \neg(\varphi \wedge \psi)) \to \Box\psi$. So by (1),

$$\vdash \Box\varphi \wedge \Box\neg(\varphi \wedge \psi) \to \Box\neg\psi$$

From this we use propositional logic to get what we want.

11.7. We prove by induction on φ that for all $U \in Th$, $\varphi \in U$ iff $U \models \varphi$. The result is clear for the atomic sentences, by the definition of the structure Th. The steps for $\neg$ and $\wedge$ are easy, since they just use general facts about maximal consistent subsets of boolean algebras.

Assume that $\Diamond\varphi \in U$; we show that $U \models \Diamond\varphi$. By Lemma 11.24, there is some $V \in Th$ so that $V \in e(U)$ and $\varphi \in V$. By induction hypothesis, $V \models \varphi$. (This is why we're proving something by induction on φ for *all* maximal consistent sets.) So $U \models \Diamond\varphi$. The converse direction is similar, and does not use Lemma 11.24.

11.8. It is easy to see that $th(\Omega)$ is closed under $\Box$. Suppose T is any maximal consistent set with this property. We show by induction on φ that $\varphi \in T \to \Omega \models \varphi$. This is true for the atomic propositions, and the induction steps for the boolean operators are easy.

Suppose that $\Diamond\varphi \in T$. We claim that φ must belong to T as well; for otherwise $\neg\varphi \in T$, so $\Box\neg\varphi \in T$, and this would contradict the consistency of T. By induction hypothesis, $\Omega \models \varphi$. Since $\Omega \in \Omega$, $\Omega \models \Diamond\varphi$. The converse is easier, so we omit it.

11.9. Let $W = \{a \in V_{afa}[A] \mid a \models \Diamond\mathsf{T}, \Box\Diamond\mathsf{T}, \Box\Box\Diamond\mathsf{T}, \Box\Box\Box\Diamond\mathsf{T}, \ldots\}$. We want to prove that $HNe = W$. In one direction, it is clear that each $a \in HNe$ satisfies $\Diamond\mathsf{T}$. Also, by Proposition 11.32.1, the theory of this class is closed under necessitation. Hence $HNe \subseteq W$. To prove the other direction, note that since the set of sentences in question is closed under necessitation, W is transitive on sets by Proposition 11.32.2. Hence we need only prove that each $a \in W$ has some set as a member. But this follows since every such a is a model of $\Diamond\mathsf{T}$.

11.10. First, assume that $a \in W$. Let $b = TC(\{a\})$ so that b is transitive on sets and $a \in b$. We claim that $\models^b T$. This follows easily from the assumption that $a \models T^\Box$ and the characterization of TC in terms of iterated unions.

Going the other way, let a be transitive on sets and such that $\models^a T$. Let $\varphi \in T$. We claim that for all n, each set $b \in a$ satisfies $\Box^n \varphi$. The proof is by induction on n. The case $n = 0$ is immediate. The induction step works easily also, using the assumption that a is transitive on sets.

11.11. First, we claim that every wellfounded a which is transitive on sets (no matter what its support) satisfies all of the sentences in $\boldsymbol{L}_\infty(A)$. For this, it is sufficient to show that every wellfounded set a satisfies each instance of (L), with φ an infinitary sentence.

Suppose that $a \models \Box(\Box\varphi \to \varphi)$. We show by $\in$-induction that for all sets b, if $b \in a$, then $b \models \varphi$. It will then follow that $a \models \Box\varphi$. Assume that all sets $c \in b$ which belong to a satisfy φ. Since a is transitive on sets, we see that $b \models \Box\varphi$. But by our assumption on a, $b \models (\Box\varphi \to \varphi)$. Therefore $b \models \varphi$.

Going the other way, fix some set $a \in V_{afa}[A]$.

We need to show that the membership relation on sets in a is transitive and wellfounded. For transitivity, take sets $c \in b \in a$. Let

$$\varphi = \bigvee\{\theta^d \mid d \in a \text{ and for all sets } e \in d, e \in a\}.$$

We show that $a \models \Box(\Box\varphi \to \varphi$. from this it follows that $a \models \Box\varphi$, and hence that $b \models \varphi$. This last fact implies that $c \in a$. So assume that $b' \in a$ and that $b' \models \Box\varphi$. We show that $b' \models \varphi$. For this, let $c' \in b'$. We need to see that $c' \in a$. But as $b' \models \Box\varphi$, this follows immediately.

Now we turn to the wellfoundedness part. Suppose towards a contradiction that this is false. Let

$$\varphi = \bigvee\{\theta^c \mid c \text{ is a non-wellfounded member of } a \cup \bigcup a\},$$

where θ^c characterizes c (see Theorem 11.14). The contrapositive of the scheme $\boldsymbol{L}_\infty$ tells us that

(1) $\qquad\qquad a \models \Diamond\varphi \to \Diamond(\varphi \wedge \Box\neg\varphi).$

Since a is non-wellfounded, it must have a non-wellfounded member. Thus $a \models \Diamond\varphi$. But then there is some $b \in a$ so that $b \models \varphi \wedge \Box\neg\varphi$. From this, b is a non-wellfounded member of a. We claim that every set $c \in b$ is wellfounded; this would give a contradiction. For each $c \in b$, $c \in \bigcup a$. So if c were non-wellfounded, $\models \theta^c \to \varphi$ and so $b \models \Diamond\varphi$. But $b \models \Box\neg\varphi$.

11.12. We only need to show that if $a \in HF^1[A]$ satisfies $\boldsymbol{L}(A)$, then a is wellfounded. Suppose not. We get a contradiction by considering a finitary version of the equation (1) just above the statement of the exercise. Note first that $a \cup \bigcup a$ is finite. List the non-wellfounded members of this set as $c_1, \ldots, c_K$. For each n, let

$$\varphi_n = \varphi_n^{c_1} \vee \cdots \vee \varphi_n^{c_K}.$$

For each n,
$$a \models \Diamond\varphi_n \to \Diamond(\varphi \wedge \Box\neg\varphi_n),$$
and also $a \models \varphi_n$. So for each n there is some $b \in a$ so that $b \models \varphi_n \wedge \Box\neg\varphi_n$. For this b, there is also some i so that $b \models \varphi_n^{c_i}$. But since a is finite, there is some b so that for infinitely many n, $b \models \varphi_n \wedge \Box\neg\varphi_n$. For this b, we can also find a fixed i so that $b \models \bigwedge_n \varphi_n^{c_i}$. This $b = c_i$. As before, every set $c \in b$ is wellfounded, and we get a contradiction.

11.13. Recall that in Example 11.5 on page 152 we built an irreflexive Kripke structure for the reflexive set Ω.

11.14. The first part is very nearly the definition of $H(\mathbf{Triv})$. For the second, let $p, q \in A$ be distinct urelements, let $a = \{p, a, b\}$, let $b = \{q, a, b\}$, and let $W = \{a, b\}$.

12.1. We should take $M = TC(a)$ and $R(\langle m_1, \ldots, m_k \rangle) = m_k$. W_I is the set of legal plays p in which for some n, $\mathrm{mv}(p, 2n)$ is either empty or an urelement.

12.2. We prove the third of these assertions, since it is the most involved. The others are a bit easier.

First, let $a \in Even_2$. Suppose towards a contradiction that $b \in a \cap Even_1$. We show that I has a winning strategy σ in $\mathcal{G}_a$. The strategy is to take $\sigma(\epsilon) = b$, and then follow a winning strategy τ for II in $\mathcal{G}_b$ which is guaranteed to have all plays finite.

This shows that every set $b \in a$ must belong to Odd or to $Even_2$. Suppose that all of the elements of a belonged to Odd. Then b would belong to $Even_1$ as follows: to get a winning strategy σ for II in $\mathcal{G}_a$, let I play some $b \in a$, and then continue by some strategy for I in $\mathcal{G}_b$. This strategy σ has the property that every play by it is decided in finitely many moves.

Finally, assume that every set $b \in a$ belongs to $Odd \cup Even_2$, and some set $b \in a$ belongs to $Even_2$. We need to see that $a \in Even_2$. For this, note that there is an infinite play: both players can choose elements of $Even_2$ at each step. So we only need to see that II has a winning strategy σ. If I plays some $b \in a \cap Odd$, then II can continue with a winning strategy for I in $\mathcal{G}_b$. If II plays some $b \in a \cap Even_2$, then $\sigma(\langle b \rangle)$ should be some $c \in b \cap Even_2$. Then II should continue like this. If at some point I plays an element of Odd, then II knows how to continue in order to win. Otherwise, the play will be an infinite sequence of elements of $Even_2$, and so II wins in this case as well.

12.3. Let C be the class of all sets of the form $\{\emptyset, \alpha\}$, as α ranges over the ordinals. Since any set containing $\emptyset$ belongs to Odd, C is a proper class of Odd sets. Further, every subset of C belongs to $Even_1$. This gives a proper class of $Even_1$ sets.

Every set in $Even_2$ is non-wellfounded. Assuming AFA, we know that Ω exists and belongs to $Even_2$. To get a proper class of such sets, we take the sets of the form $\{\{\emptyset, \alpha\}, \Omega\}$ as α ranges over the ordinals.

12.4. Let $W = \{w_1, w_2, \ldots\}$. II must arrange that the final play p has the property that r_p is different from each w_i. The key point is that in the first two moves, II can arrange that $r_p \neq w_1$. That is, if the decimal expansion of w_1 should happen to start with I's initial move, then II can alter the second digit; that is, II can play to make sure that the eventual p has $r_p \neq w_1$. Then in moves 3 and 4, II can insure that $r_p \neq w_2$. Continuing in this way, II can see to it that $w_p \notin W$.

12.5. Here is one way to do it. Instead of taking the set of positions to be $M = X \cup Y$, we take the disjoint union $N = X + Y$ (see page 13). The rest of the modification is straightforward. For example,

$$R(\epsilon) = \begin{cases} e_x + e'_y & \text{if } atoms(e, x) = atoms(e, y) \\ \emptyset & \text{if } atoms(e, x) \neq atoms(e, y) \end{cases}$$

The modifications in the proof of Theorem 12.6 are also minor.

12.6. Let $X = \{x, z, y_0, y_1, \ldots\}$. Let $\mathcal{E} = \langle X, \emptyset, e \rangle$ be the system with $e(y_0) = \emptyset$, $e(y_{n+1}) = \{y_n\}$, $e(x) = \{x\}$, and Let $e(z) = \{x\} \cup \{y_n : n \in \omega\}$. Let $\mathcal{F} = \langle X, \emptyset, e' \rangle$ be the system with e' defined just as e, except that $e'(z) = \{y_n : n \in N\}$.

The pointed systems $\langle \mathcal{E}, z \rangle$ and $\langle \mathcal{F}, z \rangle$ are not bisimilar. To see this using the bisimulation game, suppose the I starts by picking z in $\langle \mathcal{E}, z \rangle$. II must select some y_n. Then I keeps selecting z, and II must respond y_{n-1}, y_{n-2}, $\ldots$. Eventually, II will be faced with z and y_0. At that point, II loses.

We claim that II does have a winning strategy in the modified game. Suppose I declares n and then plays z. Then II can play y_{n+1} and win that play. The rest of the game is analyzed similarly.

12.7. For (1), we have two cases. If II has a winning strategy in some game $\mathcal{G} \in S$, then I has a strategy in S^+: play $\mathcal{G}$ and adopt II's strategy in that game. Otherwise, II has a winning strategy in S^+: If I plays $\mathcal{G}$ to start, then II should take a winning strategy σ for I in $\mathcal{G}$ and play according to σ from now on.

The proof of (2) is similar to this, with a twist in the second half. Suppose that I has a winning strategy in every game in S. To get a winning strategy for II in S^* we need to say what II should do if I happens to play S^*. Whenever this happens, II should play S^* to return the game back to the original position. Either the play goes on forever with S^* (and this is a win for II), or else at some point I plays one of the games in S, in which case II uses a winning strategy for that game.

13.1. (1) The only predicate which is problematic is the denotation predicate Denotes. We take that up in the next section. (2) We would need to allow function symbols in our language for the operations used in forming sentences. (3) One would need to allow variables to occur in the formation rules of sentences, as in: if x, y are either sentences or variable, so are $x \wedge y$. This is not particularly problematic as long as we make sure that our models are correct for the predicate Sent.

13.2. Let M_4 be defined as follows:

$$
\begin{aligned}
M_4 &= \langle D', L, Ext', Anti', d, c \rangle \\
D' &= \{\lambda, M_3, M_4\} \\
L &= \{\mathsf{True}\} \\
Ext' &= \{\langle \mathsf{True}, \emptyset \rangle\} \\
Anti' &= \{\langle \mathsf{True}, D' \times D' \rangle\} \\
den' &= \{\langle \mathsf{this}, \lambda \rangle, \langle \mathsf{h}, M_3 \rangle\}
\end{aligned}
$$

It is easy to check that M_4 truth-correct, total, and contains the Liar.

13.3. The proof of this is similar to the construction of M^+ earlier, except that one does not change the extension or the anti-extension of the truth predicate.

13.4. Except for the truth predicate, we proceed as before, by letting the M and M_{tot} assign the same extension to each predicate R, and assign as the anti-extension of R in M_{tot} just the complement of its extension. By Proposition 13.3, this new model satisfies (T1) and (T3), so it is still truth-correct. Thus we may suppose that, except for the truth predicate, the anti-extension of each predicate is just the complement of its extension. Now consider any pair $\langle \varphi, N \rangle$ such that $\varphi, N \in D_M$ but such that the pair $\langle \varphi, N \rangle$ is in neither the extension nor the anti-extension of True. We want to extend the model M to a model which assigns this pair to one or the other. If N is a model, $\varphi \in Def(N)$, and $N \models \varphi$, add $\langle \varphi, N \rangle$ to the extension of True. Otherwise, add it to the anti-extension. Let M' be the resulting model. Clearly M' is total and an extension of M. It is constructed so as to satisfy (T1) and (T3), hence

(T0). But does it? Notice in particular that the model M' might already be accessible in M. That is, M' might be one of the models N we considered in the construction of M' and hence $M' \in D_M$. So you should go back and read the construction again to convince yourself that in this case, things still work out correctly.

13.5. Consider the sentence $\exists x(\mathsf{Mod}(x) \land \neg \mathsf{True}_x(\mathsf{this}))$. (Treat Mod as an abbreviation for $\exists y \mathsf{True}_x(y)$.) Let M be any truth-correct model where this denotes this sentence, which contains some tautology (to make Mod work properly), and which contains no models. (How do you know such exist?) Assume $M^* = M(M^*)$ is a principal expansion of M which is truth-correct and in which the sentence in question has a truth value. It is easy to derive a contradiction by means of a Liar-like argument.

13.6. Let M_1 be defined as follows:

$$
\begin{aligned}
M_1 &= \langle D_1, L_1, Ext_1, Anti_1, den_1 \rangle \\
D_1 &= \{\tau, M_1\} \\
L_1 &= \{\mathsf{True}\} \\
Ext_1 &= \{\langle \mathsf{True}, Tr^+ \rangle\} \\
Tr^+ &= \{\langle \tau, M_1 \rangle\} \\
Anti_1 &= \{\langle \mathsf{True}, Tr^- \rangle\} \\
Tr^- &= D_1 \times D_1 - \{\langle \tau, M_1 \rangle\} \\
den_1 &= \{\langle \mathsf{this}, \tau \rangle, \langle \mathsf{h}, M_1 \rangle\}
\end{aligned}
$$

Similarly let M_2 be defined as follows:

$$
\begin{aligned}
M_2 &= \langle D_2, L_2, Ext_2, Anti_2, den_2 \rangle \\
D_2 &= \{\tau, M_2\} \\
L_2 &= \{\mathsf{True}\} \\
Ext_2 &= \{\langle \mathsf{True}, Tr^+ \rangle\} \\
Tr^+ &= \emptyset \\
Anti_2 &= \{\langle \mathsf{True}, Tr^- \rangle\} \\
Tr^- &= D_2 \times D_2 \\
den_2 &= \{\langle \mathsf{this}, \tau \rangle, \langle \mathsf{h}, M_2 \rangle\}
\end{aligned}
$$

Then M_1 and M_2 have the desired properties.

13.7. To prove (1), note that since M is total, either $M \models R(c)$ or $M \models \neg R(c)$. Suppose the former, and get a contradiction by using the same sort of reasoning as in the proof of Theorem 13.10. To prove (2), define p to be some urelement, and let M be defined as follows:

$$
\begin{aligned}
M &= \langle D, L, Ext, Anti, den \rangle \\
D &= \{p, \gamma, M\} \\
L &= \{\mathsf{R}, \mathsf{True}\} \\
Ext &= \{\langle \mathsf{R}, R^+ \rangle, \langle \mathsf{True}, Tr^+ \rangle\} \\
R^+ &= \emptyset \\
Tr^+ &= \emptyset \\
Anti &= \{\langle \mathsf{R}, R^- \rangle, \langle \mathsf{True}, Tr^- \rangle\} \\
R^- &= D \\
Tr^- &= D \times D \\
den &= \{\langle \mathsf{c}, p \rangle, \langle \mathsf{this}, \gamma \rangle, \langle \mathsf{h}, M \rangle\}
\end{aligned}
$$

Then M has the desired properties.

13.8. We first construct a denotation-correct model M_0 to satisfy the following conditions:

1. this denotes t_1 in M_0.
2. M_0 is reflexive and has a name $\ulcorner M_0 \urcorner$ for itself.
3. M_0 is total.

(As in Example 13.2, it will be some term other than this that denotes M_0.)

Let M_0 be defined by

$$
\begin{aligned}
M_0 &= \langle D, L, Ext, Anti, d, c \rangle \\
D &= \{0, 1, 2, t_1, M_0\} \\
L &= \{\mathsf{m}_0, \mathsf{Denotes}\} \\
Ext &= \{\langle \mathsf{Denotes}, De^+ \rangle\} \\
De^+ &= \emptyset \\
Anti &= \{\langle \mathsf{Denotes}, De^- \rangle\} \\
De^- &= \{\langle \mathsf{Denotes}, D \times D \times D \rangle\} \\
den &= \{\langle 0, 0 \rangle, \langle 1, 1 \rangle, \langle \mathsf{m}_0, M_0 \rangle, \langle \mathsf{this}, t_1 \rangle\}
\end{aligned}
$$

Note also that $M \models \neg \exists x \, \mathsf{Denotes}_\mathsf{h}(t_1, x)$, since h is not part of the language of M_0.

Continuing, we build a model M_1 so that

1. this denotes t_1 in M_1.
2. h denotes M_0 in M_1.
3. M_1 is total.

To do this, we only need to add M_1 to D and $\langle \mathsf{h}, M_0 \rangle$ to den.

14.1. A^∞ is not the only solution to $Z = A \times Z$. To get another solution, we go back to the proof of Theorem 14.1 and replace F, the set of functions from

N to A, with

$$F_0 \;=\; \{f \in F \mid f \text{ is eventually constant}\} \,.$$

Then the set $Z_0 = \{s_f \mid f \in F_0\}$ has the property that $Z_0 = A \times Z_0$. We could also work with $F - F_0$.

14.2. We define sets C^n by $C^0 = B$, and $C^{n+1} = A \times C^n$. Let

$$Z \;=\; A^\infty \cup C^0 \cup C^1 \cup \cdots \cup C^n \cup \cdots \,.$$

So $Z = B \cup (A \times Z)$. To show that Z is the largest such set, let $W = B \cup (A \times W)$; we'll show that $W \subseteq Z$. For this, let $D = W - \bigcup_n C^n$; we show that $D \subseteq A \times D$. Let $d \in D$. Then $d \notin B$, so $d \in A \times W$. Let $a \in A$ and $w \in W$ be such that $d = \langle a, w \rangle$. Then w must not belong to $\bigcup_n C^n$, lest d also belong to this set. Hence $w \in D$, and so $d \in A \times D$. This shows that $D \subseteq A \times D$. By Theorem 14.1, $D \subseteq A^\infty$. Thus

$$W \;=\; D \cup \bigcup_n C^n \;\subseteq\; A^\infty \cup \bigcup_n C^n \;=\; Z.$$

14.3. Assume that $Z \subseteq Z \to Z$, and further assume that $Z \neq \emptyset$. For each $z \in Z$, let x_z be a corresponding indeterminate. Consider the system

$$x_z \;=\; \{\langle x_w, x_v \rangle \mid v = z(w)\} \,,$$

where z ranges through Z. (Remember that each $z \in Z$ is a function on Z, so it makes sense to write $z(w)$.) As in the hint, we have one solution s given by $s_z = z$ for all $z \in Z$. But also, we have the solution $t_z = \Omega$, by Exercise 7.1. Therefore, for all $z \in Z$, $z = s_z = t_z$. Since $Z = \{z \mid z \in Z\}$, Z is either empty or equals $\{\Omega\} = \Omega$.

14.4. We'll prove something stronger: there are arbitrarily large sets Z such that $Z \subseteq \{a\} \cup (Z \to Z)$. Indeed, let λ be any ordinal number. By *AFA*, define functions f_α for $\alpha < \lambda$ so that $f_\alpha(a) = a$ and

$$f_\alpha(f_\beta) \;=\; \begin{cases} a & \text{if } \alpha < \beta \\ f_\beta & \text{if } \beta \leq \alpha \end{cases}$$

Recall that $a \in \mathcal{U}$, so each $f_\alpha \neq a$. Also, if $\alpha < \beta$, $f_\alpha(f_\beta) = a \neq f_\beta(f_\alpha)$. In particular $f_\alpha \neq f_\beta$. It follows that $Z = \{a\} \cup \{f_\alpha \mid \alpha < \lambda\}$ satisfies $Z \subseteq \{a\} \cup (Z \to Z)$ and has the same size as λ.

14.5. This one is a bit tricky. Let $Z = \{zip(s,t) \mid s, t \in A^\infty\}$. We must show that $Z \subseteq A \times Z$. The definition shows that $Z \subseteq A \times (A \times Z)$. Note also that $A \times Z \subseteq A \times Z$. So

$$Z \cup (A \times Z) \;\subseteq\; A \times (Z \cup (A \times Z)).$$

By coinduction, $Z \cup (A \times Z) \subseteq A^\infty$. Therefore $Z \subseteq A^\infty$. This is what we wanted to show.

14.6. Suppose we had z such that for all streams s and t,

$$z(\langle a, s \rangle, \langle b, t \rangle) \quad = \quad \langle a, \langle b, z(s,t) \rangle \rangle.$$

We would consider the set

$$\{ \langle zip(s,t), z(s,t) \rangle \mid s, t \in A^\infty \}$$
$$\cup \quad \{ \langle \langle a, zip(s,t) \rangle, \langle a, z(s,t) \rangle \rangle \mid a \in A, s, t \in A^\infty \}.$$

It is not hard to see that this is a stream bisimulation. So for all s, t, $zip(s,t) = z(s,t)$.

14.7. We take $C = A$, $G(a) = a$, and $H(a) = f(a)$. This gives us $iter_f$ so that for all $a \in A$,

$$iter_f^\cdot(a) \quad = \quad \langle a, iter_f(f(a)) \rangle.$$

14.8. In part (1), we take R to relate all pairs of the following two forms:

$$
\begin{array}{ccc}
zip(c_a, c_a) & R & c_a \\
\langle a, zip(c_a, c_a) \rangle & R & c_a
\end{array}
$$

The second components of the pairs on the top line are of the form of the bottom line, and vice-versa.

   The important point is that even though the original statement asked about $zip(c_a, c_a) = c_a$, we ended up using a bigger bisimulation. That is, if we had defined R' by

$$
\begin{array}{ccc}
zip(c_a, c_a) & R' & c_a
\end{array}
$$

then we would not have gotten a bisimulation. This is exactly parallel to what happens in proof by induction, where one often needs a stronger induction hypothesis to prove a result.

   For (2), we use R such that

$$
\begin{array}{ccc}
iter_f(a) & R & zip(iter_g(a), map_f(iter_g(a))) \\
iter_f(fa) & R & zip(iter_g(ga), map_f(iter_g(ga)))
\end{array}
$$

14.9. The key is to show by induction on n that

$$dm_f^n(s) \quad = \quad \langle f^n(1^{st}(s)), dm_f^{2n}(2^{nd}(s)) \rangle.$$

The proof is by induction on $n \geq 1$. So if $s = \langle 0, \langle 1, \langle 2, \ldots \rangle \rangle \rangle$, then

$$
\begin{array}{rcl}
dm_f(s) & = & \langle 1, dm_f^2(\langle 1, \langle 2, \ldots \rangle \rangle) \rangle \\
& = & \langle 1, \langle 3, dm_f^4(\langle 2, \ldots \rangle) \rangle \rangle \\
& = & \langle 1, \langle 3, \langle 6, \ldots \rangle \rangle \rangle
\end{array}
$$

15.1. These are all straightforward. For Γ_1, suppose that $a \subseteq b$. Since every subset of a is a subset of b, $\mathcal{P}(a) \subseteq \mathcal{P}(b)$. Γ_2 works similarly. For Γ_3, recall that $A \times a$ is the set of ordered pairs $\langle c, d \rangle$ such that $c \in A$ and $d \in a$. If $a \subseteq b$, then each such pair $\langle c, d \rangle$ also belong to $A \times b$. Concerning Γ_4, the key point is that if $a \subseteq b$, then every partial function from a to itself automatically is a partial function from b to itself. We could show the monotonicity of Γ_5 directly, but it is better to say that Γ_1 and Γ_3 are monotone and the composition of monotone functions is monotone (see Proposition 15.1). Since $\Gamma_5(a) = \Gamma_1(\Gamma_3(a))$, we're done.

15.2. Note that every subset of C is a set, so $\mathcal{P}(C) \subseteq C$. On the other hand, if p is an urelement, then $\{p\}$ would be an element of C but not a subset of C. So $C \subseteq \mathcal{P}(C)$ is false.

15.3. The forms are sets of pairs $\langle a, x \rangle$, where $a \in Act$ and $x \in X$. We have seen the fixed points already. Recall the set of canonical states over Act (see Section 3.2). The least fixed point is the set of wellfounded canonical states. The greatest fixed point is the set of all canonical states. Assuming FA, these are the same. Assuming AFA, the two fixed points differ.

15.4. Read the justification for the corresponding coinduction principle on page 218, change $\cap$ to $\cup$, and reverse the inclusions.

15.5. For part (1), define streams s_n by recursion on n so that $s_0 = s$, and $s_{n+1} = 2^{nd} s_n$. Let $b = \{s_n : n \in N\}$. (We may think of b as the set of *final segments* of the original stream s.) Then b will be a Γ_3-correct superset of $a = \{s_0\}$, since $s_{n+1} \in A \times \{s_n\}$. To see b is $\subseteq$-minimal, let b' be any Γ_3-correct superset of a, and show by induction on n that each s_n belongs to b'.

Turning to part (2), the least c will be the transitive closure of $\{b\}$.

15.6. (1) All of the examples $\Gamma_1, \ldots, \Gamma_5$ commute with all intersections. In fact, all of the examples in this book (except for the one just below) commute with all intersections.

(2) Here is a monotone operator that does not commute with binary intersections:

$$\Gamma(b) = \begin{cases} \emptyset & \text{if } b \subseteq \{\emptyset\} \\ b & \text{otherwise} \end{cases}$$

It is not hard to check that this Γ is monotone. To see that it does not commute with binary intersections, consider $b = \{\emptyset, 1\}$ and $c = \{\emptyset, 2\}$. $\Gamma(b) \cap \Gamma(c) = b \cap c = \{\emptyset\}$, but $\Gamma(b \cap c) = \Gamma(\{\emptyset\}) = \emptyset$.

For (3), let c be any set of Γ-correct sets. Then

$$\bigcap_{b \in c} b \quad \subseteq \quad \bigcap_{b \in c} \Gamma(b) \quad = \quad \Gamma(\bigcap_{b \in c} b).$$

(The equality at the end uses the commutativity assumption.) So $\bigcap_{b \in c} b$ is Γ-correct.

For (4), let $a \subseteq \Gamma^*$. We know that there is some b such that b is Γ-correct and $a \subseteq b$. Let

$$c \quad = \quad \bigcap \{b' \subseteq b \mid b' \text{ is } \Gamma\text{-correct and } a \subseteq b'\}.$$

Then by part (3), c is Γ-correct. Clearly $a \subseteq c$. To check the minimality, let d be a Γ-correct superset of a. Then so is $d \cap c$. By the definition of c, $c \subseteq d \cap c$. Thus $c \subseteq d$.

(5) Our operator Γ from part (2) gives an example of a set with no closure. Let $a = \{\emptyset\}$. Then $a \in \Gamma^*$ because $a \subseteq \{\emptyset, 1\} = \Gamma(\{\emptyset, 1\})$. However, the intersection of all Γ-correct supersets of a is a itself, and a is not Γ-correct.

15.7. (1) $\Gamma_* = \emptyset$, since $\emptyset$ is a fixed point. $\Gamma^* = \Omega$. The easiest way to see this now is to note that if $a \in \Gamma^*$, then a is non-empty, as is every element of every element of a, etc. So every element of $TC(\{a\})$ is a non-empty set. Thus in the canonical system for a, every right-hand side e_x is non-empty and contains no urelements. This implies that $a = \Omega$.

Now for (2). It is clear that the greatest fixed point is the collection HNe defined earlier. To show the second claim, let $p \in A$. We can define an injective function F from the proper class of all pure wellfounded sets into Δ_* as follows. Let $F(\emptyset) = \{p\}$ and $F(b) = \{F(x) \mid x \in b\}$ for b non-empty, pure, and wellfounded. We prove by induction on α that if b and b' are pure wellfounded sets of rank at most α, and if $b \neq b'$, then $F(b) \neq F(b')$. Then it follows that F is injective. Hence Δ_* is a proper class.

15.8. The least fixed point is the empty set. The greatest fixed point is the class *HRef* of hereditarily reflexive sets over A, as defined earlier. if $A = \emptyset$ then it is easy to see that any member of Δ^* is bisimilar to, and hence identical to, Ω.

To see that Δ^* is a proper class if $A \neq \emptyset$, assume $p \in A$. We define a transfinite sequence $\langle a_\alpha \rangle_{\alpha \in ON}$ of strictly increasing sets, each a member of *HRef*, as follows. If α is even, let $a_\alpha = \{a_\alpha\} \cup \{a_\beta \mid \beta < \alpha\}$. If α is odd, let $a_\alpha = \{p, a_\alpha\} \cup \{a_\beta \mid \beta < \alpha\}$. We are using the Solution Lemma at each stage. Limit ordinals count as being even, of course. For example, $a_0 = \Omega$,

$a_1 = \{p, a_1, a_0\}$, and $a_2 = \{a_2, a_1, a_0\}$. It is immediate that these are all reflexive sets. Moreover, the even a's and the odd a's are different, since the former do not contain p while the latter do.

Note also that for all $\alpha \geq 1$, $a_\alpha \neq a_0$, since $a_1 \neq a_0$ and $a_1 \in a_\alpha$.

We claim that for all α, and all $\beta < \alpha$, $a_\alpha \notin a_\beta$. This is vacuous for $\alpha = 0$, and it is clear for $\alpha = 1$, since $\alpha_1 \neq \Omega$.

Assume that it is true for $\alpha > 0$, and we prove it for $\alpha + 1$. Let $\beta < \alpha + 1$ be such that $a_{\alpha+1} \in a_\beta$. We know that $a_{\alpha+1} \notin a_0$, so $\beta \neq 0$. But then for some $\gamma \leq \beta$ of the same parity as $\alpha + 1$, $a_{\alpha+1} = a_\gamma$. By the parity assertion, we must have $\gamma < \alpha$. But now $a_\alpha \in a_{\alpha+1} = a_\gamma$. This contradicts the induction hypothesis, since a_α belongs to a previous a_γ.

The induction step when α is a limit is easy. In this way, we have a proper class of members of the greatest fixed point.

15.9. $\Gamma(\Delta_*) \subseteq \Delta(\Delta_*) = \Delta_*$. By the Preliminary Induction Principle for Γ_*, $\Gamma_* \subseteq \Delta_*$.

Similarly, $\Gamma^* = \Gamma(\Gamma^*) \subseteq \Delta(\Gamma^*)$. So $\Gamma^* \subseteq \Delta^*$, by the Preliminary Coinduction Principle for Δ^*.

15.10. $\Gamma^* = \Gamma(\Gamma^*) \subseteq \Delta^*$. So we're done by the Preliminary Coinduction Principle for Δ^*.

15.11. (1) We omit the subscripts. $\Gamma^* \subseteq \Delta^*$ because

$$\Gamma^* \;=\; \Gamma(\Gamma^*) \;\subseteq\; \Gamma(C \cup \Gamma^*) \;=\; \Delta(\Gamma^*).$$

Since Δ^* is the greatest fixed point of Δ, $\Gamma^* \subseteq \Delta^*$. Next, we use coinduction to show that $\Phi_* \subseteq \Delta^*$:

$$\begin{aligned}
\Phi_* \;&=\; \Phi(\Phi_*) \\
&=\; \Gamma(C \cup \Gamma^* \cup \Phi_*) \\
&\subseteq\; \Gamma(C \cup \Delta^* \cup \Phi_*) \\
&=\; \Delta(\Delta^* \cup \Phi_*).
\end{aligned}$$

Turning to (2), we first note that $\Gamma^* = A^\infty$. Second, we can write $\Delta(a)$ as $(A \times C) \cup (A \times a)$. We use Exercise 14.2 to characterize Δ^*. Let $B = A \times C$, let $C^0 = B$, and let $C^{n+1} = A \times C^n$. By Exercise 14.2, $\Delta^* = A^\infty \cup \bigcup_n C^n$. Now

$$A^\infty \;=\; \Gamma^* \;\subseteq\; \Gamma(C \cup \Gamma^* \cup \Phi_*) \;=\; \Phi(\Phi_*) \;=\; \Phi_*.$$

Further, an easy induction on n shows that $C^n \subseteq \Phi_*$. It follows that $\Delta^* \subseteq \Phi_*$. The converse was shown in part (1).

To see (3), let $b = \{x, b\}$. Then $b \in \Delta^*$ by coinduction: $\{b\} \subseteq \mathcal{P}(C \cup \{b\})$. But an easy induction shows that for all ordinals α, $b \notin \Phi_\alpha$. So $b \notin \Phi_*$.

15.12. We claim that for each urelement x, $\{x\} \in \hat{\Gamma}(\emptyset)$. This means that $\{x\} \notin \Gamma(V_{afa}[\mathcal{U}])$. But this is clear, since $\{x\}$ is not a set of sets.

15.13. For part (1), assume that $C \subseteq D$. Then $-D \subseteq -C$, so $\Gamma(-D) \subseteq \Gamma(-C)$. By taking complements again, we see that $\hat{\Gamma}(C) \subseteq \hat{\Gamma}(D)$.

For (2), recall that we are trying to show that

$$(2) \qquad \hat{\Gamma}(C) \;=\; \bigcup \{\hat{\Gamma}(a) \mid a \subseteq C \text{ is a set}\}.$$

Let $b \in \hat{\Gamma}(C)$. Thus $b \notin \Gamma(-C)$. Since $-C = \bigcap_{a \subseteq C} -a$ and Γ commutes with intersections, there is some set $a \subseteq C$ such that $b \notin \Gamma(-a)$. So $b \in \hat{\Gamma}(a)$. This proves half of (2), and the other half follows from monotonicity.

Here is a counterexample for part (3): Let $i : V_{afa}[\mathcal{U}] \to ON$ be a bijection. Define Γ by

$$\Gamma(a) \;=\; \{p \mid (\exists q \in a)\, i(p) < i(q)\}.$$

It is easy to see that Γ is monotone. Further, $\Gamma(\emptyset) = \emptyset$, so that $\hat{\Gamma}(V_{afa}[\mathcal{U}]) = -\Gamma(\emptyset) = V_{afa}[\mathcal{U}]$. But for every proper class C, $\Gamma(C) = V_{afa}[\mathcal{U}]$. (This is because $i[C]$ is a proper class, hence an unbounded class, of ordinals. So for every set b, there is some $\alpha \in i[C]$ such that $i(b) < \alpha$.) So for all sets a, $\hat{\Gamma}(a) = \emptyset$. So Γ is a counterexample to (2).

Finally we turn to (4). We first claim that if C is a fixed point of Γ, then $-C$ is a fixed point of $\hat{\Gamma}$. This is because $\hat{\Gamma}(-C) = -\Gamma(-(-C)) = -\Gamma(C) = -C$. This shows that $-\Gamma^*$ and $-\Gamma_*$ are fixed points of $\hat{\Gamma}$. By using complements once more, we can see that these are the least and greatest fixed points of $\hat{\Gamma}$.

16.1. To check the claim, one checks that $V_{afa}[X \cup A]$ really is a fixed point of Γ_X. This is because

$$\begin{aligned} \Gamma_X(V_{afa}[X \cup A]) &= \mathcal{P}(X \cup A \cup V_{afa}[X \cup A]) \\ &= V_{afa}[X \cup A] \end{aligned}$$

To finish the verification that $V_{afa}[X \cup A]$ is the greatest fixed point, we claim that every fixed point C of the operator Γ_X must consist entirely of sets whose support lies in $X \cup A$. For suppose that $C = \Gamma_X(C)$. That is, $C = \mathcal{P}(X \cup A \cup C)$. Then C is a set, so all urelements in every $c \in C$ belong to $X \cup A$. Further, every such $c \in C$ also has the property that it is a subset of $X \cup A \cup C$. So all urelements in c belong to $X \cup A$. We can keep going like this, showing that all urelements in the transitive closure of C belong to $X \cup A$. This proves that $support(C) \subseteq X \cup A$.

16.2. X is new for Γ_5 iff X is disjoint from $support(Act)$. To see this, first let X be new for Γ_5. Let s be defined on X, and let a be any set. Then $Act[s] = Act$, so $(Act \times a)[s] = Act \times a[s]$. And

$$(\mathcal{P}(Act \times a))[s] \;=\; \mathcal{P}((Act \times a)[s]) \;=\; \mathcal{P}(Act \times a[s]).$$

This means that $\Gamma_5(a)[s] = \Gamma_5(a[s])$.

Conversely, suppose that $X \cap support(Act)$ is not disjoint; let x belong to this set. Let $y \in \mathcal{U} - support(Act)$. Let s be the substitution that interchanges x and y. Let $a = \{\emptyset\}$. So $a[s] = a$, and $\mathcal{P}(Act \times a[s]) = \mathcal{P}(Act \times a)$ does not have y in its support. But $y \in support(\mathcal{P}(Act \times a)[s])$. Thus $\mathcal{P}(Act \times a[s]) \neq \mathcal{P}(Act \times a)[s]$.

16.3. You guessed it: $\mathcal{E}[r] = \mathcal{E}'$. Since $X[r] = Y$, we only have to show that $e[r] = e'$. The reason for this is that

$$
\begin{aligned}
e[r] &= \{\langle x, e_x \rangle \mid x \in X\}[r] \\
&= \{\langle x[r], e_x[r] \rangle \mid x \in X\} \\
&= \{\langle r_x, e'(r_x) \rangle \mid x \in X\}
\end{aligned}
$$

Since r is surjective, this last set of ordered pairs is the function e'.

16.4. The equation is $e' \star r = r \star e$. If r is a Γ-morphism, then both $e' \star r$ and $r \star e$ have domain X. For $x \in X$, $r_x[e'] = e_x[r]$ by the morphism condition. This proves that $e' \star r = r \star e$. Conversely, assume that this equation holds. Then as the domain of $r \star e$ is X and the domain of $e' \star r$ is $dom(r)$, we have $dom\, r = X$. For $x \in dom(r)$, the definition of $\star$ tells us that $r_x[e'] = e_x[r]$. Thus r is a Γ-morphism.

16.5. The problem is that $\mathcal{E}$ is *not* a flat Γ = coalgebra. It is based on $\{a\} \subseteq \mathcal{U}$, but this set is not new for Γ. $\mathcal{E}$ does have a solution, and this solution is Ω. We might note that it is critical that the urelements of a system be new for an operator Γ, since only then can we be sure that the solution is included in Γ^*. In other words, the newness assumption is critical for the second half of the proof of Lemma 16.2.

16.6. (1) We check that $\Gamma(b) = a$ is uniform. Let $Y = support(b)$. Let $X \subseteq \mathcal{U}$ be disjoint from Y, and let t have domain X. Then for all sets c, $\Gamma(c[t]) = a$, and $\Gamma(c)[t] = a[t] = a$. So $\Gamma(c[t]) = \Gamma(c)[t]$.

In (2), the hypothesis implies that $\Delta \circ \Gamma$ is proper and monotone. To check that the composition commutes with almost all substitutions, suppose that Y_Γ and Y_Δ are avoidance sets for Γ and Δ, respectively. Let $Y = Y_\Gamma \cup Y_\Delta$. We check that Y is an avoidance set for $\Delta \circ \Gamma$. Let X be disjoint from Y, so X is disjoint from Y_Γ and Y_Δ. Let t be a substitution defined on X. For all sets a,

$$(\Delta \circ \Gamma)(a)[t] = \Delta(\Gamma(a))[t] = \Delta(\Gamma(a)[t]) = \Delta(\Gamma(a[t])) = (\Delta \circ \Gamma)(a[t]).$$

Parts (3) and (4) are similar to (2); we use also the fact that substitution operators commute with pairing and binary unions.

16.7. The least fixed point is a certain countable set, the smallest set containing $\emptyset$ and closed under ordered pairs. Assuming the Foundation Axiom, this is the only fixed point. Assuming *AFA*, Γ^* is bigger than Γ_*: $\Gamma_* \cup \Omega \subseteq \Gamma^*$. Indeed, Γ^* is uncountable. Γ is uniform, and every set of urelements is new for it.

16.8. Note that for all sets a, and all substitutions s, $\{a\}[s] = \{a[s]\}$. This proves that the singleton operation commutes with all substitutions. But it is not monotone; for example, $\emptyset \subseteq \{\emptyset\}$, but $\{\emptyset\} \not\subseteq \{\{\emptyset\}\}$.

16.9. Suppose not. Then let $x \in support(\Gamma(a)) - X_\Gamma$ and $x \notin support(a)$. Let $y \in \mathcal{U} - (X \cup \cup support(G(a)))$ be arbitrary. Consider the substitution $s = \{\langle x, y \rangle\}$. Then $a[s] = a$, but $\Gamma(a)[s] \neq \Gamma(a)$.

16.10. Recall from page 227 that if X is new for Γ_4 then X has at most two elements. Therefore the union of the solution sets of Γ_4-coalgebras is finite (in fact it is fairly easy to write it out explicitly). On the other hand, we showed in Example 15.3 on page 214 that Γ_4^* is a proper class.

16.11. It is easy to check that Δ is monotone, and it is almost as easy to see that if $X \subseteq \mathcal{U}$ is new for Γ, then X is also new for Δ. So Δ is uniform.

Concerning least fixed points, one checks easily that $\Delta_\alpha = \Gamma_{2\alpha}$ for all ordinals α. From this and monotonicity, it follows that $\Gamma_* = \Delta_*$.

For the greatest fixed points, it is clear that $\Delta(\Gamma^*) = \Gamma^*$. Thus $\Gamma^* \subseteq \Delta^*$. To show the reverse inclusion, we use the Representation Theorem 16.6 and the Solution Lemma Lemma. Let $\mathcal{E} = \langle X, e \rangle$ be a flat Δ-coalgebra. Note that $\Delta(X) = \Gamma(\Gamma(X))$ is a set of parametric Γ-objects over X, because $\Delta(X) \subseteq \Gamma(X \cup \Delta(X))$. By the Solution Lemma Lemma, $solution\text{-}set(\mathcal{E}) \subseteq \Gamma^*$. So by Theorem 16.6, $\Delta^* \subseteq \Gamma^*$.

Finally, we want an example of a fixed point of Δ that is not a fixed point of Γ. Let $A = \{x, y\} \subseteq \mathcal{U}$, and let π be the substitution that transposes x and y. Let $\Gamma(a) = A \times (a[\pi])$. Then Γ is a uniform operator. Let C be the set of streams over A that are eventually x; that is, C least among fixed points of Γ containing $s = \langle x, s \rangle$. Similarly, let D be the streams that are eventually y. Then $C[\pi] = D$ and $D[\pi] = C$. Also $C = A \times C$ and $D = A \times D$. It follows that $\Gamma(C) = D$ and $\Gamma(D) = C$. So we see that neither C nor D is a fixed point of Γ, but each is a fixed point of the associated Δ.

16.12. We want to solve the equation $c = \{\mathcal{P}(a \cup c)\}$. Let $b = \mathcal{P}(a \cup c)$; once we solve for b, we need only take $c = \{b\}$. We have $b = \mathcal{P}(a \cup \{b\})$. This is an equation involving b alone, and we can solve it.

Let x_b be an urelement not in $support(a)$. For each $t \subseteq a$, take an urelement x_t, and consider the equation $x_t = t \cup \{x_b\}$. The idea is that the solution s to our system should satisfy $s(x_t) = s(x_b) \cup t$. So x_t should also be taken outside of $support(a)$. We take such equations for each t, and also one more equation for b:

$$x_b = \{x_t \mid t \subseteq a\} \cup \{t \mid t \subseteq A\}.$$

Once again, the first set on the right gives the subsets that contain b, and the second gives the subsets that do not. To solve this system, we use the General Solution Lemma; note that every right hand side is a parametric set over the indeterminates of the system.

16.13. For (1), consider an urelement x, and let $a = \{x\}$. Then $\bigcup a = \emptyset$. Thus $(\bigcup a)[s] = \emptyset$ for all substitutions s. But if s_x is any non-empty set, then $\bigcup(a[s]) = s_x \neq \emptyset$. Hence no non-empty set of urelements is new for f.

Turning to (2), note that for any set a, $\bigcup\{a\} = a$. It follows that $\{a\} = \{\bigcup\{a\}\}$. In other words, the singleton of any set is a solution to $x = \{\bigcup x\}$. (Conversely, every solution must be a singleton of some set.)

16.14. In part (1), let F be the operation $F(x) = \mathcal{P}(\mathcal{P}(x))$. So we have a new $\mathcal{F}$-term $F^*(x)$, and we want to solve $x = \{F^*(x)\}$. We adjoin the equation

$$\begin{aligned} F^*(x) &= F(\{F^*(x)\}) \\ &= \{\emptyset, \{\emptyset\}, \{\emptyset, \{F^*(x)\}\}, \{\{F^*(x)\}\}\} \end{aligned}$$

The Solution Lemma now applies. Incidentally, Exercise 2.2 is a verification that the solution of this associated system assigns really gives a solution to $x = \{\mathcal{P}(\mathcal{P}(x))\}$.

To see whether the two equations mentioned have the same solution, let $a = \{\mathcal{P}(\mathcal{P}(a))\}$, and let $b = \{\mathcal{P}(b)\}$. An analysis based on bisimulations shows that a has 4 elements and b has 2. So $a \neq b$. (Alternately, one can show directly that a and b are not bisimilar.)

Turning to part (2), we introduce a new indeterminate y (abbreviating $CP^*(x, \{p, q\})$). The system is

$$\begin{aligned} x &= \{y\} \\ y &= \{\langle y, p \rangle, \langle y, q \rangle\} \end{aligned}$$

The Solution Lemma now applies. To calculate the number of elements in $TC(\{x\})$, The elements of $TC(\{x\})$ are x, y, $\langle y, p \rangle$, $\{y\}$, $\{y, p\}$, p, $\langle y, q \rangle$, $\{y, q\}$, and q. An analysis based on bisimulations shows that all 9 of these are distinct.

17.1. In (1), note that a very new set of urelements is new.

For part (2), suppose that Γ and Δ are smooth, say with avoidance sets X and Y. We'll show that we can use $X \cup Y$ as the avoidance set for $\Delta \circ \Gamma$. Let Z be disjoint from $X \cup Y$. Let a be any set, and assume that s is a substitution defined on Z. Then $(\Delta\Gamma(a))[s] = \Delta\Gamma(a[s])$ by the calculation in Exercise 16.6 (b). so that $[s]_a$ is injective. Then $[s]_{\Gamma(a)}$ is also injective, as is $[s]_{\Delta(\Gamma(a))} = [s]_{\Delta\Gamma(a)}$.

Now we turn to part (3). For $\Gamma_1(a) = \mathcal{P}(a)$, suppose that b and c are distinct subsets of a. Say that $a' \in b - c$. Assuming that $[s]$ is injective on a, we see that $a'[s] \in b[s] - c[s]$. In particular, $b[s] \neq c[s]$. As a' is arbitrary, $s_{\mathcal{P}(a)}$ is injective. Γ_2 is similar to Γ_1.

For Γ_3, suppose that $\pi' = \langle a', b \rangle$ and $\pi'' = \langle a'', c \rangle$ are distinct elements of $A \times a$. Let s be any substitution whose domain is a set of new urelements. In particular, $a'[s] = a'$ and $a''[s] = a''$. Now $\pi'[s] = \langle a', b[s] \rangle$ and $\pi''[s] = \langle a'', c[s] \rangle$ If $a' \neq a''$, then clearly $\pi'[s] \neq \pi''[s]$. If $a' = a''$, then $b \neq c$. Then the fact that $[s]$ is injective on a implies that again $\pi'[s] \neq \pi''[s]$.

To see that Γ_5 is smooth, we use the result for Γ_3 and part (2) above.

17.2. To prove that Γ is uniform, we use Exercise 16.6, parts (b) and (d). Here is an example showing that it is not smooth. Every set $X \subseteq \mathcal{U}$ is new for Γ. Let $X = \{x, y\}$. Let s be defined by $s_x = \emptyset$ and $s_y = \{\emptyset\}$. Then s is injective on X. But $\{\{x\}\}$ and $\{y\}$ belong to $\Gamma(X)$, and $\{\{x\}\}[s] = \{\{\emptyset\}\} = \{y\}[s]$. (Another example may be found in the solution to Exercise 17.8.)

17.3. We take $\Gamma(a) = A \times a$, $C = A^\infty$, and $\pi(s) = \langle f(1^{st}s), 2^{nd}s \rangle$. Then φ satisfies

$$\begin{aligned} \varphi(s) &= \langle f(1^{st}s), \ulcorner 2^{nd}s \urcorner \rangle [\varphi \circ den] \\ &= \langle f(1^{st}s), \varphi(2^{nd}s) \rangle. \end{aligned}$$

In the last equality, we used the same reasoning as in Example 17.2.

17.4. We first calculate that $\overline{\pi}$ is given by $\overline{\pi}(\ulcorner s \urcorner) = \langle 1^{st}(s), \ulcorner 2^{nd}(2^{nd}(s)) \urcorner \rangle$. Then φ satisfies

$$\varphi(s) = \langle 1^{st}(s), \varphi(2^{nd}(2^{nd}(s))) \rangle.$$

This map might be called *even*, since it takes a stream s and returns the stream of "even indexed" entries of s, starting with $s_0 = 1^{st}(s)$.

17.5. Let Γ be the stream forming operator $\Gamma(a) = A \times a$. We lift the map $\pi : \Gamma^* \to \Gamma^* \times A$ given by $\pi(s) = \langle 2^{nd}(s), 1^{st}(s) \rangle$. This gives us $\varphi : \Gamma^* \to \Delta^*$ so that $\varphi(s) = \langle \varphi(2^{nd}(s)), 1^{st}(s) \rangle$. To see that φ is injective, note that the relation $\varphi(s) = \varphi(t)$ is a stream bisimulation.

However, to prove that φ is surjective is not so immediate. We first get $\psi : \Delta^* \to \Gamma^*$ by corecursion so that $\psi(a) = \langle 2^{nd}(a), \psi(1^{st}(a)) \rangle$ for $a \in \Delta^*$. Further, we can develop a theory of bisimulations for smaerts. One consequence would be that ψ is one-to-one. Also, $\varphi \circ \psi : \Delta^* \to \Delta^*$ would satisfy $\varphi \circ \psi(s) = \langle 1^{st}(s), \varphi \circ \psi(2^{nd}(s)) \rangle$. But the identity is the only function with this property (see page 202 for a similar discussion). So $\varphi \circ \psi = i_{\Delta^*}$. Now it follows that φ is surjective, since for all $a \in \Delta^*$, $\varphi(\psi(a)) = a$.

17.6. We take $C = A^\infty \times B^\infty$, and Γ to be the operator $\Gamma(a) = (A \times B) \times a$. Let $\pi : C \to \Gamma(C)$ be given by

$$\pi(s,t) \;\; = \;\; \langle\langle 1^{st}(s), 1^{st}(t)\rangle, \langle 2^{nd}(s), 2^{nd}(t)\rangle\rangle.$$

This is all we need to apply the Corecursion Theorem. We therefore get a function φ which in this problem we are writing as μ. To see what it does, fix a Γ-notation scheme $den : X \to C$. The Γ-lift of π by den is given by

$$\overline{\pi}(\ulcorner\langle s,t\rangle\urcorner) \;\; = \;\; \langle\langle 1^{st}(s), 1^{st}t\rangle, \ulcorner\langle 2^{nd}(s), 2^{nd}(t)\rangle\urcorner\rangle.$$

The urelements new for Γ are those disjoint from

$$support(A \times B) \;\; = \;\; support(A) \cup support(B).$$

So the substitution $\mu \circ den$ fixes the elements of A and B, $\ulcorner\langle s,t\rangle\urcorner[\mu \circ den] = \mu(s,t)$. So for all $s \in A^\infty$ and $t \in B^\infty$,

$$\mu(s,t) \;\; = \;\; \langle\langle 1^{st}(s), 1^{st}(t)\rangle, \mu(2^{nd}s, 2^{nd}t)\rangle.$$

17.7. (1) The Γ-lift of π should be $\overline{\pi} : X \to \Gamma(\Gamma(X))$ given by $\overline{\pi}(\ulcorner a\urcorner) = \langle f(a), \langle g(a), \ulcorner a\urcorner\rangle\rangle$. Note that $\overline{\pi}$ maps into $\Gamma(\Gamma(X))$.

For (2), $[den]_X$ is bijective, since X is very new for Γ. So as Γ is smooth, $[den]_{\Gamma(X)}$ is also bijective. Applying the condition once more, we see that $[den]_{\Gamma(\Gamma(X))}$ is bijective. Thus we get a unique lift of π by Proposition 17.2.

Applying this to the special case of this problem, note that $[s]_{\Gamma(X)}$ takes the pair $\langle b, \ulcorner a\urcorner\rangle$ to $\langle b, a\rangle$. Similarly, $[s]_{\Gamma(\Gamma(X))}$ takes $\langle b, \langle b', \ulcorner a\urcorner\rangle\rangle$ to $\langle b, \langle b', a\rangle\rangle$. Then, working through the definition of the lifted map, we see that the Γ-lift of π is given exactly by the formula in the solution to part (1).

In part (3), the set $\Gamma(\Gamma(X))$ is a set of parametric Γ-objects. So the result about s follows from the Solution Lemma Lemma.

By the Corecursion Theorem, we get a map $\varphi : A \to B^\infty$ so that for all $a \in A$, $\varphi(a) = \overline{\pi}(\ulcorner a\urcorner)[\varphi \circ den]$. As we've seen $\overline{\pi}(\ulcorner a\urcorner) = \langle f(a), \langle g(a), \ulcorner a\urcorner\rangle\rangle$. Now the elements of B are fixed by $[\varphi \circ den]$, because X was taken to be new for Γ. Thus $\varphi(a) = \langle f(a), \langle g(a), \varphi(a)\rangle\rangle$.

17.8. We take C to be $\{\emptyset\}$. So

$$\Gamma(C) \quad = \quad \{\emptyset, \{\emptyset\}, \{\emptyset, \{\emptyset\}\}, \{\{\emptyset\}\}\}.$$

For π we take $\emptyset \mapsto \{\emptyset\}$. For X we take the singleton set $\{x\}$ of any urelement. So

$$\Gamma(X) \quad = \quad \{\emptyset, \{x\}, \{\emptyset\}, \{\emptyset, \{x\}\}, \{\{x\}\}\}.$$

(Incidentally, this gives another example showing that Γ is not smooth.) One Γ-lift $\overline{\pi}$ of π is $x \mapsto \{x\}$. Another is $\overline{\pi}'$ given by $x \mapsto \{\emptyset\}$. For the first lift, the function $\emptyset \mapsto \Omega$ satisfies (5). For the second we use $\emptyset \mapsto \{\emptyset\}$.

17.9. We take $\Gamma(a) = A \times a$, and $\Delta(a) = a \times A$. Further, let $C = A$, and let $\pi(a) = \langle a, a \rangle$. The functions $\varphi : A \to \Gamma^*$ and $\psi : A \to \Delta^*$ are different since $\Gamma^* \cap \Delta^* = \emptyset$.

17.10. One satisfier of (6) is indeed $\varphi(s) = $ the minimum value reached on s. But a second one is the constant function $\varphi(s) = 0$.

17.11. By smoothness, let $A \subseteq \mathcal{U}$ be a set with the property that the complement $\mathcal{U} - A$ is very new for Γ. Let $f : B \to \mathcal{U}$ and $g : C \to \mathcal{U}$ be given by $f(b) = \text{new}(\langle 0, b \rangle, A)$. and $g(c) = \text{new}(\langle 1, c \rangle, A)$. Then f and g are injections. Let $X = f[B]$ and $Y = g[C]$ be their images. Their inverses give the bijections den_1 and den_2 that we are after. We need to see that X and Y are disjoint. But for $b \in B$ and $c \in C$, $\langle 0, b \rangle \neq \langle 1, c \rangle$. So $f(b) \neq g(c)$ by the basic injectivity property of new.

17.12. Let s be the solution of $\langle X \cup Y, \overline{\pi} \rangle$. So $s : X \cup Y \to \Gamma^*$. Define φ by $\varphi(b) = s(\ulcorner b \urcorner)$, and ψ by $\psi(c) = s(\text{'}c\text{'})$. Then

$$s \quad = \quad (\varphi \circ den_1) \cup (\psi \circ den_2).$$

Equation (8) in the definition of simultaneous corecursion follows from this observation and the definition of s.

17.13. Let R be such that conditions (1) and (2) hold. We prove that R is a Γ-bisimulation. For this, we need to specify $e' : R \to \Gamma(R)$. We define

$$e'_{\langle x,y \rangle} \quad = \quad \{\langle a, \langle x', y' \rangle \rangle \mid \langle a, x' \rangle \in e_x \text{ and } \langle a, y' \rangle \in e_y\}.$$

Assumptions (1) and (2) imply that π_1 and π_2 are Γ-morphisms.

We next turn to the converse. Suppose that R is a Γ-bisimulation according to the definitions. We show that (1) holds; (2) is similar. Let x, y, x', and a be such that $x \mathrel{R} y$, $\langle a, x' \rangle \in e_x$. Then since π_1 is a Γ_5-morphism, $e_x = e'_{\langle x,y \rangle}[\pi_1]$.

The elements of $\Gamma(R)$ are of the form $\langle a, \langle x'', y'' \rangle \rangle$ for some $x''\, R\, y''$. Since X is new for Γ_5,

$$\pi_1(\langle a, \langle x'', y'' \rangle \rangle) \quad = \quad \langle a, x'' \rangle.$$

Going back to x', we see that for some y', $\langle a, \langle x', y' \rangle \rangle \in e'_{\langle x,y \rangle}$. This gives us some y' as needed.

17.14. We follow the notation from the Lifting Lemma. Let $\epsilon : Y \to X$ be $den^{-1} \circ den'$. Let T be $S[\epsilon^{-1}]$. Since S is a bisimulation, we have $f : S^o \to \Gamma(S^o)$ so that the projections π_1 and π_2 are morphisms. We need the same thing for T^o.

Let π_1' and π_2' be the projections for T. First, using ϵ we can define a notation scheme $\sigma : T^o \to S^o$. This map has the properties that $\epsilon^{-1} \circ \pi_1 \circ \sigma = \pi_1'$, and similarly for π_2. Let $g : T^o \to \Gamma(T^o)$ be the lift of f by σ. Then we have three morphisms: $\sigma : \langle T^o, g \rangle \to \langle S^o, f \rangle$, $\pi_1 : \langle S^o, f \rangle \to \langle X, e \rangle$, and $\epsilon^{-1} : \langle X, \overline{e} \rangle \to \langle Y, \overline{e}' \rangle$. (The last is a morphism by the Lifting Lemma.) By Lemma 17.3, these compose to give a morphism $\epsilon^{-1} \circ \pi_1 \circ \sigma : \langle T^o, g \rangle \to \langle Y, \overline{e}' \rangle$. But as we have seen, the map here is π_1'. This proves that π_1' is a morphism, and the same holds for π_2'.

17.15. Suppose towards a contradiction that $X \times X$ were a bisimulation. Then there would be a Γ_{AM}-coalgebra

$$\mathcal{E}' \quad = \quad \langle X \times X, e' \rangle$$

such that π_1 and π_2 would reorganize $\mathcal{E}'$ onto $\mathcal{E}$. Consider $\langle x, y \rangle \in X \times X$. Since π_1 is a morphism, $e'_{\langle x,y \rangle}[\pi_1] = e_x = \langle x, x, y \rangle$. And since π_2 is a morphism, $e'_{\langle x,y \rangle}[\pi_2] = e_y = \langle x, y, y \rangle$. It follows that we must have

$$e'_{\langle x,y \rangle} \quad = \quad \langle \langle x, x \rangle, \langle x, y \rangle, \langle y, y \rangle \rangle.$$

But this triple does not belong to $\Gamma_{AM}(X \times X)$. This is a contradiction.

18.1. Consider $\Gamma(c) = \mathcal{P}_{fin}(A \cup c)$ and $\Delta(c) = \mathcal{P}(B \cup c)$. Then for all c, $\Gamma(c) \subseteq \Delta(c)$. The result follows from Exercise 15.9.

18.2. Let Γ be the operator $\Gamma(a) = A \times a$. We want to consider $\Delta_1(a) = \mathcal{P}_{fin}(a \cup Y)$, so that $\Delta_1^* = HF^1[Y]$. Let $s \in A^\infty$. By the Representation Theorem for Γ^*, we can find a flat Γ-coalgebra $\mathcal{E} = \langle X, e \rangle$ so that $s \in$ solution-set$(\mathcal{E})$. Now $e : X \to A \times X$, and it is not hard to verify that $A \times X \subseteq HF^1[Y \cup X] = (\Delta_1)_X^*$. So $e : X \to (\Delta_1)_X^*$. By the Solution Lemma Lemma, solution-set$(\mathcal{E}) \subseteq \Delta_1^*$. Thus $s \in \Delta_1^*$. Since s is arbitrary, we see that $A^\infty \subseteq \Delta_1^*$.

18.3. Let $C = \bigcap_n C_n$. To prove that $C \subseteq HF^1$, we show that $C \subseteq \mathcal{P}_{fin}(C)$. In fact, for $n \geq 1$, $C_n \subseteq \mathcal{P}_{fin}(C_{n-1})$. So if $x \in C$, then x is a finite set and all elements of x belong to C.

We use Theorem 18.3.2 to prove the converse. Let $a \in C$. Then $\mathcal{E}_a$, the canonical system of equations for a, has all e_x finite. Also, $\mathcal{E}_a$ uses only countably many indeterminates. Since a belongs to the solution set of $\mathcal{E}_a$, the Theorem tells us that $a \in HF^1$.

18.4. Let $\{x_n : n \in N\}$ be a family of distinct sets, and consider the system $x_n = \langle a_n, x_{n+1} \rangle$. Let s be the solution; the set we want is $s(x_0)$. By the Solution Lemma for HF^1, the solution set is a subset of HF^1. So $s(x_0) \in HF^1$.

18.5. Note that 0 and 1 belong to $HF^0 \subseteq HF^1$. There are uncountably many sequences of 0's and 1's, so the result follows from Exercise 18.4.

18.6. Let $\mathcal{E} = \langle X, A, e \rangle$ be a system which belongs to $HF^{1/2}[X \cup A]$. We need to go through the details of the proof of Theorem 16.7 in this case, and we need to check a few details. First, b can be taken to be finite. It follows that we may assume that Y, and thus $\mathcal{E}^b$ are finite. So $\mathcal{E}^b$ is a flat system of equations, and hence its solution set is transitive. Since $\mathcal{E}^b$ is finite, $\text{solution-set}(\mathcal{E}^b) \subseteq HF^{1/2}[A]$.

18.7. Note that $HF^{1/2}$ is transitive and closed under pairing and union. Since the power set of a finite set is finite, it is also closed under power set. This implies that the corresponding axioms are satisfied, and in addition the Replacement Scheme and the Axiom of Choice also hold. To check that *AFA* holds, note that the notions of being a system of equations and of being a solution are absolute, and then use Exercise 18.6. Incidentally, HF^1 is also a model of all axioms of ZFC^- except Infinity.

18.8. The case $\kappa = \omega_1$ and is entirely typical, so we restrict ourselves to it for the most part. There are two sets involved: HC^1, the largest collection X of sets with the property that every countable subset of X is an element of X; and also $HC^{1/2}$, the collection of sets whose transitive closures are countable. (We are ignoring the urelements in this solution.) Exactly as before, HC^1 is the collection of sets pictured by countably branching graphs. However, the transitive closure of all such sets is countable. This is because a countably branching accessible graph can have only a countable set of nodes. Therefore, unlike the situation with finiteness, $HC^{1/2} = HC^1$. Further, the analogue of the Solution Lemma holds. Finally HC^1 is a model of all axioms except

Power Set; it is a model of Infinity. The same fact holds for all infinite regular cardinals κ. In the case where κ is a strongly inaccessible cardinal (a regular cardinal satisfying the condition that if b has size less than κ then so does $\mathcal{P}(b)$) then H_κ^1 is closed under $\mathcal{P}$ so we also get the power set axiom to hold. Hence H_κ^1 is a model of full ZFA.

18.9. We only prove the more general statement. Assume first that $a \in \Gamma^* \cap HF^{1/2}[A]$. By the Representation Theorem for Γ^*, let $X \subseteq \mathcal{U}$ and $e : X \to \Gamma(X)$ determine a Γ-coalgebra $\mathcal{E}$ whose solution set contains a, say as $a = s_x$. We may assume that this solution s is injective as well. Let $Y \subseteq X$ be the subset of X accessible from x in the natural sense: Y is the least set such that $x \in Y$, and if $y \in Y$ and $z \in X \cap TC(e_y)$, then $z \in Y$ also. For each $y \in Y$, $s_y \in TC(\{a\})$. Since s is injective Y must be finite. But the restriction of e to Y gives a flat Γ-coalgebra, and the restriction of s to Y is its solution. This gives us the desired finite flat coalgebra.

The converse direction follows from Exercise 18.6 and the assumption that $\Gamma(X) \subseteq HF^{1/2}[A \cup X]$.

Now we turn to part (2). First we show that $\Gamma^r \subseteq \Gamma(\Gamma^r)$. Let $a \in \Gamma^r$. By part (1), let $\mathcal{E} = \langle X, e \rangle$ be a finite flat Γ-coalgebra so that for some $x \in X$, $a = s_x$. By (1) again, $X[s] \subseteq \Gamma^r$. But then $s_x = e_x[s] \in \Gamma(X)[s]$, and $\Gamma(X)[s] = \Gamma(X[s]) \subseteq \Gamma^r$.

In the other direction, note first that $\Gamma(\Gamma^r) \subseteq \Gamma(\Gamma^*) = \Gamma^*$. Further, we show $\Gamma(\Gamma^r) \subseteq HF^{1/2}[A]$. Let $a \in \Gamma(\Gamma^r)$. Let $den : X \to \Gamma^r$ be a bijection from some set Y which is new for Γ. By uniformity, $[den]_{\Gamma(Y)}$ maps $\Gamma(Y)$ onto $\Gamma(\Gamma^r)$. So we can find $c \in \Gamma(X)$ so that $c[den] = a$. Now $c \in HF^{1/2}[A \cup X]$, so we can find a finite flat system of equations in the sense of Chapter 6, say $\mathcal{E} = \langle Y, A \cup X, e \rangle$ whose solution set contains c. The transitive closure of c is thus the (finite) solution set of $\mathcal{E}$ together with a finite subset of $A \cup X$. And $TC(d) = TC(c)[den]$ the elements of $X \cap TC(d)$ with sets from $HF^{1/2}[A]$. So overall, $c \in HF^{1/2}[A]$.

18.10. Let R be such that for all t,

(3) $$lb(t) \quad R \quad rb(invert(t))$$

We claim that R is a stream bisimulation. Take a pair in R, for example the one in (3). It is not hard to check that the first component of both sides is $1^{st}(t)$. Also, $2^{nd}(lb(t)) = lb(2^{nd}(t))$, and

$$2^{nd}(rb(invert(t)) \quad = \quad rb(3^{rd}(invert(t))) \quad = \quad rb(invert(2^{nd}(t))).$$

Therefore $2^{nd}(lb(t)) \, R \, 2^{nd}(rb(invert(t))$, as needed.

18.11. The complete set is as follows:

$lb \circ copy$	$=$	i_{str}	$rb \circ invert$	$=$	lb
$rb \circ copy$	$=$	i_{str}	$lb \circ invert$	$=$	rb
$invert \circ invert$	$=$	i_{tr}	$invert \circ copy$	$=$	$copy$

We have omitted all of the laws that use trivial properties of the identities, such as $i_{tr} \circ copy = copy$ and $i_{tr} \circ i_{tr} = i_{tr}$.

It is not hard to check that all the laws listed above are valid. We discuss completeness. The basic point is that to check whether a very long equation is or is not a valid law, we can use the rules above to reduce things to a normal form. This can be illustrated by an example:

(4) $\qquad invert \circ copy \circ lb \circ copy \circ rb \quad = \quad rb \circ invert.$

The right side equals lb by one of the rules. For the left side, we can "cancel out" the part $lb \circ copy$ since it equals i_{str}. Continuing, we get

$$invert \circ copy \circ rb \quad = \quad i_{tr} \circ rb \quad = \quad rb.$$

So the question of whether (4) is a valid law boils down to the question of whether $rb = lb$ is a valid law. Of course, this last equation is not a valid law.

In more detail, every well-formed term can be reduced to one of the following forms: rb, lb, $copy \circ rb$, $copy \circ lb$, i_{str}, i_{tr}, or $invert$. This fact concerning reduction can be checked by induction on the terms. So these are our normal forms. It is easy to see that the only valid laws between normal forms are those where the left and right hand sides are identical. It follows that our list of laws above is complete.

18.12. The definition of f is by corecursion. Consider Δ_3, and let $C = Act^\infty$. We consider the coalgebra $\tau : C \to \Delta_3(C)$ given by

$$\tau(s) \quad = \quad \{s\} \quad = \quad \{\langle 1^{st}(s), 2^{nd}(s)\rangle\}.$$

By corecursion, we get a unique map $f : C \to Can(Act)$. When we work out the details concerning f, we see that for all streams s, $f(s) = \{\langle 1^{st}(s), f(2^{nd}(s))\rangle\}$.

The states which are of the form $f(s)$ are those in which exactly one action is possible, and after that action, exactly one action is possible, etc. More formally, let

$$\Delta(b) \quad = \quad \{\{a, b'\} \mid a \in Act \ \& \ b' \in b\}.$$

Then Δ^* is the set of states of the form $f(s)$.

18.13. We apply the Corecursion Theorem. Let

$$\pi : LSet(A) \to A \times \mathcal{P}(LSet(A))$$

be given by $\pi(\langle p, b \rangle) = \langle \mu(p), b \rangle$. The Corecursion Theorem gives us a map $\varphi : LSet(A) \to LSet(A)$ which satisfies a certain equation relative to any Δ_5-notation scheme for $LSet(A)$. When we write down that equation and simplify it, we see that φ satisfies the desired equation for $\overline{\mu}$.

19.1. As the table shows, for Δ_4, we take constants 0 and 1, and also operators $a :$ for all $a \in Act$. The semantics on a flat coalgebra $\langle X, e \rangle$ is given by

$$
\begin{array}{lll}
x \models 0 & \text{iff} & 1^{st}(e_x) = 0 \\
x \models 1 & \text{iff} & 1^{st}(e_x) = 1 \\
x \models a : \varphi & \text{iff} & (2^{nd}(e_x))(a) \models \varphi
\end{array}
$$

In this last expression, note that $2^{nd}(e_x)$ is a function from Act to X.

For Δ_5, we take A as the set of constants, and also $\Diamond$ as the only modal operator. The semantics is

$$
\begin{array}{lll}
x \models a & \text{iff} & 1^{st}(e_x) = a \\
x \models \Diamond\varphi & \text{iff} & (\exists y)(y \in 2^{nd}(e_x) \text{ and } y \models \varphi)
\end{array}
$$

19.2. We show that for every word w there is a sentence φ_w such that for all states s in all automata $\mathcal{A}$, $s \models \varphi_w$ iff $w \in lang_A(s)$. If $w = a_1 a_2 \cdots a_k$, then we take

$$
\varphi_w \quad = \quad a_1 : a_2 : \cdots : a_k : 1.
$$

The result on languages now follows. We should note that the sentences of finite modal depth might not be finite in this case, since Act need not be finite. That is, $\bigwedge_{a \in Act} a : 1$ is a sentence of modal depth 1, but if Act is infinite, it is not hereditarily finite.

19.3. We would like to take $A \cup B$ as the set of constants, but if $A \cap B \neq \emptyset$, the semantics will not work out. Instead we take the disjoint union $A + B$. We also use two modal operators, $\Diamond_1$ and $\Diamond_2$. The semantics is

$$
\begin{array}{lll}
x \models \langle 1, b \rangle & \text{iff} & 1^{st}(e_x) = b \\
x \models \langle 2, c \rangle & \text{iff} & 2^{nd}(e_x) = c \\
x \models \Diamond_1\varphi & \text{iff} & (\exists y)(y \in 3^{rd}(e_x) \text{ and } y \models \varphi) \\
x \models \Diamond_2\varphi & \text{iff} & (\exists y)(y \in 4^{th}(e_x) \text{ and } y \models \varphi)
\end{array}
$$

An alternative way to go would be to take $A \times B$ as the set of constants.

19.4. The statement of θ for Δ_3 is on page 289. First, suppose that $x \models_\varepsilon \theta(f, s)$. We take

$$
c \quad = \quad \{\langle a, \langle y, w \rangle \rangle \in \Gamma(\models_{s, \varepsilon}) \mid x \xrightarrow{a} y\}.
$$

Then
$$c[\pi_1] \quad = \quad \{\langle a, y \rangle \in e_x \mid (\exists w)\, \langle a, w \rangle \in f\}$$
$$c[\pi_2] \quad = \quad \{\langle a, w \rangle \in f \mid (\exists y)\, x \xrightarrow{a} y\}$$

So automatically, $c[\pi_1] \subseteq e_x$ and $c[\pi_2] \subseteq f$. Since x satisfies all of the subformulas of $\theta(f, s)$ that begin with $\square$ (one for each $a \in A$), we see that $e_x \subseteq c[\pi_1]$. And since x satisfies all of the subformulas of $\theta(f, s)$ that begin with $\lozenge$ (one for each pair $\langle a, x \rangle \in f$), $f \subseteq c[\pi_2]$.

Going the other way, suppose that $c \in \Gamma(\models_{s,\varepsilon})$ exists so that $c[\pi_1] = e_x$ and $c[\pi_2] = f$. Since $e_x \subseteq c[\pi_1]$, x satisfies the $\square$-subformulas of $\theta(f, s)$. And since $f \subseteq c[\pi_2]$, x satisfies the $\lozenge$-subformulas.

19.5. Fix a Γ-coalgebra $\mathcal{E} = \langle X, e \rangle$. Then by the definition of canonicalizing, $x \models_{\mathcal{E}} \theta(e_x, j_{\mathcal{E},s})$. This is just because $\theta(e_x, j_{\mathcal{E},s}) = \theta(e_x, j_{\mathcal{E},s})$. And the definition also tells us that if $x \models_{\mathcal{E}} \theta(f, t)$, then $\theta(f, t) = \theta(e_x, j_{\mathcal{E},s})$. So S' is indeed a separator.

19.6. We only give the argument for Δ_1, since the others are similar. Let $\mathcal{E} = \langle X, e \rangle$ be a flat Δ_1-algebra. Let r be a reorganization with domain X, so for all $x \in X$, $e(r_x) = e_x[r]$.

For Δ_1, note that since X is new for the operator, for all $a \in A$, $a[r] = a$. So $x \models a$ iff $a \in e_x$ iff $a \in e_x[r] = e(r_x)$. This translates to the reorganization condition for the atomic sentences. For the general case, we use induction on φ. The only interesting induction step is for the modal operator. Suppose that $x \models_{\mathcal{E}} \lozenge\varphi$. Then for some $y \in e_x$, $y \models_{\mathcal{E}} \varphi$. So by induction hypothesis, $r_y \models_{\mathcal{E}'} \varphi$. Since $r_y \in e_x[r] = e(r_x)$, $e(r_x) \models \lozenge\varphi$.

19.7. We have already seen part (1). If two canonical states b and b' satisfy the same finitary formulas, then the languages associated to those states are the same, by Exercise 19.2. But then b and b' are equal, by Lemma 18.8. (No results in Chapter 18 ever assumed that Act was a finite set.)

The second part is easier: $Act \subseteq TC(c)$, so c is not hereditarily finite.

19.8. For Δ_1, fix $\mathcal{E}$, f, and x. We have two cases. If $1^{st}(f) \cap A \neq e_x \cap A$, then no matter what s is, $x \not\models \theta(f, s)$. So we take some Q which falsifies the second condition. For example, we can take $Q = \{\langle y, z \rangle\}$, where $y \not\models s_z$. So we assume that $1^{st}(f) \cap A = e_x \cap A$. We then take Q to be the set of pairs $\langle y, z \rangle$ such that either y and z are both in $e_x \cap X$; or else $y = z \notin e_x \cap X$. (We need this last condition since part (2) of the strong separator condition requires that the domain of Q be all of X.) We check that if $x \models \theta(f, s)$, then the strong separator conditions both hold. On the other hand, if they do hold, then by considering the elements of $e_x \cap X$, we see that $x \models \theta(f, s)$.

Turning to L_4, we again fix $\mathcal{E}$, f, and x; we also have two cases. If $1^{st}(f) \neq 1^{st}(e_x)$, then we take Q to be the empty relation. That way, both parts of the strong separator condition fail. On the other hand, if $1^{st}(f) = 1^{st}(e_x)$, then $x \models \theta(f, s)$ iff for all $a \in Act$, $(2^{nd}(e_x))(a) \models_{\mathcal{E}} s((2^{nd}(e_x))(a))$. Let $W = 2^{nd}(e_x)[A]$; this is the set of $y \in X$ which occur in the support of e_x. Thus we let Q be the set of pairs $\langle y, z \rangle$ such that either $y = z \in W$; or else $y \notin W$ and $y \models s_z$. The fact that $2^{nd}(e_x)$ is a function insures that two separator conditions are equivalent.

19.9. Fix x and y so that $j_\omega(x) = j_\omega(y)$. By Lemma 19.7, we see that for all n, $j_n(y) = j_n(x)$. Also fix a finite-to-finite relation Q as in the strong separator condition for e_x. We show that $y \models \theta(e_x, j_\omega)$.

Let Q' be the relation given by

$$v \, Q' \, w \quad \text{iff} \quad v \, Q \, w \text{ and for all } n, v \models j_n(w).$$

We need only show that for each v there is some w such that $v \, Q' \, w$, and vice-versa. For each v and n there is some w (depending on n) so that $v \, Q \, w$ and $w \models j_n(w)$. For this v, there are only finitely many w such that $v \, Q \, w$. So some fixed w must have the property that for infinitely many n, $w \models j_n(w)$. By Lemma 19.7 again, this w has the property that for all n, $w \models j_n(w)$. The other direction is similar.

Now we know that $y \models \theta(e_x, j_\omega)$. Then, by the definition of a canonical-izing map and the fact that $y \models \theta(e_y, j_\omega)$, we have $\theta(e_x, j_\omega) = \theta(e_y, j_\omega)$, as desired.

For part (2), we argue as in Theorem 19.4. If $x \, R \, y$, then $\theta(e_x, j_\omega) = \theta(e_y, j_\omega)$. So by the equivalence condition on θ, $e_x[j_\omega] = e_y[j_\omega]$. Since Γ preserves covers, j_ω shows that R is indeed a Γ-bisimulation.

19.10. The axioms for the operators are

$$left\colon \varphi \;\to\; \neg left\colon \neg\varphi$$
$$right\colon \varphi \;\to\; \neg right\colon \neg\varphi$$
$$left\colon (\varphi \to \psi) \;\to\; ((left\colon \varphi) \to (left\colon \psi))$$
$$right\colon (\varphi \to \psi) \;\to\; ((right\colon \varphi) \to (right\colon \psi))$$

The first two express the fact that in a Δ_2-coalgebra $\langle c, e \rangle$, for each $b \in c$ there is a *unique* b_1 such that $b_1 = 2^{nd}(b)$, and a unique b_2 such that $b_2 = 3^{rd}(b)$. The last two distribution axioms are analogues of the axioms for $\Box$ from ordinary modal logic.

For (2), we need an axiom that says that exactly one $a \in A$ holds. When A is a finite set, this can be expressed in a single finitary axiom.

Turning to the rules of inference, we have two forms of necessitation: from φ, derive $left\colon \varphi$; and from φ, derive $right\colon \varphi$.

19.11. For part (1), we actually show something stronger: for each φ and each canonical ψ of rank at least the modal depth of φ, either $\vdash \psi \to \varphi$ or $\vdash \psi \to \neg\varphi$. The proof is by induction on φ. In the case of atomic φ, any canonical sentence ψ of rank at least 1 will do.

All of the boolean steps are easy. For the modal operators, assume the result for φ and consider $left\colon \varphi$. Let ψ have rank $n + 1$, so ψ is of the form $a \wedge left\colon \psi_1 \wedge right\colon \psi_2$ for some ψ_1 and ψ_2. By induction hypothesis, either $\vdash \psi_1 \to \varphi$ or $\vdash \psi_1 \to \neg\varphi$. Assume the first case, since the other is similar. By necessitation, $\vdash left\colon (\psi_1 \to \varphi)$. So by the distribution axiom and modus ponens, $\vdash left\colon \psi_1 \to left\colon \varphi$. Therefore, $\vdash \psi \to left\colon \varphi$ as well.

For part (2), we need a preliminary remark. For each maximal consistent T, consider the sets

$$
\begin{aligned}
T_1 &= \{\psi \mid left\colon \psi \in T\} \\
T_2 &= \{\psi \mid right\colon \psi \in T\}
\end{aligned}
$$

Then using the axioms for the modalities, we see that each of these is maximal consistent. Further, there is a unique $a_T \in A \cap T$. The desired map α takes a theory T to the triple $\langle a_T, T_1, T_2 \rangle$.

Finally, we prove completeness. Let T be a maximal consistent set. We use part (3) to prove that for all finitary canonical φ, $T \models_{Th} \varphi$ iff $\varphi \in T$. (Then the result holds for all φ, not just the canonical sentences, by part (1).) The proof is by induction on the rank. For rank 0, $\varphi = \mathsf{T}$ and the result is trivial. Assume the result for rank n, and let ψ be a canonical sentence of rank $n + 1$. We can write

$$\psi = a \wedge left\colon \psi_1 \wedge right\colon \psi_2,$$

where ψ_1 and ψ_2 are canonical sentences of rank n. Then

$$
\begin{aligned}
\psi \in T \quad &\text{iff} \quad a \in T, \psi_1 \in T_1, \text{ and } \psi_2 \in T_2 \\
&\text{iff} \quad a \in T, T_1 \models_{Th} \psi_1, \text{ and } T_2 \models_{Th} \psi_2 \\
&\text{iff} \quad T \models_{Th} \psi
\end{aligned}
$$

The first and third equivalences are by the definition of coalgebra map α, the second is by the induction hypothesis. This completes the proof.

20.1. This result could be done using the fact that ordered pairs are presentable. We give a different proof, just for variety. Let φ hold exactly on the pairs shown below:

$$
\begin{array}{ll}
\varphi(\langle 4, b\rangle, \langle 5, b\rangle) \text{ for all } b & \varphi(\langle 3, b\rangle, \langle 5, b\rangle) \text{ for all } b \\
\varphi(\langle 2, b\rangle, \langle 4, b\rangle) \text{ for all } b & \varphi(\langle 2, b\rangle, \langle 3, b\rangle) \text{ for all } b \\
\varphi(\langle 2, c\rangle, \langle 2, b\rangle) \text{ if } c \in b & \varphi(\langle 2, c\rangle, \langle 1, b\rangle) \text{ if } c \in f(b)
\end{array}
$$

This φ may be taken to be Σ_1. Let χ be determined from φ as in Proposition 20.1. Then an argument involving bisimulations shows that

$Ext(\ulcorner\chi\urcorner, \langle 2, b\rangle) = b$ for all b. From this, $Ext(\ulcorner\chi\urcorner, \langle 1, b\rangle) = f(b)$ for all b.
Then $Ext(\ulcorner\chi\urcorner, \langle 3, b\rangle) = \{b\}$, $Ext(\ulcorner\chi\urcorner, \langle 4, b\rangle) = \{b, f(b)\}$, and

$$Ext(\ulcorner\chi\urcorner, \langle 5, b\rangle) \quad = \quad \{\{b\}, \{b, f(b)\}\} \quad = \quad \langle b, f(b)\rangle.$$

Now there is a Σ_1 formula χ' so that for all b, $\chi'(b, c)$ iff $c = \langle\chi, \langle 5, b\rangle\rangle$. Hence
$Ext(\ulcorner\chi'\urcorner, b) = Ext(\chi, \langle 5, b\rangle) = \langle b, f(b)\rangle$. This χ' is as desired.

20.2. If the conclusion were not true, then the class of all wellfounded canonical classes would itself be a wellfounded canonical class, hence an element of itself, hence not wellfounded after all. The second sentence is proven in the same way as (1) in Corollary 20.6.

20.3. $U - \{U\} = \{x \in U \mid x \neq U\}$. By the Difference Lemma, this does not belong to U. But if U satisfied the Axiom of Separation, $U - \{U\}$ would belong to U.

Bibliography

Abramsky, Samson. 1988. A Cook's Tour of the Finitary Non-Well-Founded Sets. Unpublished manuscript.

Aczel, Peter. 1988. *Non-Well-Founded Sets*. CSLI Lecture Notes Number 14. Stanford: CSLI Publications.

Aczel, Peter, and Nax Mendler. 1989. A Final Coalgebra Theorem. In *Category Theory and Computer Science*, ed. D. H. Pitt (et al). 357–365. Lecture Notes in Computer Science. Heidelberg: Springer Verlag.

d'Agostino, G., and C. Bernardi. To appear. Translating the hypergame paradox; remarks on the set of founded elements of a relation. *Journal of Philosophical Logic*.

van Aken, J. 1986. Axioms for the set-theoretic hierarchy. *Journal of Symbolic Logic* 51(4):992–1004.

Baltag, Alexandru. To appear. Modal Characterisations for Sets and Kripke Models. Unpublished manuscript.

Bartlett, Steven J., and Peter Suber (ed.). 1987. *Self-reference: Reflections on Reflexivity*. Martinus Nijhoff Philosophy Library, Vol. 21. Dordrecht: Martinus Nijhoff Publishers.

Barwise, Jon. 1972. The Hanf Number of Second Order Logic. *Journal of Symbolic Logic* 37:588–594.

———. 1974. *Admissible Sets and Structures*. Heidelberg: Spinger-Verlag.

———. 1989. On the Model Theory of Common Knowledge. In *The Situation in Logic*. 201–220. CSLI Lecture Notes Number 17. Stanford: CSLI Publications.

Barwise, Jon, and John Etchemendy. 1987. *The Liar: An Essay in Truth and Circularity*. Oxford: Oxford University Press.

Barwise, Jon, and Lawrence S. Moss. 1991. Hypersets. *Mathematical Intelligencer* 13:31–41.

———. 1996. Modal Correspondence for Models. In *Proceedings of the Tenth Amsterdam Colloquium*, ed. Paul Dekker and Martin Stokhof. Department of Philosophy, University of Amsterdam: ILLC.

van Benthem, Johan. 1985. *Modal Logic and Classical Logic*. Naples: Bibliopolis.

van Benthem, Johan, and Jan Bergstra. 1995. Logic of Transition Systems. *Journal of Logic, Language, and Information* 3:247–284.

Bloom, Steven L., S. Ginali, and J. Rutledge. 1977. Scalar and Vector Iteration. *Journal of Computer and System Sciences* 14:251–256.

Bloom, Steven L., and Zoltán Ésik. 1993. *Iteration Theories: The equational logic of iterative processes*. EATCS Monographs on Theoretical Computer Science. Berlin: Springer-Verlag.

Boffa, Maurice. 1968. Graphes extensionneles et axiome d'universitalité. *Zeitschrift für Math. Logik und Grundlagen der Math.* 14:329–334.

Booth, David. 1990. Hereditarily finite Finsler sets. *Journal of Symbolic Logic* 55:700–706.

Clark, Herbert, and C. Marshall. 1981. Definite Reference and Mutual Knowledge. In *Elements of Discourse Understanding*, ed. A. Joshi (et al). 10–63. Cambridge, MA: Cambridge University Press.

Devlin, Keith. 1993. *The Joy of Sets*. Undergraduate Texts in Mathematics. Berlin: Springer-Verlag.

Fagin, Ronald. 1994. A Quantitative Analysis of Modal Logic. *Journal of Symbolic Logic* 59(1):209–252.

Felgner, Ulrich. 1971. *Models of ZF-Set Theory*. Lecture Notes in Mathematics, No. 223. Berlin: Springer-Verlag.

Fiore, M. P. 1993. A Coinduction Principle for Recursive Data Types Based on Bisimulation. In *Proceedings 8th LICS*, 110–119. IEEE.

Forster, T. E. 1995. *Set Theory with a Universal Set, second edition*. Oxford Logic Guides, No. 31. Oxford: Oxford Science Publications.

Forti, Marco, and Furio Honsell. 1983. Set Theory with Free Construction Principles. *Annali Scuola Normale Supeiore di Pisa, Classe di Scienze* 10:493–522.

———. 1992. Weak foundation and anti-foundation properties of positively comprehensive hyperuniverses. In *L'Antifondation en Logique et en Theéorie des Ensembles*, ed. R. Hinnion. 31–43. Cahiers du Centre de Logique.

———. To appear. A General Construction of Hyperuniverses. *Theoretical Computer Science* 156.

Forti, Marco, Furio Honsell, and Marina Lenisa. 1994. Processes and Hyperuniverses. In *Mathematical Foundations of Computer Science*, ed. I. Privara (et al). Lecture Notes in Computer Science, Vol. 841, 352–361. Berlin: Springer-Verlag.

Goguen, Joseph A. 1973. Realization is Universal. *Math. Systems Theory* 6:359–374.

Grice, P. 1966. Meaning. *Philosophical Review* 66:377–388.

Gupta, Anil, and Nuel Belnap. 1993. *The Revision Theory of Truth*. Bradford Books. Cambridge MA: MIT Press.

Hallett, Michael. 1984. *Cantorian Set Theory and Limitation of Size*. Oxford: Clarendon Press.

Harman, Gilbert. 1986. *Change in View*. Cambridge, MA: MIT Press.

Heifetz, Aviad. To appeara. Eliminating Redundancies in Partition Spaces. In *Epistemic Logic and the Theory of Games and Decisions*, ed. M. Bacharach (et al).

———. To appearb. Non-Well-Founded Type Spaces. In *Games and Economic Behavior*.

Jones, N. D., C. K. Gomard, and P. Sestoft. 1993. *Partial Evaluation and Automatic Program Generation*. Prentice Hall International.

Kripke, Saul A. 1975. Outline of a Theory of Truth. *The Journal of Philosophy* 72:53–81.

Lévy, Azriel. 1965. *A Hierarchy of Formulas in Set Theory*. Memoirs, No. 57. Providence: American Math Society.

Lewis, David. 1969. *Convention, A Philosophical Study*. Cambridge, MA: Harvard University Press.

Lismont, L. 1995. Common Knowledge: Relating anti-founded situation semantics to modal logic neighborhood semantics. *Journal of Logic, Language, and Information* 3:285–302.

Milner, Robin, and Mads Tofte. 1991. Co-induction in relational semantics. *Theoretical Computer Science* 87:209–220.

Moschovakis, Yiannis N. 1974. *Elementary Induction on Abstract Structures*. Studies in Logic and the Foundations of Mathematics. Amsterdam: North Holland.

———. 1980. *Descriptive Set Theory*. Studies in Logic and the Foundations of Mathematics. Amsterdam: North Holland.

———. 1994. *Set Theory Notes*. Undergraduate Texts in Mathematics. Heidelberg: Springer-Verlag.

Moss, Lawrence S., and Norman Danner. To appear. On the Foundations of Corecursion. Unpublished manuscript.

Newman, Donald J. 1982. *A Problem Seminar*. Berlin: Springer-Verlag.

Pitts, A. 1993. Relational Properties of Recursively Defined Domains. In *Proceedings 8th LICS*, 86–97. IEEE.

Rogers, Jr., Hartley. 1967. *Theory of Recursive Functions and Effective Computability*. New York: McGraw Hill.

Rutten, J. J. M. M. 1993. A structural co-induction theorem. In *Proceedings of the 9th International Conference on Mathematical Foundations of Programming Semantics*, ed. S. Brookes, M. Main, A. Melton, M. Mislove, and D. Schmidt, LNCS, Vol. 802, 83–102. New Orleans. Springer-Verlag.

Rutten, J. J. M. M., and D. Turi. 1993. On the Foundations of Final Semantics: nonstandard sets, metric spaces, partial orders. In *Proceedings of the REX Workshop on Semantics: Foundations and Applications*, ed. J. W. de Bakker, W.-P. de Roever, and G. Rozenberg, LNCS, Vol. 666, 477–530. Beekbergen. Springer-Verlag.

———. 1994. Initial algebra and final coalgebra semantics for concurrency. In *Proceedings of the REX School/Symposium 'A decade of concurrency'*, ed. J. W. de Bakker, W.-P. de Roever, and G. Rozenberg, LNCS, Vol. 803, 530–582. Springer-Verlag.

Sainsbury, R. M. 1988. *Paradoxes*. Cambridge: Cambridge Univeristy Press.

Searle, John. 1983. *Intentionality, An Essay in the Philosophy of Mind*. New York: Cambridge University Press.

Smullyan, Raymond M. 1983. *5000 B.C. and Other Philosophical Fantasies*. New York: St. Martin's Press.

Tarski, Alfred. 1939. On undecidable statements in enlarged systems of logic and the concept of truth. *Journal of Symbolic Logic* 4:105–112.

————. 1956. The Concept of Truth in Formalized Languages. In *Logic, Semantics, Metamathematics*. 152–277. Oxford: Clarendon Press. Reprinted from earlier versions in German and Polish.

Zwicker, William S. 1987. Playing games with games: the hypergame paradox. *American Mathematical Monthly* 94(6):507–514.

Index

A-bisimulation, 78
Abramsky, S. (1988), 157
Act, 35
*Act**, 279
Aczel, P.
  (1988), 5, 6, 67, 76, 89, 221, 242, 337
  and Mendler, N. (1989), 326
AFA, 72
  AFA_0, 128
  equivalent form, 99, 101
d'Agostino, G.
  and Bernardi, C. (to appear), 29, 46, 175
van Aken, J. (1986), 303, 304
atom, 70
$Aut(Act)$, 278
automaton, 39
  canonical, 39
  deterministic, 278
avoidance set, 233, 245
Axiom
  Anti-Foundation (*AFA*), *see AFA*
  Foundation (*FA*), 24
  Global Choice, 28
  Plenitude, 23
  Strong Plenitude, 23
  table of, 28

Baltag, A. (to appear), 142, 329
Bartlett, S. J. and Suber, P. (1987), 196
Barwise, J.

(1972), 304, 313, 331
(1989), 49
  and Etchemendy, J. (1987), 8, 125, 183, 185, 195
  and Moss, L. S. (1991), 8
  and Moss, L. S. (1996), 156, 329
Belnap, N.
  and Gupta, A. (1993), 196
van Benthem, J.
  (1985), 153
  and Bergstra J. (1995), 157
Bergstra, J.
  and van Benthem, J. (1995), 157
Bernardi, C.
  and d'Agostino, G. (to appear), 29, 46, 175
bisimulation, 78
  algorithm for, 87–89
  and modal logic, 135–137
  between labeled graphs, 129
  Γ, 258
  of deterministic automata, 278
  of lts's, 37, 276
  of pointed systems, 88
  stream, 84
Bloom, S. L.
  and Ginali, S. and Rutledge, J. (1977), 102
Bloom, S. L. and Ésik, Z. (1993), 102
Boffa, M. (1968), 337
Booth, D. (1990), 282
$b[s]$, 93

canonical invariant, 140, 294
canonical sentence, 294
canonicalizing map, 293
cardinal, 21
    regular, 140
    strongly inaccessible, 310
cartesian product, 13
Clark, H.
    and Marshall, C. (1981), 49
class, 15
    canonical, 310
        small, 312
    large and small, 15–17
    proper, 15
class notation system, 310
closure (programming semantics), 40
coalgebra, 226–228
    canonical, 230
    strongly extensional, 260
coinduction
    for greatest fixed points, 217–218
    for streams, 201–205
compiler, 44
    generator, 44
consequence, 133
corecursion, 243–258
    for streams, 200–205
    simultaneous, 255–258
corecursion conditions, 93
Corecursion Theorem, 250, 327
cover of a set, 262
cumulative conception, 302

Danner, N.
    and Moss, L. S. (to appear), 328
deg(a), 142
denotation
    in partial models, 178
descending sequence, 24
Devlin, K. (1993), 7, 242, 325
difference, 13
Difference Lemma, 85–86

E-subset, 347
equinumerous, 21
equivalence condition, 290

Ésik, Z.
    and Bloom, S. L. (1993), 102
Etchemendy, J.
    and Barwise, J. (1987), 8, 125, 183, 185, 195
expansion, 180
    principal, 181
expression map, 287–293
expressive logic, 293
extension, 178

Fagin, R. (1994), 157
Felgner, U. (1971), 7
Fiore, M. P. (1993), 242, 282
fixed point, 199, 213
    greatest, 199, 213–214, 216–219
    least, 213–216
flat system, see system of equations, flat
Forster, T. E. (1995), 175
Forti, M.
    and Honsell, F. (1983), 5, 67, 76, 115
    and Honsell, F. (1992), 321
    and Honsell, F. (to appear), 321
    and Honsell, F. and Lenisa, M. (to appear), 321
function, 12
    bijective, 12
    one-to-one (injective), 12
    onto (surjective), 12
    transition, 35

game, 161
    and quantifiers, 165–167
    bisimulation, 167–170
    closed, 162
    determined, 164
    for fixed points, 220
    membership, 162
    open, 162
    wellfounded, 162
Γ-bisimulation, 258
Γ-closed, 213
Γ-closure, 217
Γ-coalgebra, 226

flat, 227
  general, 236
Γ-corecursion, 249
Γ-correct, 213
Γ-form, 213, 288
Γ-lift, 246
Γ-logic, 287
Γ-morphism, 229
Γ-notation scheme, 246
Γ-substitution, 288
Γ_*, 215
Γ^*, 216
Ginali, S.
  and Bloom, S. L. and Rutledge, J.
    (1977), 102
Goguen, J. A. (1973), 282
Gomard, C. K.
  and Jones, N. D. and Sestoft, P.
    (1993), 46
graph, 119
  accessible pointed, 128
  canonical, 123–124
  decoration of, 120
  isomorphism, 122
  labeled, 125
    canonical, 128
  picture of a set, 120
Grice, P. (1966), 50
Gupta, A. and Belnap, N. (1993), 196

Hallett, M. (1984), 302, 303, 319
Harman, G. (1986), 49
Heifetz, A.
  (1993), 52
  (1994), 52
hereditarily $\mathcal{C}$, 152
$HF^0[A]$, 268
$HF^1[A]$, 268
$HF^{1/2}[A]$, 268
$H_\kappa[A]$, 140–142
Honsell, F.
  and Forti, M. (1983), 5, 67, 76, 115
  and Forti, M. (1992), 321
  and Forti, M. (to appear), 321
  and Forti, M. and Lenisa, M. (to ap-
    pear), 321

hypergame, 59, 172
Hypergame Paradox, 58–59, 170–175
hyperuniverse, 321

image, 12
indeterminate, 70
induction
  for least fixed points, 215–216
  for the natural numbers, 218
  for wellfounded sets, 25
interpreter, 44
intersection, 13
invariant under reorganizations, 297
isomorphism, 19
  graph, 122
iterative conception, 302

Jones, N. D.
  and Gomard, C. K. and Sestoft, P.
    (1993), 46

Kripke frame, 149
Kripke structure, 134
  canonical, 134
  finitely branching, 137
Kripke, S. A. (1975), 56, 195

labeled transition system (lts), 35–40,
  276
  canonical, 38, 276
  deterministic, 38
$\mathcal{L}_\infty[A]$, 132
$\mathcal{L}[A]$, 132
language, 279
Lenisa, M.
  and Forti, M. and Honsell, F. (to ap-
    pear), 321
Lévy, A. (1965), 313
Lewis, D. (1969), 4, 49
Liar Paradox, 55–57, 87, 187–191
  Strengthened, 87, 189
Liar sentence, 187
lift, 246
linear order, 18
Lismont, L. (1995), 49

map, 12

Marshall, C.
  and Clark, H. (1981), 49
Mendler, N.
  and Aczel, P. (1989), 326
Milner, R. and Tofte, M. (1991), 40,
    41, 46
modal depth, 143
model
  denotation-correct, 193
  partial, 178
  reflexive, 181
  total, 178
  truth-complete, 185
  truth-correct, 185
morphism, 229
  of automata, 278
  of lts's, 277
Moschovakis, Y. N.
  (1974), 221
  (1980), 165, 175
  (1994), 7, 325, 328
Moss, L. S.
  and Barwise, J. (1991), 8
  and Barwise, J. (1996), 156, 329
  and Danner, N. (to appear), 328
Myhill-Nerode Theorem, 39, 278

natural numbers, 13
necessitation, 146
new set of urelements, 205, 227
Newman, D. J. (1982), 53
notation scheme, 246
  binary, 256
notation system
  $\mathcal{F}$, 313

Ω, 72
operation, 16
  substitution-like, 97
operator, 212
  appendix on, 335–336
  cover preserving, 262, 291, 294
  dual, 219
  monotone, 199, 212
  proper, 230
  smooth, 245

uniform, 233
ordered pair, 11
ordinal, 17–20, 86
  limit, 19
  successor, 19

paradoxical terms, 57–58
parametric Γ-object, 236
partial function, 12
partial order, 18
Pitts, A. (1993), 242
play
  in accord with σ, 163
  legal, 160
  terminal, 160
  winning, 161
$\mathcal{P}(a)$, 13
$\mathcal{P}_{fin}(a)$, 14
power set, 13
preorder, 18
program
  self-applicative, 42
  self-producing, 42
projection, 258
proper class, 15
$\mathcal{PS}$, 105

Recursion Theorem, 43
  for class notation systems, 314
  for self-sufficient sets, 62
reflexive set, 57
relation, 12
  antisymmetric, 18
  circular, 24
  equivalence, 18
  finite-to-finite, 298
  irreflexive, 18
  non-wellfounded, 24
  reflexive, 18
  symmetric, 18
  transitive, 18
  wellfounded, 24
relational structure, 12
reorganization, 229
Representation Theorem for Γ^*, 235
restriction, 12